Oscar Castillo, Patricia Melin, Janusz Kacprzyk, Witold Pedrycz (Eds.)

Hybrid Intelligent Systems

Studies in Fuzziness and Soft Computing, Volume 208

Editor-in-chief
Prof. Janusz Kacprzyk
Systems Research Institute
Polish Academy of Sciences
ul. Newelska 6
01-447 Warsaw
Poland
E-mail: kacprzyk@ibspan.waw.pl

Further volumes of this series
can be found on our homepage:
springer.com

Vol. 192. Jose A. Lozano, Pedro Larrañaga, Iñaki Inza, Endika Bengoetxea (Eds.)
Towards a New Evolutionary Computation, 2006
ISBN 3-540-29006-0

Vol. 193. Ingo Glöckner
Fuzzy Quantifiers: A Computational Theory, 2006
ISBN 3-540-29634-4

Vol. 194. Dawn E. Holmes, Lakhmi C. Jain (Eds.)
Innovations in Machine Learning, 2006
ISBN 3-540-30609-9

Vol. 195. Zongmin Ma
Fuzzy Database Modeling of Imprecise and Uncertain Engineering Information, 2006
ISBN 3-540-30675-7

Vol. 196. James J. Buckley
Fuzzy Probability and Statistics, 2006
ISBN 3-540-30841-5

Vol. 197. Enrique Herrera-Viedma, Gabriella Pasi, Fabio Crestani (Eds.)
Soft Computing in Web Information Retrieval, 2006
ISBN 3-540-31588-8

Vol. 198. Hung T. Nguyen, Berlin Wu
Fundamentals of Statistics with Fuzzy Data, 2006
ISBN 3-540-31695-7

Vol. 199. Zhong Li
Fuzzy Chaotic Systems, 2006
ISBN 3-540-33220-0

Vol. 200. Kai Michels, Frank Klawonn, Rudolf Kruse, Andreas Nürnberger
Fuzzy Control, 2006
ISBN 3-540-31765-1

Vol. 201. Cengiz Kahraman (Ed.)
Fuzzy Applications in Industrial Engineering, 2006
ISBN 3-540-33516-1

Vol. 202. Patrick Doherty, Witold Łukaszewicz, Andrzej Skowron, Andrzej Szałas
Knowledge Representation Techniques: A Rough Set Approach, 2006
ISBN 3-540-33518-8

Vol. 203. Gloria Bordogna, Giuseppe Psaila (Eds.)
Flexible Databases Supporting Imprecision and Uncertainty, 2006
ISBN 3-540-33288-X

Vol. 204. Zongmin Ma (Ed.)
Soft Computing in Ontologies and Semantic Web, 2006
ISBN 3-540-33472-6

Vol. 205. Mika Sato-Ilic, Lakhmi C. Jain
Innovations in Fuzzy Clustering, 2006
ISBN 3-540-34356-3

Vol. 206. A. Sengupta (Ed.)
Chaos, Nonlinearity, Complexity, 2006
ISBN 3-540-31756-2

Vol. 207. Isabelle Guyon, Steve Gunn, Masoud Nikravesh, Lotfi A. Zadeh (Eds.)
Feature Extraction, 2006
ISBN 3-540-35487-5

Vol. 208. Oscar Castillo, Patricia Melin, Janusz Kacprzyk, Witold Pedrycz (Eds.)
Hybrid Intelligent Systems, 2007
ISBN 3-540-34719-1

Oscar Castillo
Patricia Melin
Janusz Kacprzyk
Witold Pedrycz
(Eds.)

Hybrid Intelligent Systems

Analysis and Design

Springer

Oscar Castillo
Patricia Melin
Tijuana Institute of Technology
Department of Computer Science
Tijuana, Mexico
P.O. Box 4207
Chula Vista CA 91909, USA
E-mail: ocastillo@tectijuana.mx
pmelin@tectijuana.mx

Witold Pedrycz
University of Alberta
Department of Electrical
and Computer Engineering
Edmonton, Alberta T6J 2V4, Canada
E-mail: pedrycz@ece.ualberta.ca

Janusz Kacprzyk
Polish Academy of Sciences
Systems Research Institute
Newelska 6
01-447 Warszawa Poland
E-mail: kacprzyk@ibspan.waw.pl

ISSN print edition: 1434-9922
ISSN electronic edition: 1860-0808
ISBN 978-3-642-07223-9
e-ISBN 978-3-540-37421-3

Springer is a part of Springer Science+Business Media
springer.com

Cover design: Erich Kirchner, Heidelberg

Preface

We describe in this book, new methods for analysis and design of hybrid intelligent systems using soft computing techniques. Soft Computing (SC) consists of several computing paradigms, including fuzzy logic, neural networks, and genetic algorithms, which can be used to produce powerful hybrid intelligent systems for solving problems in pattern recognition, time series prediction, intelligent control, robotics and automation. Hybrid intelligent systems that combine several SC techniques are needed due to the complexity and high dimensionality of real-world problems. Hybrid intelligent systems can have different architectures, which have an impact on the efficiency and accuracy of these systems, for this reason it is very important to optimize architecture design. The architectures can combine, in different ways, neural networks, fuzzy logic and genetic algorithms, to achieve the ultimate goal of pattern recognition, time series prediction, intelligent control, or other application areas.

This book is intended to be a major reference for scientists and engineers interested in applying new computational and mathematical tools to design hybrid intelligent systems. This book can also be used as a textbook or major reference for graduate courses like the following: soft computing, intelligent pattern recognition, computer vision, applied artificial intelligence, and similar ones. We consider that this book can also be used to get novel ideas for new lines of research, or to continue the lines of research proposed by the authors of the book.

The first contribution by Witold Pedrycz deals with hybridization schemes in architectures of Computational Intelligence. The description of the hybridization schemes is described. While the essence of Computational Intelligence hinges profoundly on the symbiotic use of their underlying technologies (viz. Neuro-computing, granular computing, and predominantly fuzzy sets, and evolutionary optimization), there are several other equally promising development avenues where a hybrid usage of the underlying technologies is worth pursuing. In this study, the author concentrates on the hybrid concepts and constructs available within the realm of Granular Computing (GC). Given the highly diversified landscape of

GC, the author discusses main directions of forming hybrid structures involving individual technologies of information granulation, elaborate on the fundamental communication, interoperability, and orthogonality issues and propose some general ways of building hybrid constructs of GC, which are of immediate interest to system modeling realized in the realm of Computational Intelligence. The author also sheds light on the central role of the concepts of information granularity, information granules and ensuing hybrid constructs. Furthermore the author emphasizes a role of hierarchical modeling that is directly supported by stratified aspect of information granules formed at nested levels of specificity. The central issue of human-centricity of such models is also highlighted.

The contribution by Oscar Montiel et al. deals with an evolutionary optimization of a Wiener model. There exists no standard method for obtaining a nonlinear input-output model using external dynamic approach. In this work, the authors are using an evolutionary optimization method for estimating the parameters of an NFIR model using the Wiener model structure. Specifically, the authors are using a Breeder Genetic Algorithm (BGA) with fuzzy recombination for performing the optimization work. The BGA was selected because it uses real parameters (it does not require any string coding), which can be manipulated directly by the recombination and mutation operators. For training the system, amplitude modulated pseudo random binary signal (APRBS) were used. The adaptive system was tested using sinusoidal signals.

The contribution by Cornelio Posadas-Castillo et al. deals with the synchronization of chaotic neural networks with a generalized Hamiltonian systems approach. In this paper, the authors describe a Generalized Hamiltonian forms approach to synchronize chaotic neural networks unidirectionally coupled. Synchronization is thus between the master and the slave networks with the slave network being given by an observer. In particular, we present two cases of study: the first is a second-order 3×4 CNN array, and the second is a CNN with delay. The chaotic CNNs are used as transmitter and receiver in encrypted information transmission.

The contribution by Oscar Montiel and Oscar Castillo deals with Mediative Fuzzy Logic, which is a novel approach for handling contradictory knowledge. In this paper, the authors propose a novel fuzzy method that can handle imperfect knowledge in a broader way than Intuitionistic Fuzzy Logic does (IFL). This fuzzy method can manage non-contradictory, doubtful, and contradictory information provided by experts, providing a mediated solution, that is why it is called Mediative Fuzzy Logic (MFL). A comparative study of the results given by MFL, IFL and traditional Fuzzy logic (FL) was performed.

The contribution by Ieroham Baruch deals with direct and indirect adaptive neural control of nonlinear systems. A comparative study of various control systems using neural networks was done. The paper proposes to use a Recurrent Trainable Neural Network (RTNN) identifier with backpropagation method of learning. Two methods of adaptive neural control with integral plus state action are applied – an indirect and a direct trajectory tracking control. The first one is the indirect Sliding Mode Control (SMC) with I-term where the SMC is resolved using states and parameters identified by RTNN. The second one is the direct adaptive control with I-term where the adaptive control is resolved by a RTNN controller. The good tracking abilities of both methods are confirmed by simulation results obtained using a MIMO mechanical plant and a 1-DOF mechanical system with friction plant model. The results show that both control schemes could compensate constant offsets and that - without I- term did not.

The contribution by Eduardo Gomez-Ramirez deals with a simple Tuning of fuzzy controllers. The number of applications in the industry using the PID controllers is bigger than fuzzy controllers. One reason is the problem of the tuning, because it implies the handling of a great quantity of variables like: the shape, number and ranges of the membership functions, the percentage of overlap among them and the design of the rule base. The problem is more complicated when it is necessary to control multivariable systems due that the number of parameters. The importance of the tuning problem implies to obtain fuzzy system that decrease the settling time of the processes in which it is applied, or in some cases, the settling time must be fixed to some specific value. In this work a very simple algorithm is presented for the tuning of a fuzzy controller using only one variable to adjust the performance of the system. The results are based on the relation that exists between the shape of the membership functions and the settling time. Some simulations are presented to exemplified the algorithm proposed.

The contribution by Nohe Cazarez et al. deals with a stability and robustness Study from type-1 to type-2 fuzzy logic control. Stability is one of the more important aspects in the traditional knowledge of Automatic Control. Type-2 Fuzzy Logic is an emerging and promising area for achieving Intelligent Control (in this case, Fuzzy Control). In this work, the authors use the Fuzzy Lyapunov Synthesis, as proposed by Margaliot, to build a Lyapunov Stable Type-1 Fuzzy Logic Control System. Then an extension from a Type-1 to a Type-2 Fuzzy Logic Control System was done, ensuring the stability on the control system and proving the robustness of the correponding fuzzy controller.

The contribution by Roberto Sepulveda and Patricia Melin deals with a comparative study of controllers using type-2 and type-1 fuzzy logic. Uncertainty is an inherent part in controllers used for real-world applications.

The use of new methods for handling incomplete information is of fundamental importance in engineering applications. This paper deals with the design of controllers using type-2 fuzzy logic for minimizing the effects of uncertainty produced by the instrumentation elements. Simulations of the type-1 and type-2 fuzzy logic controllers were done to perform a comparative analysis of the systems' response, in the presence of uncertainty.

The contribution by Oscar Castillo et al. deals with evolutionary computing for topology optimization of type-2 fuzzy controllers. In this paper, the authors describe the use of hierarchical genetic algorithms for fuzzy system optimization in intelligent control. In particular, the authors consider the problem of optimizing the number of rules and membership functions using an evolutionary approach. The hierarchical genetic algorithm enables optimization of the fuzzy system design for a particular application. The approach was illustrated with the case of intelligent control in a medical application. Simulation results for this application show that the optimal set of rules and membership functions for the fuzzy system was obtained.

The contribution by Giovanni Pazienza et al. deals with decision trees and CBR for the Navigation System of a CNN-based Autonomous Robot. In this paper, the authors present a navigation system based on decision trees and CBR (Case-Based reasoning) to guide an autonomous robot. The robot has only real-time visual feedback, and the image processing is performed by CNNs to take advantage of the parallel computation. The approach was validated by successfully testing the system on a SW simulator.

The contribution by Arnulfo Alanis et al. deals with intelligent agents in distributed fault tolerant systems. Intelligent Agents have originated a lot of discussion about what they are, and how they are different from general programs. The authors describe in this paper a new paradigm for intelligent agents. This paradigm helped us deal with failures in an independent and efficient way. The authors proposed three types of agents to treat the system in a hierarchic way. A new way to visualize fault tolerant systems (FTS) is proposed, in this paper with the incorporation of intelligent agents, which as they grow and specialized create the Multi-Agent System (MAS). The MAS contains a diversified range of agents, which depending on the perspective will be specialized or evolutionary (from our initially proposal) they will be specialized for the detection and possible solution of errors that appear in an FTS). The initial structure of the agent is proposed in [1] and it is called a reflected agent with an internal state and in the Method MeCSMA [2].

The contribution by Mahmoud Tarokh deals with an approach for genetic path planning with fuzzy logic adaptation for rovers traversing rough terrain. The paper develops a genetic algorithm approach to path planning for a mobile robot operating in rough environments. Path planning consists of a description of the environment using a fuzzy logic framework, and a two-stage planner. A global planner determines the path that optimizes a combination of terrain roughness and path curvature. A local planner uses sensory information, and in case of detection of previously unknown and unaccounted for obstacles, performs an on-line planning to get around the newly discovered obstacle. The adaptation of the genetic operators is achieved by adjusting the probabilities of the genetic operators based on a diversity measure of the population and traversability measure of the path. Path planning for an articulate rover in a rugged Mars terrain is presented to demonstrate the effectiveness of the proposed path planner.

The contribution by Ricardo Guerra et al. describes chattering attenuation using linear-in-the-parameter neural nets in variable structure control of robot manipulators with friction. Variable structure control is a recognized method to stabilize mechanical systems with friction. Friction produces non-linear phenomena, such as tracking errors, limit cycles, and undesired stick-slip motion, degrading the performance of the closed-loop system. The main drawback of variable structure control is the presence of chattering, which is not suitable in mechanical systems. In this paper, the authors design a variable structure controller complemented with Linear-in-the-Parameter neural nets to attenuate chattering. Experimental validation applied to a three degree of freedom robot mechanical manipulator is shown to support the results.

The contribution by Selene Cardenas et al. describes tracking control for a unicycle mobile robot using a fuzzy logic controller. The authors develop a tracking controller for the dynamic model of unicycle mobile robot by integrating a kinematic controller and a torque controller based on Fuzzy Logic Theory. Computer simulations are presented confirming the performance of the tracking controller and its application to different navigation problems.

The contribution by Julian Garibaldi et al. describes intelligent control and planning of autonomous mobile robots using fuzzy logic and genetic algorithms. This paper describes the use of a Genetic Algorithm (GA) for the problem of offline point-to-point autonomous mobile robot Path Planning. The problem consist of generating "valid" paths or trajectories, for an holonomic robot to use to move from a starting position to a destination across a flat map of a terrain, represented by a two dimensional grid, with obstacles and dangerous ground that the Robot must evade. This means that the GA optimizes possible paths based on two criteria: length and difficulty.

The contribution by Pilar Gomez-Gil describes the role of neural networks in the interpretation of antique handwritten documents. The need for accessing information through the web and other kind of distributed media makes it mandatory to convert almost every kind of document to a digital representation. However, there are many documents that were created long time ago and currently, in the best cases, only scanned images of them are available, when a digital transcription of their content is needed. For such reason, libraries across the world are looking for automatic OCR systems able to transcript that kind of documents. In this work the authors describe how Artificial Neural Networks can be useful in the design of an Optical Character Recognizer able to transcript handwritten and printed old documents. The properties of Neural Networks allow this OCR to have the ability to adapt to the styles of handwritten or antique fonts. Advances with two prototype parts of such OCR are presented.

The contribution by Thompson Sarkodie-Gyan describes object recognition using fuzzy inferential reasoning. This paper introduces a vision-based pattern recognition scheme for the identification of very high tolerances of manufactured industrial objects. An image-forming device is developed for the generation and the capture of images/silhouettes of the components. A simple but effective feature extraction algorithm is employed to produce distinguishable features of the components in question. Radial basis function (RBF) based membership functions are used as classifiers for the pattern classification. For the decision making process, a fuzzy logic based inferential reasoning algorithm is implemented for the approximate reasoning scheme.

The contribution by Olivia Mendoza and Patricia Melin describes the fuzzy Sugeno integral as a decision operator in the recognition of images with modular neural networks. The authors describe the implementation of the Fuzzy Sugeno Integral formulas for integration of responses in modular neural networks. In this work the authors illustrate their approach with modular neural networks for image recognition, using images divided in parts. The Fuzzy Sugeno Integral was used to make a final decision. Simulation results show that the approach has potential application.

The contribution by Patricia Melin et al. describes modular neural networks and fuzzy Sugeno integral for pattern recognition for the case of human face and fingerprint. The authors describe in this paper a new approach for pattern recognition using modular neural networks with a fuzzy logic method for response integration. A new architecture for modular neural networks for achieving pattern recognition in the particular case of human faces and fingerprints is proposed. Also, the method for achieving response integration is based on the fuzzy Sugeno integral with some modifications. Response integration is required to combine the outputs of all the modules in the modular network. The authors applied the new approach for fingerprint and face recognition with a real database from students of our institution.

The contribution by J. A. Ruz-Hernandez et al. describes optimal training for associative memories with application to fault diagnosis in fossil electric power plants. In this contribution, the authors discuss a new synthesis approach to train associative memories, based on recurrent neural networks. They propose to update the weight vector as the optimal solution of a linear combination of support patterns. The proposed training algorithm maximizes the margin between the training patterns and the decision boundary. This algorithm is applied to the synthesis of an associative memory, for fault diagnosis in fossil electric power plants. The scheme is evaluated via a full scale simulator to diagnose the main faults occurred in this kind of power plants.

The contribution by F. Rivero-Angeles and Eduardo Gomez-Ramirez describes the acceleration output prediction of buildings using polynomial artificial neural networks. Severe earthquake motions could make civil structures to undergo hysteretic cycles and crack or yield their resistant elements. The present research proposes the use of a polynomial artificial neural network to identify and predict, on-line, the behavior of such non-linear systems. Predictions are carried out first on theoretical hysteretic models and later using two real seismic records acquired on a 24-story concrete building in Mexico City. Only two cycles of movement are needed for the identification process and the results show fair prediction of the acceleration output.

The contribution by Ileana Leal and Patricia Melin describes time series forecasting of tomato prices in Mexico using modular neural networks and Parallel Processing. In this paper, the authors give a brief explanation of the concepts of Time Series, the Neural Networks, the Modular Neural Networks, and Parallelism is given. Modular Neural Networks and Parallel Processing are used for Time Series Forecasting of the Tomato Price in Mexico. The modular neural network was implemented in a parallel architecture for improving the accuracy and efficiency of the obtained results.

The contribution by Patricia Melin et al. describes modular neural networks with fuzzy Sugeno integration applied to time series prediction. The authors describe in this paper the application of several neural network architectures to the problem of simulating and predicting the dynamic behavior of complex economic time series. The authors use several neural network models and training algorithms to compare the results and decide at the end, which one is best for this application. The authors also compare the simulation results with the traditional approach of using a statistical model. In this case, real time series of prices of consumer goods were used to test the models. Real prices of tomato and green onion in the U.S. show complex fluctuations in time and are very complicated to predict with traditional statistical approaches.

The contribution of Kacprzyk, Zadrożny and Wilbik presents the use of the Sugeno integral as a means for a fuzzy linguistic quantifier based aggregation to the extraction of trends in time series data. The authors start with the use of the well-known Sklansky and Gonzalez algorithms to derive a piecewise linear approximation of time series data. Then, the concept of classic Yager's linguistic summary of a data(base) is employed to derive linguistic description of trends in time series data. Two basic types of linguistic descriptions (summaries) are proposed that refer to the frequency of occurrence and duration of trends. As opposed to the classic Zadeh's fuzzy logic based calculus of linguistically quantified propositions employed in Yager's approach, the authors propose here the use of the Sugeno integral which provides more intuitively appealing results.

We end this preface of the book by giving thanks to all the people who have help or encourage us during the making of this book. We would like to thank our colleagues working in Soft Computing, which are too many to mention each by their name. Of course, we need to thank our supporting agencies for their help during this project. We have to thank our institutions for always supporting our projects. Finally, we thank our families for their continuous support during the time that we spend in this project.

Tijuana, Mexico	*Oscar Castillo*
Tijuana, Mexico	*Patricia Melin*
Warsaw, Poland	*Janusz Kacprżyk*
Calgary, Canada	*Witold Pedrycz*
July 2006	

Contents

Part I Theory

Part II Intelligent Control Applications

Part IV Pattern Recognition Applications

Part V Time Series and Diagnosis

Part I Theory

Hybridization Schemes in Architectures of Computational Intelligence

Witold Pedrycz

Department of Electrical & Computer Engineering, University of Alberta, Edmonton, Canada, & Systems Research Institute, Polish Academy of Sciences, Warsaw, Poland
[pedrycz@ee.ualberta.ca]

Abstract. While the essence of Computational Intelligence hinges profoundly on the symbiotic use of their underlying technologies (viz. neurocomputing, granular computing, and predominantly fuzzy sets, and evolutionary optimization), there are several other equally promising development avenues where a hybrid usage of the underlying technologies is worth pursuing. In this study, we concentrate on the hybrid concepts and constructs available within the realm of Granular Computing (GC). Given the highly diversified landscape of GC, we discuss main directions of forming hybrid structures involving individual technologies of information granulation, elaborate on the fundamental communication, interoperability, and orthogonality issues and propose some general ways of building hybrid constructs of GC which are of immediate interest to system modeling realized in the realm of Computational Intelligence. We also shed light on the central role of the concepts of information granularity, information granules and ensuing hybrid constructs. Furthermore we emphasize a role of hierarchical modeling that is directly supported by stratified aspect of information granules formed at nested levels of specificity. The central issue of human-centricity of such models is also highlighted.

1. Introductory Comments

With the rapidly growing complexity of systems encountered today in various disciplines, it becomes evident that new developments need to address the important and commonly present issues of efficient human-centricity of pursuits emerging within a specific domain under investigation. The important facet of human-centricity and bi-directional efficient human-system communication comes hand in hand with the omnipresent concept of abstraction and information granularity.

W. Pedrycz: *Hybridization Schemes in Architectures of Computational Intelligence*, StudFuzz **208**, 3–20 (2007)
www.springerlink.com

Computational Intelligence dwells on symbiotic links between fuzzy sets (or information granules, in general), neurocomputing and evolutionary optimization. This facet of synergy is well known, investigated in depth and carefully documented. We have encountered numerous architectures of CI that take full advantage of the fully orchestrated usage of the contributing technologies of the CI. The reader may refer to various neuro-fuzzy systems that are commonly present in control, classification, data mining and other domains. Interestingly, another equally interesting facet of synergy within the CI realm remains far less explored and documented. It concerns the knowledge-oriented aspects of the CI that are inherently associated with Granular Computing. Fuzzy sets or rough sets are regarded to be the key frameworks within which information granules are constructed and processed.

In this study, we investigate the key features of information granules and elaborate on the underlying processes of information granulation. Fundamental formalisms (including fuzzy sets, interval analysis and rough sets) applied there are presented in a succinct manner. A careful comparative analysis is offered along with a presentation of some mechanisms of interoperability and communication. It is shown how the mechanisms of knowledge-based clustering give rise to a suite of algorithms aimed at the design of information granules. We stress how such algorithms help capture the nature of data and incorporate any domain knowledge that becomes available in the context of the given problem. We also deliver a number of observations about the development of granular models and discuss them in the context of complexity handling as being supported by constructs of information granules.

The paper is organized in the following manner. First, in Section 2, we focus on the human-centric orientation which becomes an evident trend in intelligent systems and highlight its main points. In the sequel, we elaborate on the main formal models of information granules with a particular emphasis on shadowed sets which form a interesting and operationally justifiable bridge between fuzzy sets and interval analysis (Section 3 and 4). Communication and interoperability mechanisms are covered in Section 5. Then we show the role played by fuzzy clustering and knowledge-based clustering in the buildup of information granules. It is stressed how various formats of domain knowledge could be captured through a suitable refinement of the underlying objective function guiding the clustering mechanisms. Modeling aspects, both architectural and conceptual are investigated in Section 7 and 8.

2. Human-Centric Orientation in Intelligent Systems

The rapidly growing complexity of systems and processes we encounter every day and intend to model in an efficient way brings new challenges. One of the very much visible tendencies we encounter today comes with high expectations and a genuine need for making systems predominantly human centric. In essence, this implies that one expects to communicate with the system in a seamless manner, formulate requests and receive interpretable and meaningful results. Typically, the higher the complexity of the domain in which the system functions, the higher the expectations with regard to efficient communication. In the areas such as management, finances, decision-making, etc., this factor of human centricity becomes very visible. Human-centricity is a highly desirable quality yet its operational definition and the subsequent realization requires a plethora of conceptual developments and a suite of supporting algorithms. The effective two-way communication is a key to the success of constructs of Computational Intelligence, in particular if we are concerned how computing activities become invoked. For instance, the mechanisms of relevance feedback that become more visible in various interactive systems hinge upon the well-established and effective human-centric schemes of processing. It is expected that they could effectively accept user hints and directives and afterwards release results in a highly comprehensible format.

Abstraction and information granularity along with information granules (being the ultimate constructs arising within this setting) are the key components that vitally support a realization of the human-centric aspects of the systems. This has led us to the development of various frameworks of information granules.

3. Selected Formal Models of Information Granules

Let us briefly present the major formalisms used in information granulation. Our objective is to emphasize their key features and demonstrate the genuine diversity that exists in the area of granular computing. Further on, we pay more attention to the buildup of shadowed sets illustrating the tendency of offering new interpretation capabilities to fuzzy sets.

Sets and interval analysis The two-valued world of sets and interval analysis [7][8][14] ultimately dwells upon a collection of intervals of real numbers, say [a,b], [c,d],...etc or their Cartesian products. Conceptually, sets are rooted in a two-valued logic with its fundamental predicate of membership ($\in$). We emphasize here an important isomorphism between the structure of two-valued logic endowed with truth values (false-true) and set theory along with its characteristic functions. The interval analysis is a cornerstone of reliable computing that is ultimately associated with digital computing where any variable is expressed at some finite level of accuracy (being implied by the fixed number of bits that are available to represent numbers). The Boolean character of the constructs (sets, intervals, etc.) helps capture possible ranges of values of system variables yet makes the elements that belong to some set completely indistinguishable.

Fuzzy sets Fuzzy sets offer an important and conceptually different formalism of representing information granules [15][16]. It is offered in the language of constructs with partial membership so that we can discriminate between elements that are "typical" to the concept and those of the borderline character. Information granules such as *high* speed, *warm* weather, *fast* car are examples of information granules falling under this category. We cannot specify a single, well-defined element that forms a solid border between full belongingness and full exclusion. Fuzzy sets with their soft transition boundaries are an ideal vehicle to capture the notion of partial membership. In this way they realize a non-Aristotelian view of reality emphasized even before the inception of fuzzy sets, just to mention Max Black and Alfred Korzybski. Obviously, the 3-valued and multivalued Lukasiewicz logics build the solid foundations of the non-binary concepts. When looking at fuzzy sets from the computational perspective, the smoothness and non-disruptive character of membership functions are highly beneficial in forming and solving various optimization problems.

Shadowed sets Fuzzy sets help describe and quantify concepts with continuous boundaries. By introducing a certain α-cut, we convert a fuzzy set into a set. By choosing the threshold level (α) that is high enough, we admit elements whose membership grades are sought meaningful (as being viewed from the standpoint of the imposed threshold). The use of a certain α-cut transforming a fuzzy set into some set leads to a fairly misleading impression that any fuzzy set could be made equivalent to some set. This point of view is highly deceptive. In essence, by building any α-cut we elevate some membership grades to 1 (full membership) and reduce others to 0 (total exclusion). Surprisingly, no account is taken for the distribution of elements with partial membership so that this effect cannot be quantified in the resulting construct. The idea of shadowed sets [11][13], see also [4][5] is aimed at alleviating this problem by forming regions of complete ignorance about membership grades. In essence, a shadowed set $A^{\sim}$ induced by given fuzzy set A defined in **X** is an interval-valued set in **X**

which maps elements of the space into 0, 1, and the unit interval [0,1]. Formally, A˜ is a mapping A˜ : **X** → { 0, 1, [0,1]. 0 denotes complete exclusion from A˜, 1 stands for complete inclusion in A˜. A˜(x) that is equal to [0,1] represents a complete ignorance – nothing is known about membership of x in A˜: we *neither* confirm its belongingness to A˜ *nor* commit to its exclusion. An example of a shadowed set is illustrated in Figure 1.

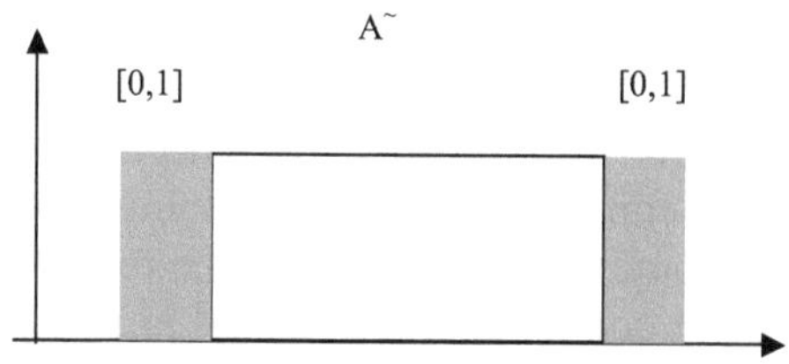

Fig. 1. An example of a shadowed set A˜; observe "shadows" produced at the edges of the characteristic function

Rough sets The fundamental concept represented and quantified by rough sets [9] is the one concerning a description of a given concept in the language of a certain collection (vocabulary) of generic terms being agreed upon in advance to be regarded as the generic ones. Depending upon this collection relative to the concept, we can encounter situations where it cannot be possible to fully and uniquely describe the concept. This situation may give rise to an approximate, or better to say, a rough description of the concept. An example of the discrepancy of the description yielding the lower and upper bound of the description is displayed in Figure 2.

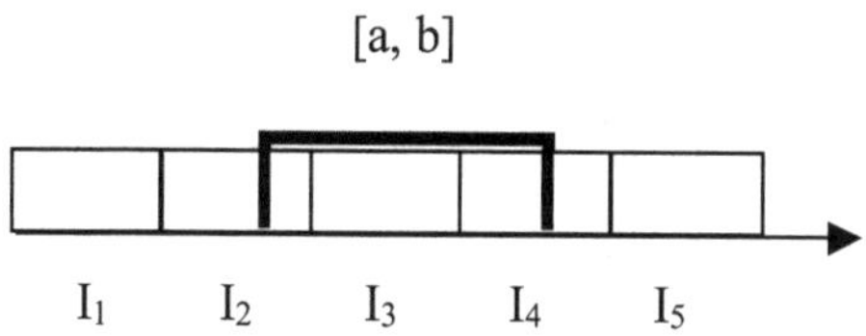

Fig. 2. Concept (set) [a, b] represented in the language of uniformly distributed intervals; note the emergence of the upper and lower bound contributing to their representation

Given the fact that shadowed sets form an interesting and algorithmically appealing bridging mechanism between fuzzy sets and interval analysis, we focus on a way in which they are constructed.

4. The Development of Shadowed Sets

Accepting the point of view that shadowed sets are algorithmically implied (induced) by some fuzzy sets, we are interested in the transformation mechanisms translating fuzzy sets into the corresponding shadowed sets. The underlying concept is the one of uncertainty condensation or "localization". While in fuzzy sets we encounter intermediate membership grades located in-between 0 and 1 and distributed practically across the entire space, in shadowed sets we "localize" the uncertainty effect by building constrained and fairly compact shadows. By doing so we could remove (or better to say, re-distribute) uncertainty from the rest of the universe of discourse by bringing the corresponding low and high membership grades to zero and one and then compensating these changes by allowing for the emergence of uncertainty regions. This transformation could lead to a certain optimization process in which we complete a total balance of uncertainty.

To illustrate this optimization, let us start with a continuous, symmetric, unimodal, and normal membership function A. In this case we can split the problem into two tasks by considering separately the increasing and decreasing portion of the membership function, Figure 3.

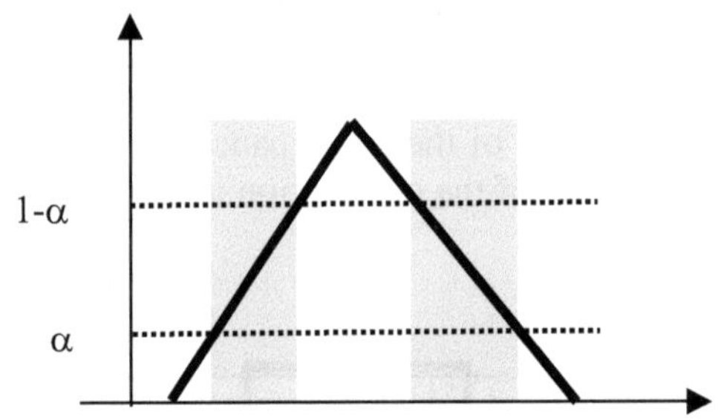

Fig. 3. The concept of a shadowed set induced by some fuzzy set; note the range of membership grades (located between α and 1-α) generating a shadow

For the increasing portion of the membership function, we reduce low membership grades to zero, elevate high membership grades to one and compensate these changes (which in essence lead to an elimination of partial membership grades) by allowing for a region of the shadow where there are no specific membership values assigned but we admit the entire unit interval as feasible membership grades. Computationally, we form the following balance of uncertainty preservation that could be symbolically expressed as

Reduction of membership + Elevation of membership = shadow (1)

Again referring to Figure 3 and given the membership grades below α and above $1-\alpha$, $\alpha \in (0, ½)$, we express the components of the above relationship in the form (we assume that all integrals do exist)
Reduction of membership (low membership grades are reduced to zero)

$$\int_{x:A(x)\le\alpha} A(x)dx \tag{2}$$

Elevation of membership (high membership grades elevated to 1)

$$\int_{x:A(x)\ge 1-\alpha} (1-A(x))dx \tag{3}$$

Shadow

$$\int_{x:\alpha<A(x)<1-\alpha} dx \tag{4}$$

The minimization of the absolute difference

$$V(\alpha) = \left| \int_{x:A(x)\le\alpha} A(x)dx + \int_{x:A(x)\ge 1-\alpha} (1-A(x))dx - \int_{x:\alpha<A(x)<1-\alpha} dx \right| \tag{5}$$

completed with respect to α is given in the form of the following optimization problem

$$\alpha_{opt} = \arg\min_{\alpha} V(\alpha) \tag{6}$$

where $\alpha \in (0, ½)$. For instance, when dealing with triangular membership function (and it appears that the result does not require the symmetry requirement), the optimal value of α is equal to $\sqrt{2}-1 \approx 0.4142$ [11]. For the parabolic membership functions, the optimization leads to the value of α equal to 0.405.

5. Communication and Interoperability Mechanisms in Granular Computing

Information granules come with a significant level of diversity indicating a wealth of conceptual and algorithmic approaches available in the area. An example of this diversity is illustrated in Figure 4. Let us note that there are a number of links that help associate various granular constructs. Some of them are quite often used. Say, we typically use α-cuts to articulate in some selective manner fuzzy sets in the language of sets (intervals). The choice of the specific value of α is quite critical to capture the essence of the fuzzy set. Various generalizations of fuzzy sets are worth stressing; those include type-2 fuzzy sets and fuzzy sets of higher order. We can appreciate interesting linkages and ensuing generalization emerging at the junction of fuzzy sets and rough sets (in the form of fuzzy-rough and rough-fuzzy sets).

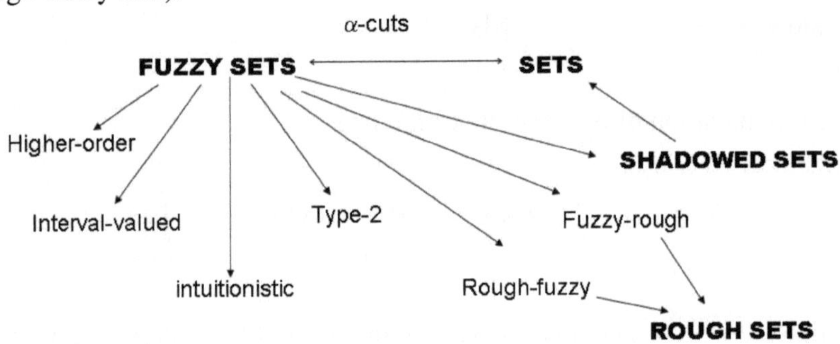

Fig. 4. A roadmap of linkages in various formal platforms of Granular Computing

The apparent diversity present within the realm of Granular Computing becomes crucial when it comes to various interactions and establishing collaborative linkages between the autonomous systems exploiting various mechanisms of information granules (say, fuzzy systems, rough models, fuzzy-rough models, interval models and alike), refer to Figure 5. Likewise one may encounter various users with their specific preferences as to the usage of specific machinery of granular computing.

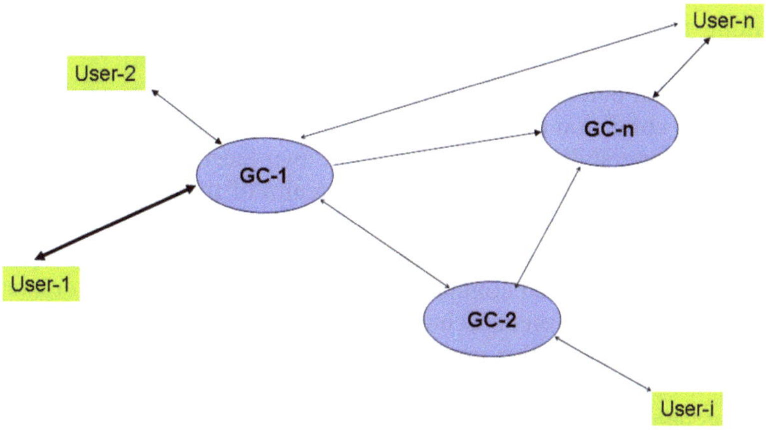

Fig. 5. Collaborative linkages between granular models established within the frameworks of various formalisms of Granular Computing (GC-1, GC-2, etc) and communities of users

Different formal models of information granules support unique points of view when it comes to knowledge representation, organization and its utilization. It is very likely that in problem solving we could be faced with several fairly heterogeneous environments for which we would like to establish a certain level of interoperability and communication. In essence, one could distinguish between the two essential and orthogonal coordinates of the space in which the communication between various models has to be realized, see Figure 6.

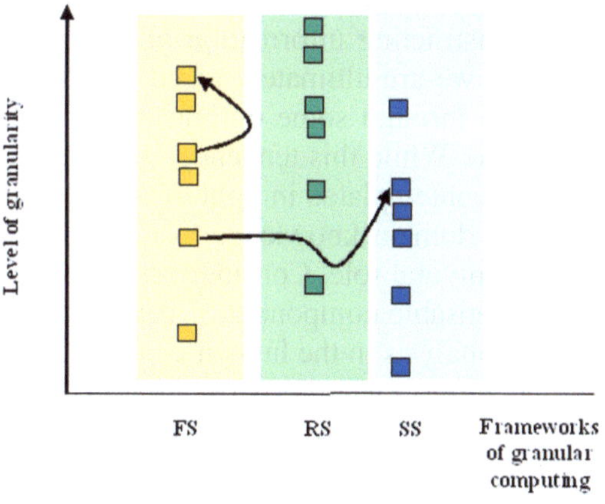

Fig. 6. Two fundamental and orthogonal coordinates of interoperability of information granules (formal frameworks of information granulation and levels of granularity–abstraction)

Along the y- coordinate, we encounter different levels of granularity (depending upon the formalism of granulation, it could be quantified through sigma-count, cardinality, etc). The x- coordinate of this graph identifies a collection of formal frameworks of granulation. When designing the interoperability mechanisms, these two facets of the problem need to be considered. When staying within a realm of the same formal framework and moving across different levels of granularity this does not require any transformation yet one has to consider a mechanism to quantify the effectiveness of such communication. When forming the interoperability links between different technologies of granular information, we have to proceed carefully as this implies a need to follow the principles of transformation of information granules between the frameworks so that they could be deemed equivalent. For instance, a fuzzy set transformed into a construct in the formal setting of sets (interval) calls for the determination of its optimal α-cut.

6. The Development of Information Granules as Human-Centric Constructs

Information granules are reflective of some abstraction processes using which we attempt to hide some unnecessary details and focus our modeling or decision-making activities on the most essential and dominant facets of the problem. Thus building information granules that are reflective of the existing sources of knowledge and available data is of paramount importance. Clustering and fuzzy clustering are well known algorithmic tools that are aimed at constructing information granules [2][3][12][13]. In all clustering algorithms we are ultimately faced with the problem of revealing structures in data through some optimization driven by the available experimental evidence. While this tendency is evident, a shift of this data-oriented paradigm is contemplated in light of the fact that not only the data are essential but any domain knowledge available from users, designers, experts has to play a pivotal role. Considering domain knowledge to be an important and indispensable component of data analysis, it becomes clear that it positions data analysis in the human-centric perspective. To be more descriptive, we may refer to pursuits carried out in this way as knowledge-based clustering [13]. There are two fundamental issues that need to be addressed in the setting of the knowledge-based clustering: (a) what type of knowledge-based hints could be envisioned, and (b) how they could be incorporated as a part of the optimization scheme. More specifically, what needs to be done with regard to the possible augmentation of the objective function and how the ensuing optimization scheme has to be augmented to efficiently cope with the modified objective function.

6. 1. Fuzzy Clustering and Mechanisms of Human-Oriented Guidance

In what follows, we highlight several commonly encountered alternatives that emerge when dealing with domain knowledge and building formal mechanisms, which reformulate the underlying objective function. We focus on two quite commonly encountered formats of domain knowledge being available in this setting that is labeling of some selected data points and assessments of proximity of some pairs of data.

To make a consistent exposure of the overall material and assure linkages with the ensuing optimization developments, we confine ourselves to the Fuzzy C-Means (FCM) [2] governed by the following objective function

$$Q = \sum_{i=1}^{c} \sum_{k=1}^{N} u_{ik}^{m} \, \| \mathbf{x}_k - \mathbf{v}_i \|^2 \qquad (7)$$

where $\mathbf{x}_k$ denotes an multidimensional data point (pattern) , $\mathbf{v}_i$ is an i-th prototype and $U=[u_{ik}]$, i=1, 2, …, c; k=1, 2,…,N stands for a partition matrix. Moreover $\|.\|$ denotes a certain distance function and "m" is a fuzzification coefficient; m>1.0. The minimization of (1) is realized with respect to the partition matrix and the prototypes. The optimization scheme and all specific features of the minimization of Q are well reported in the literature.

6.2. Mechanisms of Partial Supervision

The effect of partial supervision involves a subset of labeled data, which come with their pre-assigned values of class membership [1][3][10][13]. These knowledge-based hints have to be included into the objective function and reflect that some patterns have been labeled. In the optimization, we expect that the structure to be discovered conforms to the membership grades already provided for these selected patterns. More descriptively, we can treat the labeled patterns to form a grid of "anchor" (navigation) points using which we attempt to discover the entire structure in the data set. Put it differently, such labeled data should help us navigate a process of revealing the structure. The generic objective function shown in the form (7) has to be revisited and expanded so that the structural information (labeled data points) is taken into consideration. While there could be different alternatives possible with this regard, we consider the following additive expansion of the objective function, cf. also [10]

$$Q = \sum_{i=1}^{c} \sum_{k=1}^{N} u_{ik}^{2} \| \mathbf{x}_k - \mathbf{v}_i \|^2 + \alpha \sum_{i=1}^{c} \sum_{k=1}^{N} (u_{ik} - f_{ik} b_k)^2 \| \mathbf{x}_k - \mathbf{v}_i \|^2 \tag{8}$$

The first term is aimed at the discovery of the structure in data and is the same as in the standard FCM. The second term (weighted by some positive scaling factor α) addresses the effect of partial supervision. It requires careful attention because of the way in which it has been introduced into the objective function and the role it plays during its optimization. There are two essential data structures containing information about the initial labeling process (labeled data points)

- the vector of labels, denoted by $\mathbf{b}=[b_1\ b_2 \ldots b_N]^T$. Each pattern $\mathbf{x}_k$ comes with a Boolean indicator: we assign b_k equal to1 if the pattern has been already labeled and $b_k = 0$ otherwise.
- The partition matrix $F = [f_{ik}]$, i=1, 2,...,c; k=1,2,...N which contains membership grades assigned to the selected patterns (already identified by the nonzero values of b). If $b_k = 1$ then the corresponding column shows the provided membership grades. If $b_k = 0$ then the entries of the corresponding k-th column of F do not matter; technically we could set them up to zero.

The nonnegative weight factor (α) helps set up a suitable balance between the supervised and unsupervised mode of learning. When $\alpha = 0$ then we end up with the standard FCM.

6.3. Clustering with Proximity Hints

The concept of proximity is one of the fundamental notions useful in assessing the mutual dependency between membership values occurring for two patterns. Consider two patterns with their corresponding columns in the partition matrix denoted by "k" and "l", that is $\mathbf{u}_k$ and $\mathbf{u}_l$, respectively. The proximity between them, Prox($\mathbf{u}_k$, $\mathbf{u}_l$), is defined in the form [2][13]

$$\text{Prox}(\mathbf{u}_k, \mathbf{u}_l) = \sum_{i=1}^{c} \min(u_{ik}, u_{il}) \tag{9}$$

Note that the proximity function is symmetric and returns 1 for the same pattern (k=l). In virtue of the properties of any fuzzy partition matrix we obtain

$$\mathrm{Prox}(\mathbf{u}_k, \mathbf{u}_l) = \sum_{i=1}^{c} \min(u_{ik} u_{il}) = \mathrm{Prox}(\mathbf{u}_l, \mathbf{u}_k)$$
$$\mathrm{Prox}(\mathbf{u}_k, \mathbf{u}_k) = \sum_{i=1}^{c} \min(u_{ik} u_{ik}) = 1 \tag{10}$$

The incorporation of the proximity-based knowledge hints leads to the two-optimization processes. The first one is the same as used to optimize the original objective function. In the second one, we reconcile the proximity hints with the proximity values induced by the partition matrix generated by the generic FCM. Denote the proximity values delivered by the user as $\mathrm{Prox}[k_1, k_2]$ where k_1 and k_2 are the indexes of the data points for which the proximity value is provided. Obviously these hints are given for some pairs of data so to emphasize that we introduce a Boolean predicate $B[k_1, k_2]$

$$B[k_1,k_2] = \begin{cases} 1, & \text{if the value of } \mathrm{Prox}[k_1,k_2] \text{ has} \\ & \text{been specified for the pair } (k_1,k_2) \\ 0, & \text{otherwise} \end{cases} \tag{11}$$

Note that for any pair of data, the corresponding induced level of proximity that is associated with the partition matrix produced by the FCM. We request that the proximity knowledge-based hints brought in by the designer coincide with the induced proximity values implied by the structure revealed by the FCM on the basis of numeric data. Computationally, we express this requirement by forming the expression (which is a sum of distances between the corresponding values of the proximity values) [13]

$$\sum_{k_1} \sum_{k_2} \| \mathrm{Prox}[k_1,k_2] - \sum_{i=1}^{c} \min(u_{ik_1}, u_{ik_2}) \|^2 B[k_1,k_2] \tag{12}$$

By making changes to the entries of the partition matrix U, we minimize the value of the expression given above thus arriving at some agreement between the data and the domain knowledge. The optimization activities are then organized into two processes exchanging results as outlined in Figure 7. There are two interacting optimization processes. The first one, being driven by data produces some partition matrix. The values of this matrix are communicated to the second optimization process driven by the proximity-based knowledge hints. At this stage, the proximity values induced by the partition matrix are compared with the proximities coming as knowledge hints and (11) is minimized thus giving rise to the new values of the partition matrix U which in turn is communicated to the data driven optimization phase. At this point, this "revised" partition matrix is used to further minimize the objective function following the iterative scheme of the FCM.

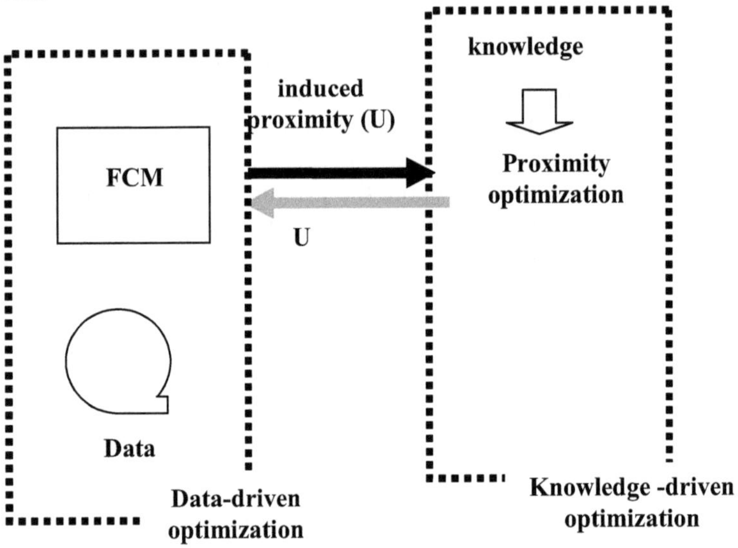

Fig. 7. The optimization data – and knowledge-driven processes of proximity-based fuzzy clustering

7. Modeling with Information Granules

Given the current developments in granular computing and its individual technologies, they are fully reflected in the diversity of models exploiting the individual formalisms of information granules. One could easily point at the growing variety of architectures and related learning schemes found in these areas. Information granules offer an interesting opportunity to form models at different levels of abstraction depending upon the needs of the modeling process, specificity (granularity) of available information, internal format of the model associated with the topology of model and available estimation algorithms. An overall scheme of such modeling showing the main functional modules of the architecture is illustrated in Figure 8.

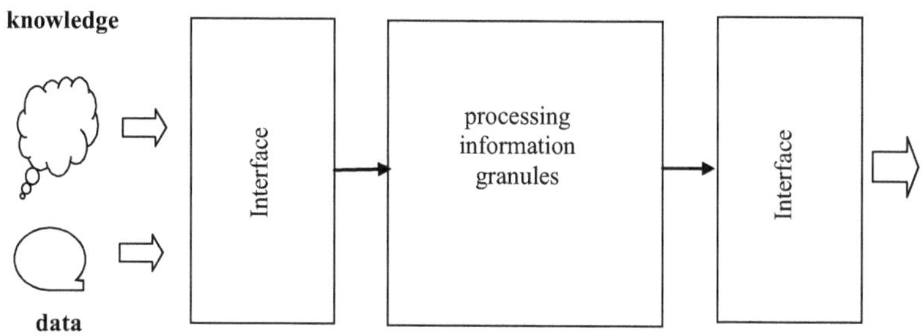

Fig. 8. A general structure of modeling realized in the setting of granular information

The role of interfaces is quite profound in the overall scheme. The input interface supports the development of information granules and could include a number of mechanisms of knowledge-based clustering given the fact that one is provided with numeric data and various knowledge hints coming from domain experts, designers, and potential users. The core processing part situated in the central part of the scheme is focused on processing of information granules and be realized in terms of rule-based computing, logic neurocomputing or any other architecture that is geared towards handling granular information. The output interface brings the results of modeling to the users in the format that is considered as the most suitable for them. In a very special case the interface may convert information granules to some numeric representatives if the numeric character of results is required. The complexity of systems can be alleviated by considering models formed at the higher levels of abstraction (so a lot of details are intentionally left out). The models built at different levels of granularity give rise to their conceptual hierarchy. Information granules offer an effective and algorithmically sound way of forming abstractions. Hence they can be considered as a viable conceptual and computational environment of system modeling.

8. Granular Multimodels

Being fully cognizant of the role of information granules, the diversity of formal frameworks (where this diversity manifests both in terms of the existing formal environments as well as the levels of specificity or generality we can consider several general architectural scenarios. Those are schematically visualized in Figure 9. As evident the two criteria being used here concern homogeneity (or heterogeneity) applied to the formalisms of information granulation or the level of granularity itself.

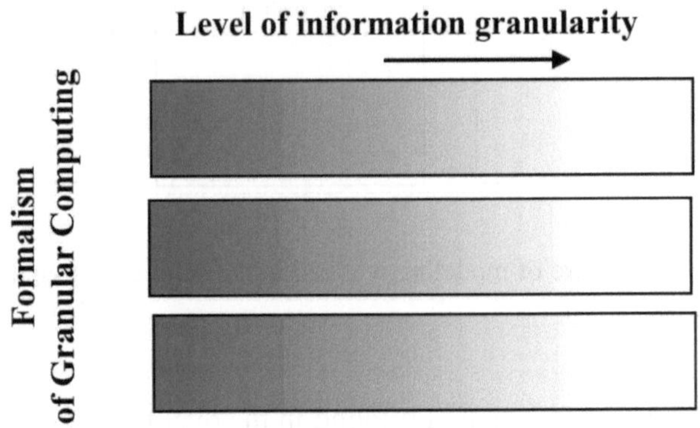

Fig. 9. Generic scenarios of heterogeneity–homogeneity in information granularity and a collection of formalisms of granular computing

While the taxonomy presented above exhibits two dimensions, the quantification of heterogeneity along the line of the formalisms of granular computing. Noticeably the coordinate of the variable level of homogeneity is definitely continuous and there is no well delineated boundary. Let us elaborate on some scenarios in more detail.

<u>Homogeneity of formal frameworks of information granularity – heterogeneity of information granules</u> Formation of a family of models depending upon the already assumed levels of granularity of information. At each level we maintain the same formal framework of Granular Computing. For instance, at each level we may consider the use of fuzzy sets. Given the variable level of granularity, the specificity of the each model in the hierarchy, Figure 10, could be very different. The aggregation of outcomes of these individual models could be completed in several different ways. For instance, one could consider an intersection of the corresponding fuzzy sets formed by the models. The other alternative would be to form higher order constructs such as type-2 fuzzy sets.

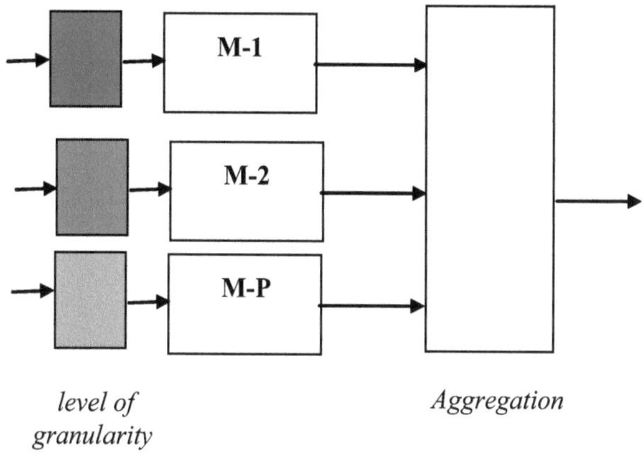

Fig. 10. A hierarchy of granular models guided by sources of data available at different levels of specificity (granularity) and aggregation of their results

Heterogeneity of formal frameworks of information granularity This scenario becomes present when we encounter models built on a basis of different fundamental frameworks in which information granules are formed. The aggregation of the outputs of the model becomes more challenging. Sometimes it becomes more suitable to consider a transformation of information granules from one framework to another so that all models are homogenous in terms of the formalism being used.

9. Conclusions

In this study, we discussed the role of information granules in the setting of Computational Intelligence, outlined the main formal frameworks of granular computing and demonstrated its role in system modeling. The issue of dealing with the heterogeneity of sources of data and knowledge was identified and along this line we presented a new category of knowledge-based fuzzy clustering demonstrating how they support the development of information granules. While we focused on some fundamental concepts and architectures, the detailed algorithmic considerations still deserve careful attention.

Acknowledgments

Support from the Natural Sciences and Engineering Research Council of Canada (NSERC) and Canada Research Chair (CRC) program is gratefully acknowledged.

References

1. A. M. Bensaid, L. O. Hall, J. C. Bezdek, L. P. Clarke, Partially supervised clustering for image segmentation, *Pattern Recognition,* 29, 5, 1996, 859-871.
2. J. C. Bezdek, Pattern Recognition with Fuzzy Objective Function Algorithms, Plenum Press, NY, 1981.
3. J.C.,Bezdek, J. Keller, R. Krishnapuram, N.R. Pal, *Fuzzy Models and Algorithms for Pattern Recognition and Image Processing*, Kluwer Academic Publishers, Boston, 1999.
4. G. Cattaneo and D. Ciucci, An algebraic approach to shadowed sets , *Electronic Notes in Theoretical Computer Science,* vol 82, 2001, no. 4, 1-12.
5. E. Gürkan, I Erkmen, A. M. Erkmen, Two-way fuzzy adaptive identification and control of a flexible-joint robot arm, *Information Sciences,* vol.145, no. 1-2, 13-43.
6. F. Hoppner, F. Klawonn, R. Kruse, T. Runkler, *Fuzzy Cluster Analysis: Methods for Classification, Data Analysis and Image Recognition*, John Wiley, New York,1999.
7. L. Jaulin, M. Kieffer, O. Didrit, E. Walter, *Applied Interval Analysis*, Springer, London, 2001.
8. R. Moore, *Interval Analysis*, Prentice Hall, Englewood Cliffs, NJ, 1966.
9. Z. Pawlak, *Rough Sets. Theoretical Aspects of Reasoning About Data*, Kluwer Academic Publishers, Dordercht, 1991.
10. W. Pedrycz, Algorithms of fuzzy clustering with partial supervision, *Pattern Recognition Letters*, 3, 1985, 13-20.
11. W. Pedrycz, Shadowed sets: representing and processing fuzzy sets, *IEEE Trans. on Systems, Man, and Cybernetics, part B*, 28, 1998, 103-109.
12. W. Pedrycz, Fuzzy clustering with a knowledge-based guidance, *Pattern Recognition Letters,* vol. 25, no. 4, 2004, 469-480.
13. W. Pedrycz, Knowledge-Based Clustering: From Data to Information Granules, J. Wiley, N. York, 2005.
14. M. Warmus, Calculus of approximations, *Bulletin de l'Academie Polonaise des Sciences*, 4, 5, 1956, 253-259.
15. L.A. Zadeh, Fuzzy logic = computing with words, *IEEE Trans. on Fuzzy Systems*, 4, 1996, 103-111.
16. L. A. Zadeh, Toward a generalized theory of uncertainty (GTU) – an outline, *Information Sciences*, 172, 1-2, 2005, 1-40.

Boltzmann Machines Learning Using High Order Decimation

Enric Farguell[1], Ferran Mazzanti[1] and Eduardo Gomez-Ramirez[2]

[1]Dept. d'Electrònica, EALS La Salle, Universitat Ramon Llull, Quatre Camins 2, 08022 Barcelona, Spain {efarguell, mazzanti}@salleurl.edu

[2]Postgrado e Investigación, Universidad La Salle México, Benjamín Franklin 47, Col. Hipódromo Condesa 06140 México D.F., México, egr@ci.ulsa.mx

Abstract. Boltzmann Machines are recurrent and stochastic neural networks that can learn and reproduce probability distributions. This feature has a serious drawback in the exhaustive computational cost involved. In this context, decimation was introduced as a way to overcome this problem, as it builds a smaller network that is able to reproduce exactly the quantities required to update the weights during learning. Decimation techniques developed can only be used in sparsely connected Boltzmann Machines with stringent constraints on the connections between the units. In this work, decimation is extended to any Boltzmann Machine with no restrictions on connections or topology. This is achieved introducing high order weights, which incorporate additional degrees of freedom.

1 Introduction

The Boltzmann Machine (BM) [0, 2] is a stochastic and recurrent neural network based on the Hopfield [3] model that can be taken as a parallel version of the Simulated Annealing Algorithm [4]. However, its stochastic properties give the neural network the great ability of learning and extrapolating probability distributions. Furthermore, it has a strong analogy with the physical magnetic spin glass model, which provides a very strong formal definition of the neural network's dynamics. On the other hand, the Boltzmann Machine has an important drawback that has prevented further usage, this is, learning stage is computationally exhaustive. One alternative to this problem is the Decimation [5, 6] that was introduced as a way to reduce the number of calculations needed at each step of the learning process. Decimatable Boltzmann Machines can be reduced to simpler struc-

E. Fargell et al.: *Boltzmann Machines Learning Using High Order Decimation*, StudFuzz **208**, 21–42 (2007)
www.springerlink.com

tures where analytical expressions for correlations between units, needed to update weights during learning, can be found. However, this method can only be applied to sparsely connected Boltzmann Machines with important constraints on their connections. In this work, we propose a way to decimate any Boltzmann Machine regardless of its topology and number of units. That is accomplished by transforming the original network into a High Order Boltzmann Machine (HOBM) [7].

The structure of the paper is the following: in section 2 we briefly describe the Boltzmann Machine and its advanced model, the High Order Boltzmann Machine, and introduce the general notation used along this work. In section 3 the traditional decimation method is introduced, just to explain the new procedure on section 4, which will be thereafter referred to as *High Order Decimation*. Section 5 is used to provide a graphical example of decimation over a typical structure. Section 6 offers some results with an example on how this method works with the N-bit parity or XOR problem. Finally, section 7 provides some conclusions about the previous results.

2 The Boltzmann Machine

2.1 The Simulated Annealing Algorithm

The Simulated Annealing (SA) algorithm [8] is a powerful, global optimization algorithm which may be considered a numerical implementation of the physical process known as annealing. This process consists on heating up any material until it arrives at a state with a maximum entropy and energy values associated. Once it reaches equilibrium at such temperature, it will be slowly cooled until achieving absolute zero. When this happens, the system will be on its minimum energy and entropy values.

The Simulated Annealing algorithm numerically makes an equivalence between a physical system and a cost function to be optimized, where its possible energetic states are no more than different values for the variables of the cost function. On the other hand, the temperature is introduced as a control parameter T, which stands for a stochastic noise measure. The SA is implemented by fixing a high temperature value where it starts searching over the function and a final temperature value where it is expected to converge on its the global minima of the function. The algorithm performs a stochastic search, starting at a random point and proposing new different points, which will be rejected or accepted according to a Metropolis criteria. Each temperature is associated to a number of iterations that the algorithm will have to be run in order to reach equilibrium. This set of temperatures and iterations is known as *annealing schedule.*

So, for a standard N variables $f(\vec{x}) = f(x_0, x_1, \ldots, x_{N-1})$ function, the algorithm itself is implemented as follows:

- Fix the initial temperature from the annealing schedule.
- Selecting a random starting point $\vec{x}_\alpha = (x_0^\alpha, x_1^\alpha, \ldots, x_{N-1}^\alpha)$.
- Selecting another random point $\vec{x}_\gamma = (x_0^\gamma, x_1^\gamma, \ldots, x_{N-1}^\gamma)$.
- Evaluate the quantity

$$\Delta E = f(\vec{x}_\gamma) - f(\vec{x}_\alpha) , \tag{1}$$

which is the energy gain between the first and the second state.

- Then, if $\Delta E \leq 0$ or

$$e^{-\Delta E / T_k} \geq p , \tag{2}$$

$\vec{x}_\gamma$ is selected as the new searching point, p being a realization of a uniform probability distribution and T the corresponding temperature from the cooling schedule.

- This evaluation is performed as many times as necessary to make the system reach equilibrium, or as stated in the annealing schedule.
- Finally, T_{k+1} is selected and this process is repeated until the annealing schedule ends.

So, if this process is correctly performed, the global minima of the function will be finally set. However, it is not only difficult to find out a suitable annealing schedule, but the algorithm itself is quite time consuming. Hence, there are problems where it is not worth using, given its computational exhaustive requirements.

2.2 Boltzmann Machine Dynamics

The neural network known as Boltzmann Machine has the same dynamics as the Simulated Annealing optimization algorithm. However, the neural network has an own free energy functional associated, known as the Helmholtz free energy

$$F = E - TS , \tag{3}$$

where T is the control parameter usually known as temperature, S represents the entropy of the system and E is the energy functional related to the system. This energy term is easily written in terms of unit states $S_i = \{-1,+1\}$, symmetric weights $w_{ij} = w_{ji}$ and external bias h_i in the form

$$E = -\frac{1}{2}\sum_{i,j} w_{ij} S_i S_j - \sum_i h_i S_i \ , \tag{4}$$

although biases are often replaced by two-body weights w_{ij} (weights which connect any pair of units) connecting every unit in the network to an external one clamped to 1 [0]. This functional has to be worked to fit any cost function that one wishes to optimize, as explained on Ref. [9]. However, when the Boltzmann Machine is working as a neural network this energy functional is evaluated with the Simulated Annealing algorithm until the neural network reaches equilibrium at a given final T which is not zero (as it would be if implementing the SA). At this moment, the Helmholtz free energy becomes zero and the neural network reaches a stationary probability distribution, which is known as Boltzmann probability distribution

$$p_\alpha = \frac{e^{-E_\alpha/T}}{Z} , \tag{5}$$

Z is a normalization term called the *partition function*

$$Z = \sum_\gamma e^{-E_\gamma/T} \ , \tag{6}$$

which ensures that the total probability sum over all possible energy states is one. On the other hand

$$E_\alpha = -\frac{1}{2}\sum_{i,j} w_{ij} S_i^\alpha S_j^\alpha - \sum_i h_i S_i^\alpha \ . \tag{7}$$

The previous equation is the energy of a generic energy state α characterized by the binary value of the different S_i units in the network.

To sum up, the process of bringing the system to equilibrium is computationally exhaustive as it has to be done with the help of the Simulated Annealing algorithm, where a sequence of decreasing temperatures is used to cool the system down to a desired final temperature T. In this algorithm, the equilibrium has to be reached at each intermediate temperature and the number of iterations required grows significantly with the number of units in the net.

2.3 Learning in Boltzmann Machines

Learning in a BM is done by minimization of the Kullback-Leibler distance [11] between the actual probability distribution and the one to be learnt. For a BM with different input and output units this magnitude is expressed as

$$G = \sum_{\gamma} P_{\gamma} \sum_{\alpha} R_{\alpha|\gamma} \ln \frac{R_{\alpha|\gamma}}{P_{\alpha|\gamma}} , \tag{8}$$

where $P_{\alpha|\gamma}$ is the Boltzmann probability distribution of finding a state α in the output units when a state γ has been set in the input units and, in consequence, it depends on the weights and biases of the neural network given by Eq. (4). $R_{\alpha|\gamma}$ is the desired probability distribution to be learnt for the same input and output, and is given by the user as an input-output vectors set. P_{γ} is the probability of having the input clamped to state γ regardless of the output and, finally, G is positive whenever $P_{\alpha|\gamma} \neq R_{\alpha|\gamma}$ and zero when $P_{\alpha|\gamma} = R_{\alpha|\gamma}$. This means that we can use gradient descent on G to find the update rule for both weights and biases [0]

$$\Delta w_{ij} = -\eta \frac{\partial G}{\partial w_{ij}} = \eta \sum_{\gamma} P_{\gamma} \sum_{\alpha} \frac{R_{\alpha|\gamma}}{P_{\alpha|\gamma}} \frac{\partial P_{\alpha|\gamma}}{\partial w_{ij}}$$

$$= \eta \sum_{\gamma} P_{\gamma} \sum_{\alpha} \frac{R_{\alpha|\gamma}}{P_{\alpha|\gamma}} \frac{\frac{\partial e^{-E_{\alpha|\gamma}}}{\partial w_{ij}} Z - e^{-E_{\alpha|\gamma}} \frac{\partial Z}{\partial w_{ij}}}{Z^2} , \tag{9}$$

so we need the following derivatives

$$\frac{\partial e^{-E_{\alpha|\gamma}}}{\partial w_{ij}} = -\frac{\partial E_{\alpha|\gamma}}{\partial w_{ij}} e^{-E_{\alpha|\gamma}} = S_i^{\alpha|\gamma} S_j^{\alpha|\gamma} e^{-E_{\alpha|\gamma}} , \tag{10}$$

and

$$\frac{\partial Z}{\partial w_{ij}} = \sum S_i^{\gamma} S_j^{\gamma} e^{-E_{\alpha|\gamma}} = \frac{\sum S_i^{\gamma} S_j^{\gamma} e^{-E_{\alpha|\gamma}}}{Z} Z = \langle S_i S_j \rangle^{\gamma} Z . \tag{11}$$

Finally,

$$\Delta w_{ij} = \frac{\eta}{T}\sum_\gamma P_\gamma \sum_\alpha \frac{R_{\alpha|\gamma}}{P_{\alpha|\gamma}} \frac{S_i^{\alpha|\gamma} S_j^{\alpha|\gamma} e^{-E_{\alpha|\gamma}} Z - e^{-E_{\alpha|\gamma}} Z \langle S_i S_j \rangle^\gamma}{Z^2}$$

$$= \frac{\eta}{T}\sum_\gamma P_\gamma \sum_\alpha \frac{R_{\alpha|\gamma}}{P_{\alpha|\gamma}} \frac{S_i^{\alpha|\gamma} S_j^{\alpha|\gamma} e^{-E_{\alpha|\gamma}} - e^{-E_{\alpha|\gamma}} \langle S_i S_j \rangle^\gamma}{Z}$$

$$= \frac{\eta}{T}\sum_\gamma P_\gamma \sum_\alpha R_{\alpha|\gamma} \left(S_i^{\alpha|\gamma} S_j^{\alpha|\gamma} - \langle S_i S_j \rangle^\gamma \right)$$

$$= \frac{\eta}{T}\sum_\gamma P_\gamma \left(\sum_\alpha R_{\alpha|\gamma} S_i^{\alpha|\gamma} S_j^{\alpha|\gamma} - \sum_\alpha R_{\alpha|\gamma} \langle S_i S_j \rangle^\gamma \right)$$

$$= \frac{\eta}{T}\sum_\gamma P_\gamma \left(\sum_\alpha R_{\alpha|\gamma} S_i^{\alpha|\gamma} S_j^{\alpha|\gamma} - \langle S_i S_j \rangle^\gamma \sum_\alpha R_{\alpha|\gamma} \right). \tag{12}$$

Since

$$\sum_\alpha R_{\alpha|\gamma} = 1 , \tag{13}$$

we get the following expression

$$\Delta w_{ij} = \frac{\eta}{T}\sum_\gamma P_\gamma \left(\langle S_i S_j \rangle^{\alpha|\gamma} - \langle S_i S_j \rangle^\gamma \right). \tag{14}$$

Finally, and since the derivative for the bias terms is similar, we get the following expressions

$$\Delta w_{ij} = -\eta \frac{\partial G}{\partial w_{ij}} = \frac{\eta}{T}\left(\langle S_i S_j \rangle^* - \langle S_i S_j \rangle \right), \tag{15}$$

$$\Delta h_i = -\eta \frac{\partial G}{\partial h_i} = \frac{\eta}{T}\left(\langle S_i \rangle^* - \langle S_i \rangle \right), \tag{16}$$

where η is a convergence parameter that is tuned at will. On the other hand, $\langle \ldots \rangle$ are average values for fixed input units at a given input vector and $\langle \ldots \rangle^*$ are average values for both fixed input and output units at the same given vector but with its corresponding output value. This expression is calculated at a final annealing schedule equilibrium temperature T, and those correlation values are to be statistically evaluated for each different example from the training set for each algorithm iteration.

2.4 The High Order Boltzmann Machine

The High Order Boltzmann Machine [7, 10] is an extension to the original model where weights may connect more than two units. So, in an M-th Order Boltzmann Machine weights can connect a total of up to M units simultaneously and, from now on, we will consider the $M = 2$ case as the standard situation. According to this, we will change the energy functional notation by

$$E = -\frac{1}{2}\sum_{i,j} w_{ij}^2 S_i S_j - \sum_i w_i^1 S_i \ , \tag{17}$$

where w_{ij}^2 stands for the typical weight connecting two different units and w_i^1 stands for the bias terms. High Order Boltzmann Machines still keep the stochastic properties of their simpler model: such networks have an associated energy functional [7, 10] of the form

$$E = -\sum_{m=1}^{M} \frac{1}{m!} \sum_{\lambda} w_{\lambda}^m S_{i_0} S_{i_1} \ldots S_{i_\lambda} \ , \tag{18}$$

w_{λ}^m is the generic m-th order weight connecting m given binary units, assuming that a bias is a first order term. This functional is evaluated each iteration of the Simulated Annealing algorithm in order to bring the network to equilibrium at every temperature of the cooling schedule. In the end, the system follows a Boltzmann distribution at the lowest T. M-th order weights can be graphically represented by a line connecting groups of M units. As an example, Fig. (1a) shows a second order weight linking units S_i and S_j, while Fig. (1b) shows a third order weight linking units S_i, S_j and S_k.

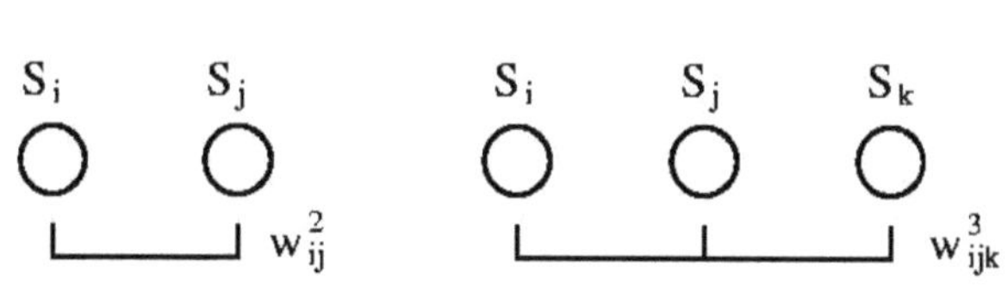

Fig. 1. Simplified notation for 2nd(a) and 3rd(b) order connections

Learning in a HOBM is carried out as in a standard BM, so it the Kullback-Leibler distance [11] between the actual probability distribution and the one to be learnt is minimized. For a HOBM with different input and output units this last quantity is still used

$$G = \sum_{\gamma} P_{\gamma} \sum_{\alpha} R_{\alpha|\gamma} \ln \frac{R_{\alpha|\gamma}}{P_{\alpha|\gamma}} , \tag{19}$$

just as in the simpler model of the neural network. Hence, we can also use gradient descent on G to find the update rule

$$\Delta w_{\lambda}^{m} = -\eta \frac{\partial G}{\partial w_{\lambda}^{m}} = \frac{\eta}{T} \left(\left\langle \prod_{\lambda} S_{i_{\lambda}} \right\rangle^{*} - \left\langle \prod_{\lambda} S_{i_{\lambda}} \right\rangle \right) , \tag{20}$$

though now the correlations are calculated along all the units that the weights connect. However, as the order of the neural network grows, it becomes harder to calculate these quantities by Monte Carlo simulations [10].

3 Standard Decimation of Boltzmann Machines

3.1 Introduction

Decimation is a standard technique in statistical mechanics [12] that is used in this context to build a new network with $m < N$ units from a N units neural network, that is able to reproduce exactly the m-unit correlations of Eq. (20). This is possible because units that do not appear explicitly as arguments in the expectation values enter only in the Boltzmann factors $\exp(-E/T)$. So, for every such unit S one can look for renormalized bias and weights corresponding to an equivalent BM with one less neuron where S has been removed. In terms of a new set of weights [5, 6]

$$J_{\lambda}^{m} = \frac{1}{T} w_{\lambda}^{m} , \tag{21}$$

the master equation for this process is born when the Boltzmann marginal probability distribution for the rest of the units is to remain constant for any given energy state E_α. So

$$P(E'_\alpha) = P\left(E_\alpha\big|_{S=-1}\right) + P\left(E_\alpha\big|_{S=+1}\right). \tag{22}$$

where E'_α takes into account the rest of the units from the neural network. Using (5) the Boltzmann probability distribution can be computed:

$$\frac{e^{-E'_\alpha/T}}{Z'} = \frac{e^{-E_\alpha|_{S=-1}/T}}{Z} + \frac{e^{-E_\alpha|_{S=+1}/T}}{Z}. \tag{23}$$

If E_α is separated in terms of weights connecting units S_i to unit S and the rest of the weights, we will now get an expression as

$$\frac{e^{-\widetilde{E}_\alpha/T + \sum_i S_i J_i^1 + \sum_{j<i} S_i S_j J_{ij}^2}}{Z'} = \frac{e^{-\widetilde{E}_\alpha/T + J^1 + \sum_i S_i J_i^2}}{Z} + \frac{e^{-\widetilde{E}_\alpha/T - J^1 - \sum_i S_i J_i^2}}{Z}$$

$$\frac{e^{-\widetilde{E}_\alpha/T} e^{\sum_i S_i J_i^1 + \sum_{j<i} S_i S_j J_{ij}^2}}{Z'} = \frac{e^{-\widetilde{E}_\alpha/T} \sum_{S=-1}^{+1} e^{S\left(J^1 + \sum_i S_i J_i^2\right)}}{Z}, \tag{24}$$

$\widetilde{E}_\alpha$ being the part of the energy which is not represented by unit S, but by all the rest of the neural network, linked by elements J^1 and J_i^2. If the previous expression is worked out, we will get the decimation expression proposed by Ref. [5]

$$\sum_{S=-1}^{+1} e^{S\left(J^1 + \sum_i S_i J_i^2\right)} = \sqrt{C} e^{\sum_i S_i J_i^1 + \sum_{j<i} S_i S_j J_{ij}^2}. \tag{25}$$

3.2 Decimation Equations

If Eq. (25) is worked, we obtain the generic decimation equation

$$\sum_{S=-1}^{+1} e^{S\left(J^1+\sum_i S_i J_i^2\right)} = \sqrt{C} e^{\sum_i S_i J_i^1 + \sum_{j<i} S_i S_j J_{ij}^2}$$

$$\frac{e^{J^1+\sum_i S_i J_i^2} + e^{-J^1-\sum_i S_i J_i^2}}{2} = \frac{\sqrt{C}}{2} e^{\sum_i S_i J_i^1 + \sum_{j<i} S_i S_j J_{ij}^2}$$

$$\cosh\left(J^1 + \sum_i S_i J_i^2\right) = e^{\ln\left(\sqrt{C}/2\right) + \sum_i S_i J_i^1 + \sum_{j<i} S_i S_j J_{ij}^2}$$

$$\ln\left(\cosh\left(J^1 + \sum_i S_i J_i^2\right)\right) = J^0 + \sum_i S_i J_i^1 + \sum_{j<i} S_i S_j J_{ij}^2 \ , \qquad (26)$$

J^0 being a normalization constant. Once these equations are solved one is left with an equivalent network with one less unit and different bias and weights. The star-triangle transformation of Ref. [6] is a special case where a unit connected to three other units is simplified as shown in Fig. (2c).

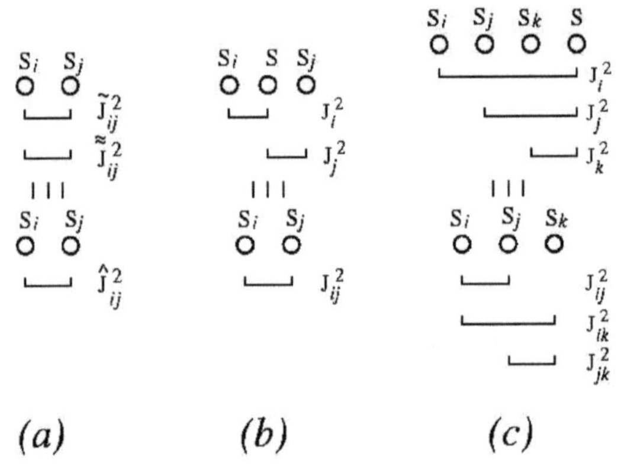

Fig. 2. Parallel (a), serial (b) and star-triangle (c) associations

Equations for the star-triangle transformation are born when binary $S = \{-1,+1\}$ are given to units S_i, S_j and S_k to this equation

$$\ln\left(\cosh\left(S_i J_i^2 + S_j J_j^2 + S_k J_k^2\right)\right) = J^0 + S_i S_j J_{ij}^2 + S_i S_k J_{ik}^2 + S_j S_k J_{jk}^2 \qquad (27)$$

which takes into account connections on Fig. (2c). Thus, we get

$$\ln\left(\cosh\left(-J_i^2 - J_j^2 - J_k^2\right)\right) = J^0 + J_{ij}^2 + J_{ik}^2 + J_{jk}^2 \qquad (28)$$

$$\ln\left(\cosh\left(-J_i^2 - J_j^2 + J_k^2\right)\right) = J^0 + J_{ij}^2 - J_{ik}^2 - J_{jk}^2 \qquad (29)$$

$$\ln\left(\cosh\left(-J_i^2+J_j^2-J_k^2\right)\right)=J^0-J_{ij}^2+J_{ik}^2-J_{jk}^2 \tag{30}$$

$$\ln\left(\cosh\left(-J_i^2+J_j^2+J_k^2\right)\right)=J^0-J_{ij}^2-J_{ik}^2+J_{jk}^2 \tag{31}$$

$$\ln\left(\cosh\left(J_i^2-J_j^2-J_k^2\right)\right)=J^0-J_{ij}^2-J_{ik}^2+J_{jk}^2 \tag{32}$$

$$\ln\left(\cosh\left(J_i^2-J_j^2+J_k^2\right)\right)=J^0-J_{ij}^2+J_{ik}^2-J_{jk}^2 \tag{33}$$

$$\ln\left(\cosh\left(J_i^2+J_j^2-J_k^2\right)\right)=J^0+J_{ij}^2-J_{ik}^2-J_{jk}^2 \tag{34}$$

$$\ln\left(\cosh\left(J_i^2+J_j^2+J_k^2\right)\right)=J^0+J_{ij}^2+J_{ik}^2+J_{jk}^2 \tag{35}$$

This set of equations is reduced to a four equation and four incognita system due to the hyperbolic cosinus symmetrical properties. So, we are left with Eq. (32) to Eq. (35). Lineal combination of this equations leads to

$$J_{ij}^2=\frac{1}{4}\ln\left(\frac{\cosh\left(J_i^2+J_j^2+J_k^2\right)\cosh\left(J_i^2+J_j^2-J_k^2\right)}{\cosh\left(J_i^2-J_j^2-J_k^2\right)\cosh\left(J_i^2-J_j^2+J_k^2\right)}\right), \tag{36}$$

$$J_{ik}^2=\frac{1}{4}\ln\left(\frac{\cosh\left(J_i^2+J_j^2+J_k^2\right)\cosh\left(J_i^2-J_j^2+J_k^2\right)}{\cosh\left(J_i^2+J_j^2-J_k^2\right)\cosh\left(J_i^2-J_j^2-J_k^2\right)}\right), \tag{37}$$

$$J_{jk}^2=\frac{1}{4}\ln\left(\frac{\cosh\left(J_i^2+J_j^2+J_k^2\right)\cosh\left(J_i^2-J_j^2-J_k^2\right)}{\cosh\left(J_i^2+J_j^2-J_k^2\right)\cosh\left(J_i^2-J_j^2+J_k^2\right)}\right). \tag{38}$$

When $J_k^2=0$ these equations are reduced to the serial association of Ref. [5] represented in Fig. (2b)

$$J_{ij}^2=\frac{1}{2}\ln\left(\frac{\cosh\left(J_i^2+J_j^2\right)}{\cosh\left(J_i^2-J_j^2\right)}\right). \tag{39}$$

The last and easiest transformation is called parallel association and is also shown in Fig. (2a). The resulting weight is immediately seen to be the sum of the weights being associated

$$\hat{J}_{ij}^2 = \tilde{J}_{ij}^2 + \tilde{\tilde{J}}_{ij}^2 \,. \tag{40}$$

4 High Order Decimation

4.1 Introduction

Decimation, as stated in the previous section, can not be used to simplify a unit connected to more than three other units [6]. Hence, a structure like this can not be handled by the traditional equations

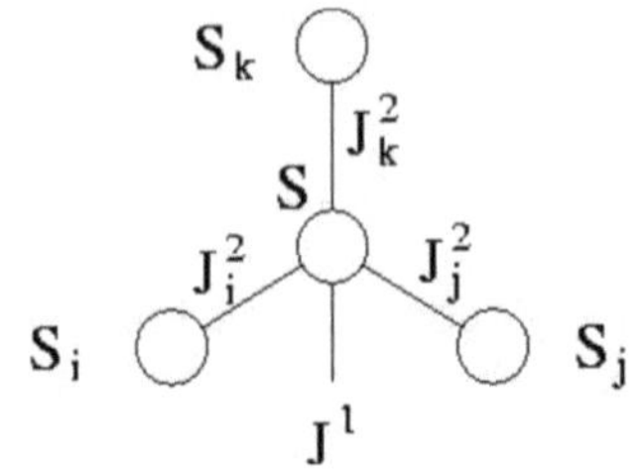

Fig. 3. Non-decimatable structure.

In this work, we propose a way to decimate such kind of structures, by making an extension to the original decimation equations.

One can shed some light on the reason why decimation fails to simplify these structures by working out a simple example, the three points star with a central biased unit, where the central unit is the one to be decimated. Using Eq. (27) one finds this set of equations

$$\ln\left(\cosh\left(J^1 - J_i^2 - J_j^2 - J_k^2\right)\right) = J^0 + J_{ij}^2 + J_{ik}^2 + J_{jk}^2 \,,$$

$$\ln\left(\cosh\left(J^1 - J_i^2 - J_j^2 + J_k^2\right)\right) = J^0 + J_{ij}^2 - J_{ik}^2 - J_{jk}^2 \,,$$

$$\ln\left(\cosh\left(J^1 - J_i^2 + J_j^2 - J_k^2\right)\right) = J^0 - J_{ij}^2 + J_{ik}^2 - J_{jk}^2 \,,$$

$$\ln\left(\cosh\left(J^1 - J_i^2 + J_j^2 + J_k^2\right)\right) = J^0 - J_{ij}^2 - J_{ik}^2 + J_{jk}^2 \,,$$

$$\ln\left(\cosh\left(J^1 + J_i^2 - J_j^2 - J_k^2\right)\right) = J^0 - J_{ij}^2 - J_{ik}^2 + J_{jk}^2 \,,$$

$$\ln\left(\cosh\left(J^1 + J_i^2 - J_j^2 + J_k^2\right)\right) = J^0 - J_{ij}^2 + J_{ik}^2 - J_{jk}^2 ,$$

$$\ln\left(\cosh\left(J^1 + J_i^2 + J_j^2 - J_k^2\right)\right) = J^0 + J_{ij}^2 - J_{ik}^2 - J_{jk}^2 ,$$

$$\ln\left(\cosh\left(J^1 + J_i^2 + J_j^2 + J_k^2\right)\right) = J^0 + J_{ij}^2 + J_{ik}^2 + J_{jk}^2 . \tag{41}$$

As it can be readily seen, the first and last equations are not compatible for arbitrary values of J^1, J_i^2, J_j^2 and J_k^2. The same happens with other pairs of equations. Actually and as stated, the problem has 4 unknown variables and 8 equations that are not compatible. In any case, the number of equations equals the number of binary states a system of 3 external units can take, and can not be reduced even if adding the bias terms units *i, j, k* would need. Alternatively, one can introduce four new variables that make the system compatible, and this is the procedure adopted here.

4.2 High Order Decimation Equations

Three new variables are easily introduced by including the missing bias terms in the resulting system. The last one is a third order term connecting the three external units together. Taking all this into account one arrives to the following set of decimation equations

$$\ln\left(\cosh\left(J^1 + \sum_i S_i J_i^2\right)\right) = J^0 + \sum_i S_i J_i^1 + \sum_{j<i} S_i S_j J_{ij}^2 + S_i S_j S_k J_{ijk}^3 , \tag{42}$$

which can be written in matrix form to realize that the coefficient matrix is of the Hadamard type [13, 14, 15]. Now, Hadamard matrices are made out of orthogonal row and column vectors, meaning that the system is compatible and has a unique solution. In summary, decimation can be carried out in this system when all possible sets of weight connecting three units are considered.

This very same strategy can be used to decimate the general structure of Fig. (4a). In this case one can write 2^N equations corresponding to the 2^N binary states units $\{S_0, S_1, \ldots, S_{N-1}\}$ can take, so once again 2^N unknown variables are required. This can be achieved by including all possible *m* order weights, with *m* spanning the range *1* to *N*. The resulting high order structure is fully connected and is shown in Fig. (4b).

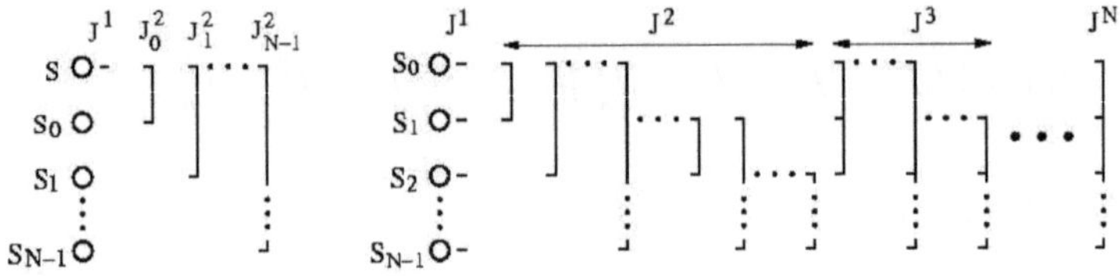

(a) (b)

Fig. 4. Initial (a) and resulting (b) structure

Counting the total number of variables is easy, since there are

- $\binom{N}{0} = 1$ normalization constants J^0,
- $\binom{N}{1}$ biases J^1,
- $\binom{N}{2}$ two body weights J^2,
- $\binom{N}{3}$ three body terms J^3,

...

- $\binom{N}{i}$ *i* body terms J^i,

...

- $\binom{N}{N} = 1$ N body weight J^N,

while we know that the sum of all these combinatorial numbers equals 2^N, that is, equals the number of decimation equations.

The complete system of equations becomes then

$$\ln\left(\cosh\left(J^1 + \sum_i S_i J_i^2\right)\right) = J^0 + \sum_i S_i J_i^1 + \sum_{j<i} S_i S_j J_{ij}^2 + \sum_{k<j<i} S_i S_j S_k J_{ijk}^3 + \ldots + S_0 S_1 S_2 \cdots S_{N-1} J_{012\cdots N-1}^N \quad , \tag{43}$$

and is still of the Hadamard type, meaning that the system can be inverted and that there is a solution that is unique. The general structure shown in Fig. (4a) can therefore be decimated. Furthermore, higher order structures can be decimated in the same way, since the coefficient matrix of the system remains unchanged and as only the independent terms are modified according to the expression

$$\ln\left(\cosh\left(J^1+\sum_i S_i J_i^2+\sum_{i<j} S_i S_j J_{ij}^3+\ldots\right)\right)= \qquad (44)$$
$$J^0+\sum_i S_i J_i^1+\sum_{j<i} S_i S_j J_{ij}^2+\sum_{k<j<i} S_i S_j S_k J_{ijk}^3+\ldots+S_0 S_1 S_2\cdots S_{N-1} J_{012\cdots N-1}^N \quad .$$

In fact, these equations are to be used iteratively, decimating one unit in each step until one is only left with the m neurons $\left\{S_{i_1}, S_{i_2}, \ldots, S_{i_m}\right\}$ required to update w_λ^m as shown in Eq. (20).

As an example, equations to decimate Fig. (3) would stand for

$$\begin{aligned}&\ln\cosh\left(J^1-J_i^2-J_j^2-J_k^2\right)=\\&=J^0-J_i^1-J_j^1-J_k^1+J_{ij}^2+J_{ik}^2+J_{jk}^2-J_{ijk}^3\end{aligned}$$

$$\begin{aligned}&\ln\cosh\left(J^1-J_i^2-J_j^2+J_k^2\right)=\\&=J^0-J_i^1-J_j^1+J_k^1+J_{ij}^2-J_{ik}^2-J_{jk}^2+J_{ijk}^3\end{aligned}$$

$$\begin{aligned}&\ln\cosh\left(J^1-J_i^2+J_j^2-J_k^2\right)=\\&=J^0-J_i^1+J_j^1-J_k^1-J_{ij}^2+J_{ik}^2-J_{jk}^2+J_{ijk}^3\end{aligned}$$

$$\begin{aligned}&\ln\cosh\left(J^1-J_i^2+J_j^2+J_k^2\right)=\\&=J^0-J_i^1+J_j^1+J_k^1-J_{ij}^2-J_{ik}^2+J_{jk}^2-J_{ijk}^3\end{aligned}$$

$$\begin{aligned}&\ln\cosh\left(J^1+J_i^2-J_j^2-J_k^2\right)=\\&=J^0+J_i^1-J_j^1-J_k^1-J_{ij}^2-J_{ik}^2+J_{jk}^2+J_{ijk}^3\end{aligned}$$

$$\begin{aligned}&\ln\cosh\left(J^1+J_i^2-J_j^2+J_k^2\right)=\\&=J^0+J_i^1-J_j^1+J_k^1-J_{ij}^2+J_{ik}^2-J_{jk}^2-J_{ijk}^3\end{aligned}$$

$$\ln\cosh\left(J^1 + J_i^2 + J_j^2 - J_k^2\right) =$$
$$= J^0 + J_i^1 + J_j^1 - J_k^1 + J_{ij}^2 - J_{ik}^2 - J_{jk}^2 - J_{ijk}^3$$

$$\ln\cosh\left(J^1 + J_i^2 + J_j^2 + J_k^2\right) =$$
$$= J^0 + J_i^1 + J_j^1 + J_k^1 + J_{ij}^2 + J_{ik}^2 + J_{jk}^2 + J_{ijk}^3 \, . \qquad (45)$$

5 High Order Decimation Method Example

In order to improve the High Order Decimation process understanding, we will provide a graphical example on how the method would work on a fully connected neural network with six free units. For this example, we will assume that input units have been parallel associated with bias terms. Therefore, there is no need of representing them.

Additionally, we would like to remark that the number of free units from the neural network will vary depending on the learning stage. Hence, Fig. (5a) represents with two circles outputs units at a learning stage where inputs have been clamped. On the other hand, Fig. (5b) has both clamped input and output units, so there are only hidden units left.

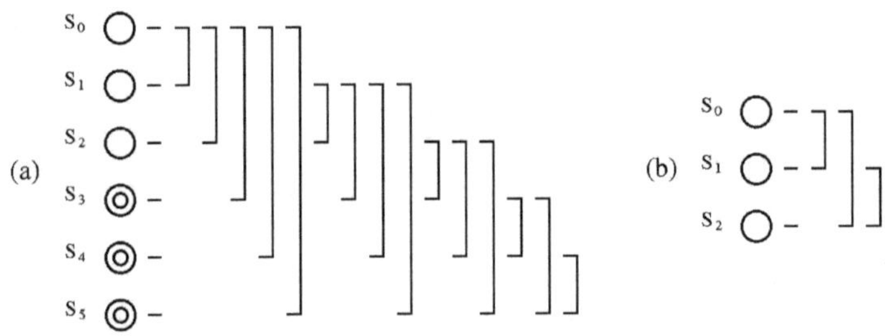

Fig. 5. Fully connected 3 inputs – 3 outputs mall, both with free (a) and clamped (b) output units

Hence, we will perform our example over Fig. (5a). For instance, we are interested on finding correlation for units S_0 and S_1 and expected value for S_0, so we are going to decimate the rest of the units following this order: S_5, S_4, S_3 and S_2. Decimating unit S_5 results on this mall

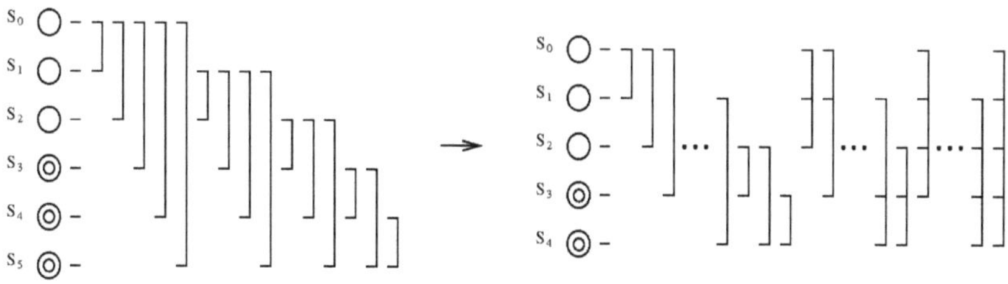

Fig. 6. Decimation process to obtain a fully connected High Order 5 units mall

Working on the other units can be seen as a process with the following stages

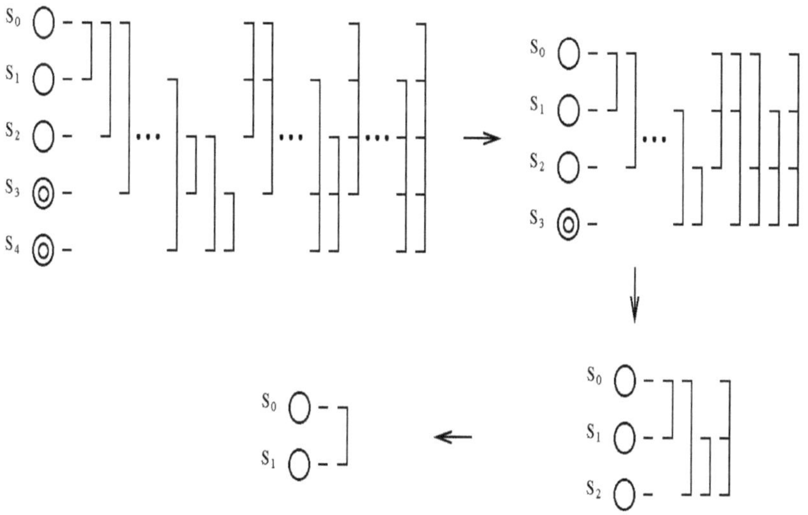

Fig. 7. Decimation process to obtain two units

Now that we have arrived to the pair S_0, S_1, correlation is analytically computed as in Ref. [5]

$$\langle S_0 S_1 \rangle = \sum_{S_0,S_1} S_0 S_1 \frac{e^{S_0 S_1 J_{01}^2 + S_0 J_0^1 + S_1 J_1^1}}{Z}$$

$$= \frac{e^{J_{01}^2 - J_0^1 - J_1^1}}{Z} - \frac{e^{-J_{01}^2 - J_0^1 + J_1^1}}{Z} - \frac{e^{-J_{01}^2 + J_0^1 - J_1^1}}{Z} + \frac{e^{J_{01}^2 + J_0^1 + J_1^1}}{Z}$$

$$= \frac{e^{J_{01}^2 + J_0^1 + J_1^1} + e^{J_{01}^2 - J_0^1 - J_1^1} - e^{-J_{01}^2 + J_0^1 - J_1^1} - e^{-J_{01}^2 - J_0^1 + J_1^1}}{e^{J_{01}^2 + J_0^1 + J_1^1} + e^{J_{01}^2 - J_0^1 - J_1^1} + e^{-J_{01}^2 + J_0^1 - J_1^1} + e^{-J_{01}^2 - J_0^1 + J_1^1}}$$

$$= \frac{e^{J_{01}^2}\left(e^{J_0^1 + J_1^1} + e^{-J_0^1 - J_1^1}\right) - e^{-J_{01}^2}\left(e^{J_0^1 - J_1^1} + e^{-J_0^1 + J_1^1}\right)}{e^{J_{01}^2}\left(e^{J_0^1 + J_1^1} + e^{-J_0^1 - J_1^1}\right) + e^{-J_{01}^2}\left(e^{J_0^1 - J_1^1} + e^{-J_0^1 + J_1^1}\right)}$$

$$\langle S_0 S_1 \rangle = \frac{e^{J_{01}^2} \cosh\left(J_0^1 + J_1^1\right) - e^{-J_{01}^2} \cosh\left(J_0^1 - J_1^1\right)}{e^{J_{01}^2} \cosh\left(J_0^1 + J_1^1\right) + e^{-J_{01}^2} \cosh\left(J_0^1 - J_1^1\right)} . \tag{46}$$

Finally, unit S_0 is isolated by performing a simple serial decimation operation, to compute its expected value

$$\langle S_0 \rangle = \sum_{S_0} S_0 \frac{e^{S_0 J_0^1}}{Z} = \frac{e^{J_0^1}}{Z} - \frac{e^{-J_0^1}}{Z} = \frac{e^{J_0^1} - e^{-J_0^1}}{e^{J_0^1} + e^{-J_0^1}} , \tag{47}$$

and, then

$$\langle S_0 \rangle = \tanh\left(J_0^1\right) , \tag{48}$$

where J_0^1 is the new bias term.

In order to calculate correlations for each pair of connected units, the neural network has to be decimated until such two units are left and, then, repeat the process.

6 Simulations and Results

In this section we are going to explain how we have used the tools described in this work to solve a typical benchmarking problem. In order to speed up the learning process the conjugate descent gradient method has been used as proposed on Ref. [16]. So, the Boltzmann learning expression

$$\Delta w_\lambda^m = \frac{\eta}{T}\left(\left\langle \prod_\lambda S_{i_\lambda} \right\rangle^* - \left\langle \prod_\lambda S_{i_\lambda} \right\rangle\right), \tag{49}$$

is now changed into

$$\Delta w_\lambda^{m(i)} = (1-\alpha)\left(\frac{\eta}{T}\left(\left\langle \prod_\lambda S_{i_\lambda} \right\rangle^* - \left\langle \prod_\lambda S_{i_\lambda} \right\rangle\right)\right) + \alpha \Delta w_\lambda^{m(i-1)}, \tag{50}$$

where i is the current algorithm iteration.

6.1 Parity Problem

The High Order Decimation procedure has been first tested on the N-bit parity or XOR problem [17]. A recurrent and fully connected network has been chosen to solve the problem and the conjugate gradient descent rule from Eq. (50) has been adopted to update the weights at each iteration of the learning process.

Preliminary results for different instances of the problem and network topologies are reported in table 1 and compared with data obtained using the Boltzmann Trees and the perceptrons trained with the SCG learning algorithm of Refs. [5] and [18], respectively. The first column in the table indicates the topology of the network in the X-Y-Z format, where X, Y and Z stand for the number of input, hidden and output units, respectively. The next three columns report the efficiency, average and maximum number of iterations allowed in the training algorithm. The procedure used to obtain these values is the following: for each architecture one performs a total of 2000 trainings. In each training the weights are allowed to be updated at most $n_{\max}$ times. A training instance is considered to be a success when the network has been trained with less than $n_{\max}$ weight updates and the predicted value for any given input coincides with the actual XOR of the inputs with a probability equal or larger than 0.9 over a batch of 1000 runs of the trained network. n_{avg} is the average number of weight updates required in each successful training. The Boltzmann Tree results of Ref. [5] are reported in columns 5, 6 and 7. Finally the perceptron results of Ref. [18] are shown in the last column, where the total number of iterations required to solve the XOR problem with an average quadratic error no larger than 10^{-6} is reported.

As it can be seen from the table, the performance of the High Order Decimation algorithm is comparable if not better than the standard decimation technique used on a Boltzmann Tree, both in efficiency and number of iterations, although more data would be required to establish a better comparison. Only in the 5-4-1 case the Boltzmann Tree achieves a higher efficiency with less iterations, but we have not been able to reproduce these figures with our Boltzmann Tree simulator.

An improvement on the number of iterations needed to solve the problem when computed with the perceptron of Ref. [18] is also apparent, although one should bear in mind the different criteria adopted to consider a learning instance successful in each case. Additionally and for fixed number of input units, the High Order Decimation method seems to be useful in this context to determine a suitable minimum number of hidden neurons required to solve the *N*-bit parity problem with a Boltzmann Machine. This number turns out to be 3, 4 and 5 for N=3, 4 and 5, which seems to indicate that the XOR problem is best solved when the number of input and output units is equal.

Table 1. Parity problem efficiency

Architecture	Efficiency (%)	n_{avg}	$n_{\max}$	Efficiency (%)	n_{avg}	$n_{\max}$	n_{avg}
2-1-1	97.2	25.0	50	97.2	25.0	50	-
3-1-1	96.2	40.0	250	96.1	42.1	250	-
3-3-1	100.0	16.3	250	-	-	-	154
4-2-1	99.5	289.4	1000	-	-	-	-
4-3-1	100.0	184.9	1000	95.1	281.1	1000	-

7 Conclusions

Previous works on decimation applied to Boltzmann Machines introduced the concepts of parallel and serial decimation firstly and, later, the original star-triangle conversion with an unbiased unit. More complex transformations were not intended to be possible due to the lack of freedom degrees on the system equations, as stated on Ref. [6]. In this work we have shown that all sort of Boltzmann Machines can be decimated when high order weights are allowed to appear in the resulting network.

Decimation is of capital interest in Boltzmann Machine learning as it allows finding update values for weights in a gradient descent exploration without resorting to the use of the Simulated Annealing algorithm. This speeds up learning considerably, but at the expense of solving a system of $2^N \times 2^N$ linear equations where N is the total number of hidden and output units. In this sense and in the context of Boltzmann Machine learning, decimation as opposed to Simulated Annealing turns exhaustive time-consuming calculations into high memory requirements. Compared with the standard decimation procedures used up to the date, high order decimation introduces many additional degrees of freedom in the form of new weights, and therefore less units are expected to be required to solve the problem at hand. This turns out to be of great advantage once the network has been trained and is operated to obtain a probability distribution with the help of the Simulated Annealing algorithm.

References

1. Ackley, D.H., Hinton, G.E. and Sejnowsky, T.J., A learning algorithm for Boltzmann Machines, Cognitive Science, vol. IX, pp. 147-169, 1985.
2. Hinton, G.E., and Sejnowski, T.J., Learning and relearning in Boltzmann Machines, Parallel Distributed Processing, vol. I, Cambridge, M.A., U.S.A., MIT Press, pp. 282-317, 1986.
3. J. J. Hopfield; Neural networks and physical systems with emergent collective computational abilities, Neurocomputing: foundations of research, Cambridge, M.A., U.S.A., MIT Press, pp. 457-464, 1988.
4. Kirkpatrick, S., Gelatt, C.D. Jr and Vecchi, M.P., Optimization by Simulated Annealing, Science 220, pp. 671-680, 1983.
5. Saul, L., and Jordan, M.I., Learning in Boltzmann Trees, Neural Computation, vol. 6, num. 6, pp. 1174-1184, 1994.
6. Rüger, S.M., Decimatable Boltzmann Machines for Diagnosis: Efficient Learning and Inference, World Congress on Scientific Computation, Modelling and Applied Mathematics, Berlin, Deutschland, vol. 4, pp. 319-324, 1997.
7. Sejnowski, T.J., High-Order Boltzmann Machines, AIP Conference Proceedings 151 on Neural Networks for Computing, Snowbird, Utah, U.S.A., pp. 398 - 403, 1987.
8. Kirkpatrick, S., Gelatt, C.D. Jr. and Vecchi M.P., Optimization by simulated annealing, Neurocomputing: foundations of research, MIT Press Cambridge, MA, USA, pp. 551-567, 1988.
9. Aarts, E., and Korst, J., Simulated Annealing and Boltzmann Machines: a stochastic Approach to Combinatorial Optimization and Neural Computing, 3rd. ed., Great Britain, John Wiley and Sons, 1997.
10. Albizuri, F.X., et al., The High-Order Boltzmann Machine: Learned distribution and Topology, IEEE Transactions on Neural Networks, vol. 6, num. 3, pp. 767-770, 1995.
11. Kullback, S., Information theory and statistics, 2nd ed, New York, U.S.A., New York: Willey, 1959.
12. Itzykson, C. and Drouffe, J., Statistical field theory, Cambridge, Cambridge University Press, 1991.

13. Álvarez, V., et al., Matrices cocílicas de Hadamard sobre productos semidirectos, III Jornadas de Matemàtica Discreta y Algorítmica 3JMDA, Sevilla, Spain, pp. 155-158, 2002.
14. Assmus Jr., E. F. and Key, J. D., Hadamard matrices and their designs: a coding theorethic approach, Trans. Amer. Soc., to appear.
15. Park, C. H., Song, H. Y. and Park, K. T., Existence and classification of Hadamard matrices, Signal Processing Proceedings, 1998. ICSP '98, Fourth International Conference on, vol. 1, pp. 117-121, 1998.
16. Duda, R.O., Hart, P.E. and Stork D.G., Pattern Classification, 2nd Ed., U.S.A., Wiley-Interscience Publication, John Wiley & Sons, INC., 2001.
17. Li, D. et al., Studying the effects of multiplication neurons for parity problem, Proceedings of the 41st SICE Annual Conference, SICE 2002, vol.: 5, pp. 2678-2681, August 2002.
18. Møller, M.F., A scaled conjugate gradient algorithm for fast supervised learning, Neural Networks, vol. 6, pp. 525-533, 1993.

Evolutionary Optimization of a Wiener Model

Oscar Montiel. CITEDI-IPN. 2498 Roll Dr. 286. San Diego, CA 92154, USA.
oross@citedi.mx.

Oscar Castillo. Department of Computer Science, Tijuana Institute of Technology. P.O. Box 4207, Chula Vista CA 91909, USA.
ocastillo@tectijuana.mx.

Patricia Melin. Department of Computer Science, Tijuana Institute of Technology. P.O. Box 4207, Chula Vista CA 91909, USA.
pmelin@tectijuana.mx.

Roberto Sepúlveda. CITEDI-IPN. Av. del Parque #1310, Mesa de Otay, Tijuana, B. C., México. rsepulve@citedi.mx

Abstract. There exists no standard method for obtaining a nonlinear input-output model using external dynamic approach. In this work, we are using an evolutionary optimization method for estimating the parameters of an NFIR model using the Wiener model structure. Specifically we are using a Breeder Genetic Algorithm (BGA) with fuzzy recombination for performing the optimization work. We selected the BGA since it uses real parameters (it does not require any string coding), which can be manipulated directly by the recombination and mutation operators. For training the system we used amplitude modulated pseudo random binary signal (APRBS). The adaptive system was tested using sinusoidal signals.

1. Introduction

A system is a human conception of a group of independent but interrelated elements comprising a unified whole (Severance, 2001). The key task of system identification (modeling) is to find a best suitable mathematical model between the inputs, outputs and disturbances of a physical system (Ljung, 1999). Nowadays, identifying linear systems has become a routine task and there are available several successful methods for solving the problem in the time or in the frequency domain, using iterative and non-iterative optimization methods for estimating the parameters. Real systems are nonlinear and their properties may change with time, obtaining models for nonlinear systems are more complex than for linear, since any difference in the dynamic behavior of these models can be extremely significant. It is a common practice to treat real systems as linear to some extent, and

O. Montiel et al.: *Evolutionary Optimization of a Wiener Model*, StudFuzz **208**, 43–58 (2007)
www.springerlink.com

as a natural consequence to consider a linear model as first choice. If the linear model does not fulfill the expectative, it is necessary to analyze the whole process to search an explanation and solution. A possibility is to change from a linear to a nonlinear model, but sometimes it can be conflicting since if the nonlinear model chosen is not flexible enough its performance can be worst. Nonlinear system identification of a dynamic process is a challenging task and it has received special attention during the past decade (Ikonen, 1999).

In this work, for modeling a nonlinear system we used an input-output model that fits in the class of models known as "external dynamics models", this name stems from the fact that it can be separated in two well defined parts: an external dynamic filter bank and a nonlinear static approximator (Nelles, 2001). This concept is illustrated in Fig. 1. In the external dynamic approach the model is conceptualized as a dynamic filter bank and a nonlinear static approximator. In principle, it does not matter the model architecture or the static nonlinear approximator.

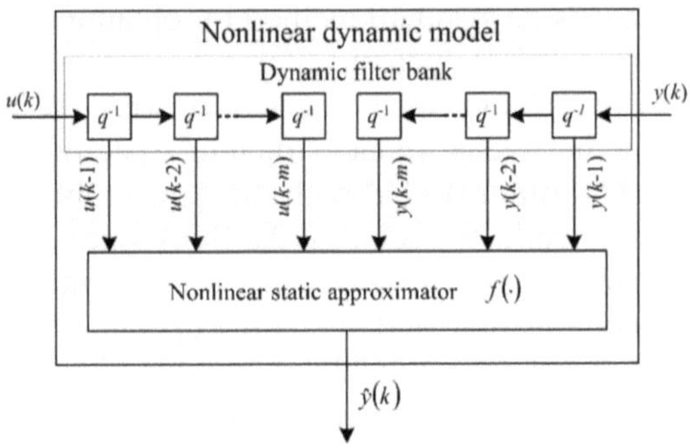

Fig. 1. In the external dynamic approach the model is conceptualized as a dynamic filter bank and a nonlinear static approximator. In principle, it does not matter the model architecture or the static nonlinear approximator.

There are several nonlinear optimization methods for estimating the parameters of an external dynamic model, most of them perform local searches and they might get trapped in local optima. Global optimization methods can perform global searches for the global optimum, although it is well known the huge computational demand that these methods require. Since nonlinear, local and global, as linear optimization methods have their own advantages and drawbacks, it is a good practice to combine these techniques.

In the field of external dynamic modeling we found some interesting works like: "An iterative method for the identification of nonlinear systems using a Hammerstein model" (Narendra, 1996), here it was proposed the traditional iterative algorithm; in "A new Identification Method for Wiener and Hammerstein Systems" (Guo, 2004), was developed a unified new recursive identification method in the prediction error; in "Worst-Case Identification of Nonlinear Fading Memory Systems" (Dahleh, 1995) was studied the problem of asymptotic identification for fading memory in the presence of bounded noise; in "Identification of Multivariable Hammerstein Systems using Rational Orthonormal Bases" (Gomez, 2004), it is presented a non iterative algorithm for the simultaneous identification of the linear and nonlinear parts of multivariable Hammerstein systems.

For the purpose of putting into context this work, we want to mention two previous works that deal with estimating parameters using evolutionary computation, they are: "The evolutionary learning rule for system identification" (Montiel et al., 2004a) and "Asynchronous hybrid architecture for parametric system identification using fuzzy real coded evolutionary algorithm" (Montiel et al., 2004b). In both papers were shown results of parameter estimation for the Finite Impulse Response (FIR) filter using the BGA.

A different approach is to use genetic programming (GP) mainly focusing on generating algebraic expression for describing a physical system instead of estimating parameters of a model structure. In this branch there are some interesting works, such as, "Finding an Impulse Response Function Using Genetic Programming" (Keane Martin A. et al., 1993) where GP was applied to obtain a symbolic expression for a linear time-invariant system (LTI); two representative works applied to nonlinear systems are: "Multiobejctive Genetic Programming: A Nonlinear System Identification Application" (Rodríguez et al., 1997), and "Identifying Nonlinear Model Structures Using Genetic Programming Techniques" (Winkler S., et al., 2004).

2 System Description

In Fig. 2. System identification with noise presence.

We have a digital signal input $u(k)$ that is fed to the unknown system and to the adaptive system at the same time, in this figure the unknown system is enclosed by dashed lines in a "black box" (Sjoberg, 1995), its output is called the desired response signal and it is represented by $y(k)$. The adaptive system, i.e. the Wiener model will compute a corresponding output signal sample $\hat{y}(k)$ at time k. Both signals, $y(k)$ and $\hat{y}(k)$ are compared subtracting the two samples at time k, to obtain the error signal, $e(k)$,

$$e(k) = y(k) - \hat{y}(k) \tag{1}$$

The adaptive system has the task of representing accurately the signal $y(k)$ at its output, i.e, $y(k) = \hat{y}(k)$. At the unknown system side, we have an additive noisy signal known as the observation noise signal because it corrupts the observation of the signal at the output of the unknown system (Vijay, 1997), then

$$y(k) = y_u(k) + \eta(k) \tag{2}$$

The unknown system can be any system, a simple or a complex system. We used a nonlinear autoregressive with exogenous input (NARX) first order Wiener model given by (Nelles, 2001),

$$y(k) = \arctan(0.1 * u(k-1) + 0.9 * \tan(y(k-1))) \tag{3}$$

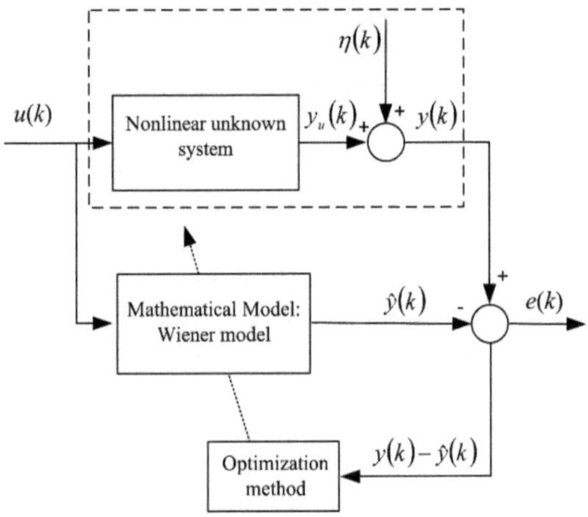

Fig. 2. System identification with noise presence.

The Wiener model structure consists of a linear dynamic block followed by a nonlinear static block. We selected an NFIR Wiener model structure described by (Nelles, 2001)(Gomez Juan C., 2004),

$$\hat{y}(k) = \arctan\left(\sum_{i=0}^{L-1} h_i(k) x(k-i) \right) \tag{4}$$

or in vectorial form

$$\hat{y}(k) = \arctan\left(H^T(k) X(k) \right) \tag{5}$$

where the coefficient vector $H(k)$ is

$$H(k) = [h_0(k)\ h_1(k)\ \ldots\ h_{L-1}(k)] \tag{6}$$

and the input signal vector is given in vectorial form by,

$$X(k) = [x(k)\ x(k-1)\ \ldots\ x(k-L+1)] \quad (7)$$

In real world problems, the identification is successful if we meet with some criterion in the error value. Moreover, in real world problems, the output of the unknown system $y_u(k)$ is contaminated with noise $\eta(k)$. Generally, we do not have direct access to the uncorrupted output $y_u(k)$ of the plant; instead we have a noisy measurement of it. In this case the output is given by equation (2). Then, we can say that the adaptive filter has reached the optimum if we find a value $y(k) = \hat{y}(k)$ this is achieved when we find an optimum weight vector's parameter *H*, such as

$$H(k) = H_{OPT}(k) \quad (8)$$

3 Evolutionary Optimization Technique

System identification uses a supervised learning method (Jang, 1997), for estimating the optimum parametric vector *H*(*k*) it is necessary to use an optimization technique. We selected the evolutionary algorithm known as BGA (Mühlenbein, 1994), and we tested it using the fuzzy recombination (FR) operator (Voigt, 1995). The BGA allows us to represent in a direct way floating point numbers, so the encoding and decoding of each variable is transparent for the user (Deb Kalyanmov, 2002). This algorithm uses a deterministic selection mechanism implemented using truncation selection, only a percent of the best individuals of the whole population is selected for recombining, in this way the survivor of the best individuals is guaranteed, and the extinction of the worst individuals is also guaranteed. The BGA is defined as an eight tuple,

$$BGA = \left(H^0, N, T\%, \Gamma, \Delta, HC, F, term\right) \quad (9)$$

where H^0 is the initial population of size *N, T is* the truncation threshold commonly referred as *T%*, Γ represents the recombination operator, Δ is the mutation operator, *HC* is a hill climbing method (for example: the LS algorithm), *F* is the fitness function and *term* is the termination criterion. In FR for obtaining an offspring it is necessary to recombine the individuals of the population, say $X=(x_1,\ldots,x_n)$, and $Y=(y_1,\ldots,y_n)$ to obtain $Z=(z_1,\ldots,z_n)$ (Voigt, 1995), the offspring z_i is obtained using triangular membership functions, where u_i and y_i are the modes. Equation (10) is the membership function of a normalized triangular fuzzy number, where *m* is the mode and s represents the spread of the fuzzy number, for example for the offspring x_i, we have

$$\mu(x_i)_T\{s,m\} = 1 - \frac{2\,|\,m - x_i\,|}{s} \tag{10}$$

and the corresponding probability distribution function (pdf) is

$$\Pr(x_i)_T\{s,m\} = \begin{cases} 0 & x_i < m - s \\ \frac{1}{s^2}(x_i + s - m) & m - s \le x_i \le m \\ \frac{1}{s^2}(-x_i + s + m) & m \le x_i \le m + s \\ 0 & x_i > m + s \end{cases} \tag{11}$$

Equation (11) is a unimodal pdf of the allele x_i. For generating the allele successor z_i, we will need a bimodal pdf, so it will be necessary to obtain a unimodal pdf for each allele of each parent. Equation (12) represents the bimodal pdf for the allele z_i,

$$\Pr(z_i)_{BT}\{s_1,m_1,s_2,m_2\} = \frac{1}{2}\left(\Pr(x_i)_T\{s_1,m_1\} + \Pr(y_i)_T\{s_2,m_2\}\right) \tag{12}$$

where the range for each triangular membership function is given in equations (13) and (14)

$$x_i - a \le x_i \le x_i + a \tag{13}$$

$$y_i - a \le y_i \le y_i + a \tag{14}$$

the offspring z_i can lie in one or both of the intervals, the variable a is given by

$$a = e\left|y_i - x_i\right|, \qquad e > 0 \tag{15}$$

Using equations (14), (15) and (16) we have that

$$\Pr(z_i) = \Pr(z_i)_{BT}\left\{e\left|y_i - x_i\right|, x_i, e\left|y_i - x_i\right|, y_i\right\} \tag{16}$$

in equation (15), the variable e is the fuzzy spread of the fuzzy numbers, generally e is selected to be 0.5. The mutation operator is applied to each offspring, and the resulting individuals are inserted in the new population $H_r(n)$. The process is repeated until a termination criterion is met.

The goal of the mutation operator Δ is to modify one or more parameters of z_i, the modified objects (i.e., the offsprings) appear in the landscape within a certain distance of the unmodified objects (i.e., the parents). In this way, an offspring Z', where $Z'=(z_1,\ldots,z_n)$ is given by (Mühlenbein Heinz and Schilierkamp-Voosen, 1993)(De Falco et al., 1997),

$$z_i' = z_i \pm range_i \cdot \delta \tag{17}$$

where $range_i$ defines the mutation range, and is calculated as ($\lambda \cdot searchinterval_i$). The variable *searchinterval* is defined into the domain of the variable to be mutated, in this case $[-1, 1]$. In the Discrete Mutation operator (DM) λ is normally set to 0.1 or 0.2 and is very efficient in some functions, but also we can set λ to 1. The sign + or – is chosen with probability of 0.5. The variable δ is computed by

$$\delta = \sum_{i=0}^{J-1} \varphi_i 2^{-i} \qquad \varphi_i \in 0,1 \tag{18}$$

Before mutation we set each φ_i equals to 0, then each φ_i is mutated to 1 with probability $p = 1/J$, and only $_j$=1 contributes to the sum. On the average there will be just one φ_i with value 1, say φ_j. Then is given by:

$$\delta = 2^{-i} \tag{19}$$

In formula (18), J is a parameter originally related to the machine precision, i.e. the number of bits used to represent a real variable in the machine we are working with, traditionally J used values of 8 and 16. In practice, however, the value of J is related to the expected value of mutation steps, in other words, the higher J is, the more fine-grained is the resultant mutation operator (De Falco, 1997).

4 Training Signal Generation

For identifying a process adequately is necessary to use an appropriated excitation signal. Nonlinear processes require excitation of their dynamic and static properties in all relevant points, so it will be necessary to use a sequence that combines excitation of both parts, static and dynamic. For these reasons it is common to use an amplitude modulated pseudo random binary signal (APRBS). Fig. 3 shows an APRBS signal (solid line), it is the input to both systems, and it is used for training the adaptive model, in this figure is magnified the minimal hold time (T_h). This time is the minimal step size of the signal, i.e., it is the shortest period of time for which the signal stays constant. For a given signal length, T_h will have a direct influence in the number of steps and the frequency characteristics. The selection of T_h is different for linear and nonlinear SI, in linear SI this time is selected equal to the sampling time, at the other hand for nonlinear SI T_h should be selected neither too small nor too large. If we select T_h too small then the system will not have time to settle, only operating conditions around $y_0 \approx (u_{\max} + u_{\min})/2$ will be covered (Nelles, 2001). If T_h is too large, then for a given signal only a reduced set of operating conditions will be covered, the amount of points mainly depends on the number of steps with different amplitude. As a consequence, if we do not modulate in amplitude the signal and we only use pseudo random signals (PRBS), we will have only two operating conditions, one for each signal's value. In Fig. 3 also we have a dashed line which is the output of the unknown system. Fig. 4. shows the data distribution of the training data of Fig. 3, there are some holes and they are random located, depending on the amplitude levels. Nevertheless, the holes trend to disappear or at least they will become smaller as the length of the training signal increases.

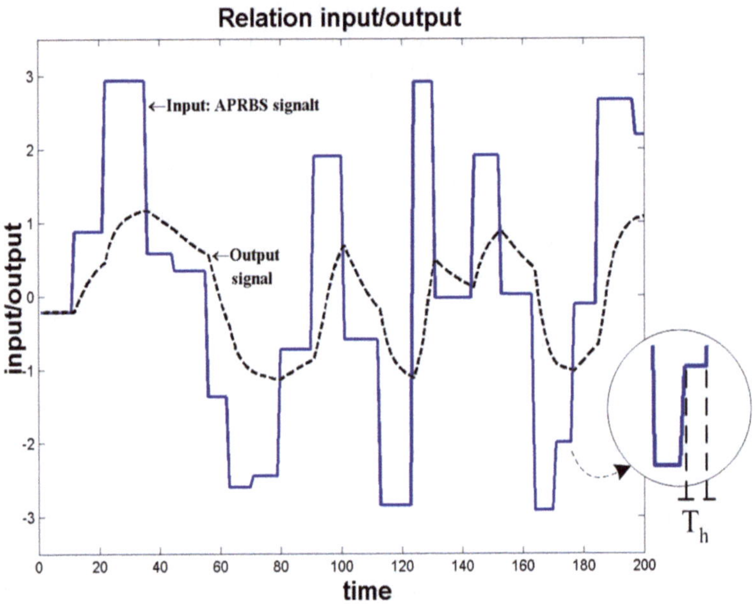

Fig. 3. At the input of both systems (unknown and adaptive model structure) we applied an APRBS signal (solid). This APRBS signal consists of 200 samples, its amplitude is in the interval of [-3,3]. The minimal hold time T_h is 5 samples and the maximum hold time Tm_h is 15 samples. Here, we are showing the unknown system's output with a dashed line.

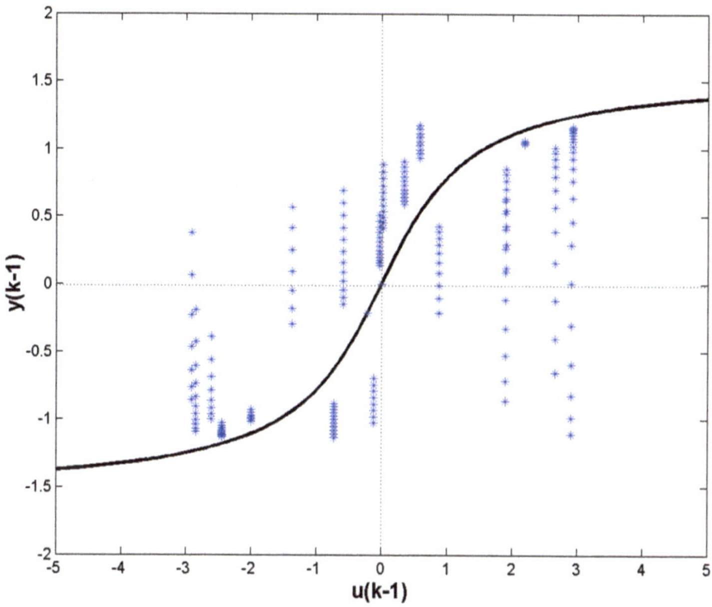

Fig. 4. Data distribution of the training signal shown in Fig. 3.

5 Experimental Results

In Fig. 5., we are showing an outline sketch of the software implementation that we used for identifying the unknown system represented by equation (3). This software uses the evolutionary floating point algorithm known as BGA explained in section 3. We used a population size of 1,000 individuals floating point coded.

We implemented the fuzzy recombination (FR) operator described by equations 10 to 16, where we used for the variable *e* a value of 0.5.

For the mutation operator, we used the discrete mutation operator (DM) with a λ value of 0.1 and a range value of [-1,1] for the variable *searchinterval*. We calculated equation 18 using $J = 16$. The new mutated offspring was calculated using equation 17, using the specifications mentioned above.

```
gen=0    %Initial generation
Initialize population of size N called H^gen, for the initial generation we have H^0.
Generate an APRBS.
Evaluate H^gen. Each row of H^gen contains the coefficients of the adaptive model; i.e.,
    each row is representing one individual. For evaluating the system, apply an
    APRBS to the unknown system, and to the adaptive model at the same time.
    Calculate the average error using an objective function (OF) for each proposal
    model. The fitness value FV is the reciprocal value of (OF).
While termination criterion is not fulfill do{
    %Selection procedure and generation of the new population
    Generate a new APRBS and apply it to the unknown system for obtaining
        the desired response sequence.
    Create a subpopulation by selecting a percentage T% of the best
        individuals in the population H^gen .
    Generate succesors {   %Create new population
        Select two individuals from the subpopulation for recombining them
            according the rules of BGA.
        Apply recombination operator for generating an individual (offspring).
        Apply the selected mutation operator to the individual.
        Evaluate successor. Apply the same APRBS obtained at this time
            in this loop for calculating the system response of the adaptive
            model, then calculate the FV for this individual.
    }
    Substitute the old population with the new one.
    gen=gen+1    %Generation counter
}   %end do
End
```

Fig. 5. Outline of the software implementation for solving the nonlinear system identification problem. We tested the software for obtaining models of the NFIR type model structure. This software can be applied for solving the generic nonlinear system identification problem known as nonlinear autoregressive moving average with exogenous input (NARMAX) type.

The BGA uses a deterministic selection mechanism in the sense that only the best individuals are selected for creating the offspring by means of applying the recombination and mutation operators, moreover, we saved the best individual, i.e. the individual with the highest fitness value through generations, this is shown in Fig. 6. with small circles 'o'. The average fitness value of the whole population at each generation is shown in Fig. 6. with 'x'.

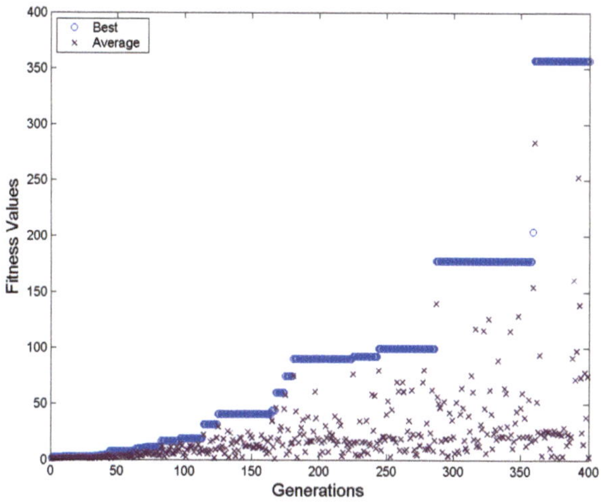

Fig. 6. Here we are showing the best fitness value at each generation, we preserved the best individual at each generation. At generation 360 we found the best fitness values of 357.1027. The average square error is 0.0028, we used 80 samples for computing this value. The fittest individual at each generation is plotted using a small 'o', and the average fitness value of the population at each generation is shown with 'x'. Note that although the selection procedure is deterministic and we are including the fittest individual through generation, the algorithm still have good explorative characteristics, preventing the population to fully converge.

At the adaptive side, we used a Wiener model structure for modeling the system, the parameters were estimated using the BGA with the above mentioned characteristics. For training the adaptive system, we made in Fig. 2, $\eta(k)=0$; but for testing the system we used random noise with normal distribution with mean equals to zero and variance $\sigma^2=0.1$. We ran the algorithm 400 generations, at each generation we applied an APRBS to the systems: the unknown, and the adaptive model. This signal (APRBS) was generated with the next characteristics: 200 samples, with random amplitude in the interval of [-3,3], the minimal hold time T_h is 5 samples, moreover, we used maximum hold time Tm_h consisting of 15 samples.

For testing the system we applied to the unknown system a sinusoidal signal, this signal and its corresponding output is shown in Fig. 7. In Fig. 8 we have the data distribution of this signal at the unknown system side.

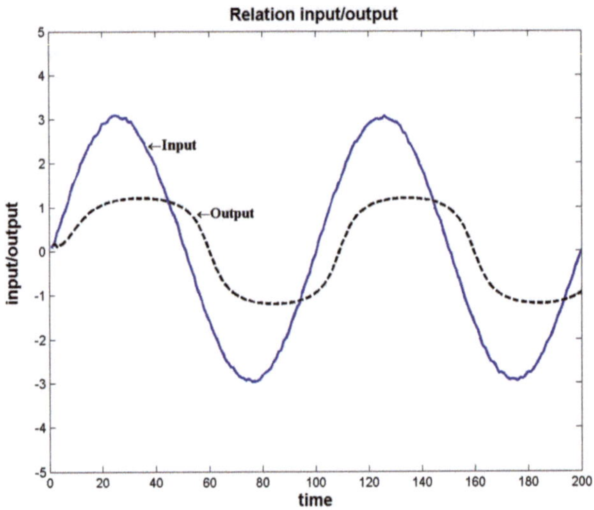

Fig. 7. We used a sinusoidal sequence for testing the system (solid line), the system's output is shown with a dashed line.

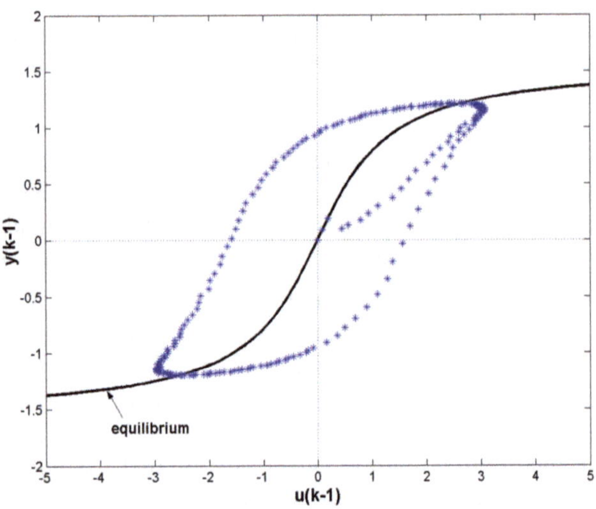

Fig. 8. Unknown system data distribution for the sinusoidal signal of Fig. 7.

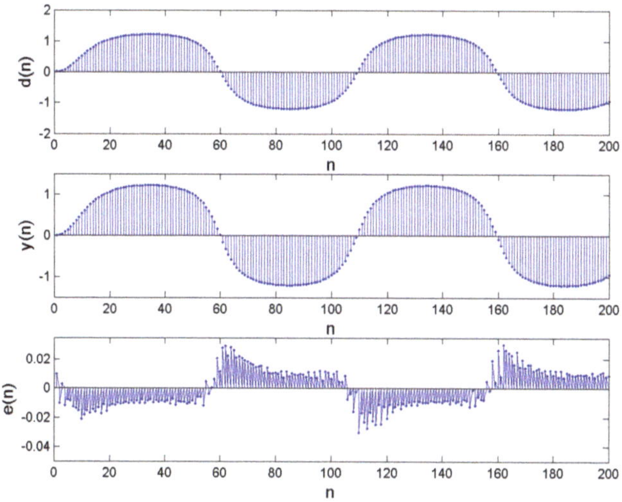

Fig. 9. In this graphic we have three plots; they were obtained using the best parametric vector found for the NFIR model structure. The upper one is the unknown system output; at the middle, we have the NFIR model output; and at the bottom we have the error sequence obtained subtracting the two preceding plots. For the error signal we found that the maximum value of this sequence is 0.02996, the mean value is 0.0002821, the median value is -0.003728, and the standard deviation value is 0.01324. These values were obtained considering 200 samples.

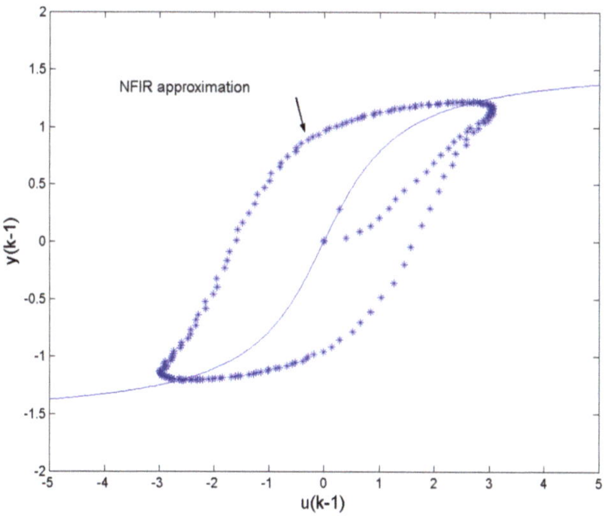

Fig. 10. Data distribution at the adaptive system side. This signal was obtained when we applied a sinusoidal excitation (**Fig. 7.**) at the NFIR model once it was trained; i.e., with the optimal parameters.

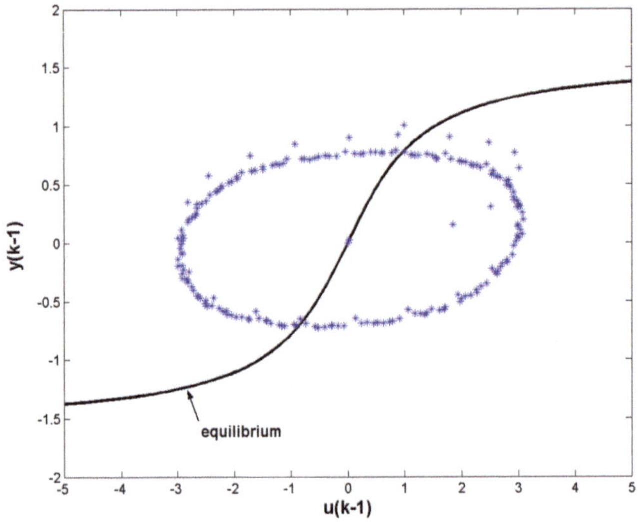

Fig. 11. This figure was obtained when we increase the sinusoidal frequency almost two times.

6 Conclusions

Global search capabilities of evolutionary algorithms can be exploited for estimating parameters in nonlinear input-output models structures such as in a Wiener model structure. This is an off-line option for system identification since it is highly computational time demanding, but it could be implemented in on-line application working in a second plane, searching for better models. Using an NFIR model structure gives us stability since it does not have feedback, but the price to pay is that we will need higher dynamic order for describing the process dynamics properly. The NFIR model can represent an unstable model for the first L samples, where L is the filter's order. For improving the performance of this method it is necessary to implement faster evolutionary strategies and better fitness functions.

Acknowledgment

The authors thank "Comisión de Operación y Fomento de Actividades Académicas del I.P.N.", and "Instituto Tecnológico de Tijuana" for supporting our research activities.

References

Dahleh M. A., E. D. Sontag, D. N. C. Tse, and J. N. Tsitsiklis, 1995, "Worst-case identification of nonlinear fading memory systems", Automatica, vol. 31, pp. 303–308. http://citeseer.ist.psu.edu/dahleh95worstcase.html

De Falco, I., Della Cioppa, A., Natale, P., Tarantino, E. (1997), "Artificial Neural Networks Optimization by means of Evolutionary Algorithms", http://citeseer.nj.nec.com/defalco97artificial.html

Deb Kalyanmoy (2002), "Multi-Objective Optimization using Evolutionary Algorithms", John Wiley & Sons, LTD, New York, USA.

Gomez Juan and Baeyens Enrique (2004), "Identification of Multivariable Hammerstein Systems using Rational Orthonormal Bases", http://citeseer.ist.psu.edu/421047.html.

Gómez Juan C., Enrique Baeyens (2004), "Identification of Nonlinear Systems using Orthonormal Bases", http://citeseer.ist.psu.edu/596443.html

Guo Fen (2004), "A New Identification Method for Wiener and Hammerstein Systems", Institut für Angewandte Informatik, http://bibliothek.fzk.de/zb/berichte/FZKA6955.pdf

Ikonen E., Najim K. (1999), "Learning control and modelling of complex industrial processes, Overview report of our activities within the European Science Foundation's programme on Control of Complex Systems (COSY) Theme 3: Learning control". http://cc.oulu.fi/~iko/lccs.html

Keane Martin A., Koza John R., Rice James P. (1993). Finding an impulse response function using genetic programming. In Proceedings of the 1993 American Control Conference, volume 3, pages 2345–2350, San Francisco, CA, 2.-4. June 1993. IEEE, New York. http://citeseer.ist.psu.edu/keane93finding.html

Ljung Lennart (1999), "System Identification. Theory for the User. Second Edition", Prentice Hall PTR, USA.

Nelles Oliver (2001), "Nonlinear System Identification. From Classical Approaches to Neural networks and Fuzzy Models", Springer-Verlag Berlin Heidelberg. Germany. 2001. pp. 15, 457-511.

Montiel R. Oscar, Oscar Castillo, Roberto Sepúlveda, Patricia Melin (2004a), "The evolutionary learning rule for system identification", Applied Soft Computing Journal. Special issue: Soft Computing for Control of Non-Linear Dynamical Systems. Volume 3, Issue 4. December 2003. pp. 343-352

Montiel Oscar, Oscar Castillo, Patricia Melin, Roberto Sepúlveda (2004b), "Asynchronous hybrid architecture for parametric system identification using fuzzy real coded evolutionary algorithm", Nonlinear Studies, Volume 11, Number 1.

Mühlenbein Heinz, Dirk Schlierkamp-Voosen (1994), "The science of breeding and its application to the breeder genetic algorithm BGA". Evolutionary Computation, 1(4):335-360.

Mühlenbein Heinz, Evolutionary Algorithms: Theory and Applications, http://citeseer.ist.psu.edu/110687.html.

Mühlenbein Heinz and Schilierkamp-Voosen (1993), "Predictive Model for Breeder Genetic Algorithm", Evolutionary Computation. 1(1): 25-49.

Narendra K. S. and P. G. Gallman (1996), "An iterative method for the identification of nonlinear systems using a Hammerstein model". IEEE Transactions on Automatic Control, AC-11:546-550. July 1966.

Rodríguez Katya Vázquez, Fonseca Carlos M. Fleming Peter J. (1997). Multiobjective Genetic Programming: A Nonlinear System Identification Application. Late Breaking Papers at the Genetic Programming 1997 Conference, Editor John R. Koza, Standford Bookstore, USA, pp. 207-212.

Severance Frank L. (2001), "System Modeling and Simulation. An Introduction", John Wiley & Sons Ltd., UK.

Sjoberg, J., Q. Zhang, L. Ljung, A. Benveniste, B. Delyon, P.-Y. Glorennec, H. Hjalmarsson and A. Juditsky (1995). "Nonlinear black-box modeling in system identification: a unified overview". Automatica 31(12), 1691--1724. http://citeseer.nj.nec.com/sjoberg95nonlinear.html

Vijay K. Madisetti, Douglas B. Williams (1997). "The Digital Signal Processing Handbook", A CRC Handbook Published in Cooperation with IEEE Press, pp. 15-1, 18-1, 18-12, 20-1, 20-4.

Jang J.-S.R., C.-T. Sung, E. Mizutani (1997), "Neuro-Fuzzy and Soft Computing. A Computational Approach to Learning and Machine Intelligence". Prentice Hall. NJ, USA.

Voigt H.M., Mühlenbein, D. Cvetkovic (1995), "Fuzzy Recombination for the Breeder Genetic Algorithm", Proceedings of the Sixth International Conference on Genetic Algorithms, published by Morgan Kaufmann, pp. 1104-111.

Winkler S., Affenzeller M., Wagner S. (2004), "Identifying Nonlinear Model Structures Using Genetic Programming Techniques". *Cybernetics and Systems 2004*, pp. 689-694. Austrian Society for Cybernetic Studies, 2004.

Synchronization of Chaotic Neural Networks: A Generalized Hamiltonian Systems Approach

C. Posadas-Castillo[1], C. Cruz-Hernández[2], D. López-Mancilla[3]

[1] Engineering Faculty, Baja California Autonomous University (UABC) and FIME University Autonomous of Nuevo León (UANL), México.
E-mail:cposadas@fime.uanl.mx,

[2] Telmatics Direction, Scientific Research and Advanced Studies of Ensenada (CICESE), Km. 107 Carretera Tijuana-Ensenada, 22860 Ensenada, B.C., México.
E-mail:ccruz@cicese.mx,

[3] Engineering Faculty, Baja California Autonomous University (UABC), México.
E-mail: didier@uabc.mx

Abstract. In this paper, we use a Generalized Hamiltonian forms approach to synchronize chaotic neural networks unidirectionally coupled. Synchronization is thus between the master and the slave networks with the slave network being given by an observer. In particular, we present two cases of study: the first is a second-order 3×4 **CNN** array, and the second is a **CNN** with delay. The chaotic **CNNs** are used as transmitter and receiver in encrypted information transmission.

1 Introduction

In recent years many complex network structures have been observed in diverse fields as physics, biology, economics, ecology, electronics and computer science. In particular, Cellular Neural Networks (**CNNs**) constitute an important example in such cases. **CNN** is a nonlinear system defined by coupling only identical simple dynamical systems called cells located within a prescribed sphere of influence, such as nearest neighbors [3]. **CNN** has broad applications in image and video signal processing, robotic and biological visions [30], and higher brain functions [18]. Many proceedings of workshop and special issues see e.g., [23]; [24]; [25]; [26] have been devoted to **CNNs.**

On the other hand, recently synchronization of complex dynamics (chaotic and hyperchaotic) has become a field of active research see e.g., [20];

C. Posadas-Castillo et al.: *Synchronization of Chaotic Neural Networks: A Generalized Hamiltonian Systems Approach*, StudFuzz **208**, 59–73 (2007)
www.springerlink.com

[17]; [27]; [28]; [29]; [10]; [4]; [5]; [22]; [21]; [1]; [7]; [11]; [12]; [13]; and references therein. Data encryption using chaotic dynamics was reported in the early 1990's as a new approach for signal encoding which differs from the conventional methods using numerical algorithms as the encryption key. One of the motivations for synchronization is the possibility of sending confidential information through chaotic signals for secure communications. The idea is use two highly dynamic nonlinear systems (as transmitter and receiver). So, the confidential information is imbedded into the transmitted chaotic signal by direct modulation, masking or another method. At the receiver end, if chaos synchronization can be achieved, then it is possible to recover the original information. The communication schemes based on chaos synchronization can be broadly categorized into three approaches. They include the chaotic masking scheme [8], the chaotic shift keying scheme [19]; [8]; [9], and the chaotic modulation scheme [31].

The main goal of this paper is to synchronize chaotic neural networks. This objective is achieved by using Generalized Hamiltonian forms and observer approach developed in [22]. Moreover, we proceed to illustrate this synchrony to transmit encrypted confidential information using a modified chaos-based communication scheme [16]; [14]. The synchronization method presents the following advantages: i) it is systematic, ii) it is useful to synchronize several well-known chaotic and hyperchaotic oscillators, iii) it does not require the computation of any Lyapunov exponent, and iv) it does not require initial conditions belonging to the same basin of attraction.

The paper is organized as follows: In Section 2, we give a brief review on chaos synchronization via Generalized Hamiltonian forms and observer approach. In Section 3, we apply this approach to synchronize chaotic neural networks using two numerical examples; a second-order 3×4 **CNN** array and a **CNN** with delay. In Section 4, we present the stability analysis related to the synchronization process. In Section 5, we apply the synchronization of chaotic neural networks to confidential communication for transmission and recovering of audio messages. Finally, in Section 6, we give some concluding remarks.

2 Review of Chaos Synchronization via Hamiltonian Forms and Observer Approach

Consider the following n-dimensional autonomous system

$$\dot{x} = f(x(t)), \qquad x \in \Re^n, \tag{1}$$

which provides an example of complex oscillator, whit f a nonlinear function of the state x. Following the approach provided in [22], many **CNN** models described by Eq. (1) can be written in the following "Generalized Hamiltonian" canonical form,

$$\dot{x} = J(x)\frac{\partial H}{\partial x} + S(x)\frac{\partial H}{\partial x} + F(x) \tag{2}$$

where *H(x)* denotes a smooth *energy function* which is globally positive definite in $\Re^n$. The column *gradient vector* of *H*, denoted by $\partial H/\partial x$, is assumed to exist everywhere. We use quadratic energy function $H(x) = 1/2\, x^T Mx$ with M being a, constant, symmetric positive definite matrix. In such a case, $\partial H/\partial x = Mx$. The square matrices, $J(x)$ and $S(x)$ satisfy, for all $x \in \Re^n$, the following properties, which clearly depict the *energy managing* structure of the system, $J(x) + J^T(x) = 0$ and $S(x) = S^T(x)$. The vector field $J(x)\,\partial H/\partial x$ exhibits the *conservative* part of the system and it is also referred to as the *workless* part, or *workless* forces of the system; and $S(x)$ depicting the *working* or *nonconservative* part of the system. For certain systems, $S(x)$ is *negative definite* or *negative semidefinite*. In such cases, the vector field is addressed to as the *dissipative* part of the system. If, on the other hand, $S(x)$ is positive definite, positive semidefinite, or indefinite, it clearly represents, respectively, the global, semiglobal and local *destabilizing* part of the system. In the last case, we can always (although nonuniquely) descompose such an indefinite symmetric matrix into the sum of a symmetric negative semidefinite matrix $R(x)$ and a symmetric positive semidefinite matrix $N(x)$. And where $F(x)$represents a *locally destabilizing* vector field.

We consider a special class of Generalized Hamiltonian systems given by

$$\begin{aligned} \dot{x} &= J(y)\frac{\partial H}{\partial x} + (I + S)\frac{\partial H}{\partial x} + F(y), && x \in \Re^n \\ y &= C\frac{\partial H}{\partial x}, && y \in \Re^m \end{aligned} \tag{3}$$

where S is a constant symmetric matrix, not necessarily of definite sign. The matrix I is a constant skew symmetric matrix. The vector variable $y(t)$ is referred to as the system *output*. The matrix C is a constant matrix. The destabilizing vector field $F(y)$.

We denote the **estimate** of the state vector $x(t)$ by $\xi(t)$, and consider the Hamiltonian energy function $H(\xi)$ to be the particularization of H in terms of $\xi(t)$. Similarly, we denote by $\eta(t)$ the estimated output, computed in terms of the estimated state $\xi(t)$. The gradient vector $\partial H(\xi)/\partial \xi$ is, naturally, of the form $M\xi$ with M being a, constant, symmetric positive definite matrix.

A dynamic nonlinear **state observer** for the special class of Generalized Hamiltonian forms (3) is readily obtained as

$$\dot{\xi} = J(y)\frac{\partial H}{\partial \xi} + (I+S)\frac{\partial H}{\partial \xi} + F(y) + K(y-\eta), \qquad (4)$$
$$\eta = C\frac{\partial H}{\partial \xi},$$

where K is a constant matrix, known as the *observer gain*. The **state estimation error**, defined as $e(t) = x(t) - \xi(t)$ and the output estimation error, defined as $e_y(t) = y(t) - \eta(t)$, are governed by

$$\dot{e} = J(y)\frac{\partial H}{\partial e} + (I+S-KC)\frac{\partial H}{\partial e}, \quad e \in \Re^n \qquad (5)$$
$$e_y = C\frac{\partial H}{\partial e}, \quad e_y \in \Re^m$$

where the vector, $\partial H/\partial e$ actually stands, with some abuse of notation, for the gradient vector of the *modified* energy function, $\partial H(e)/\partial e = \partial H/\partial x - \partial H/\partial \xi = M(x-\xi) = Me$. We set, when needed, $I + S = W$.

Definition 1 (Complete synchronization problem) We say that the slave system (4) synchronizes with the master system (3), if

$$\lim_{t\to\infty} \|x(t) - \xi(t)\| = 0,$$

no matter which initial conditions $x(0)$ and $\xi(0)$ have. Where the state estimation error $e(t) = x(t) - \xi(t)$ represents the synchronization error.

3 Synchronization of Chaotic Neural Networks: Examples

In this section, we present two numerical examples of synchronization of chaotic neural networks, to this purpose, let us first briefly give a suitable material on CNN.

Definition 2 (**CNN**) A **CNN** is any spatial arrangement of **locally coupled cells**, where each cell is a dynamical system which has an **input**, and a **state** evolving according to some prescribed dynamical laws [3].

In three-dimensional lattice **CNN** architecture, mathematically each cell C_{ijk} at location (i,j,k) is a dynamical system whose states evolve according to some prescribed state equations, whose dynamics are **coupled** only among the neighboring cells lying within some prescribed **sphere of influence** S_{ijk}, centered at (i,j,k). In two-dimensional case, using a double subscript, the variables for an **isolated** cell are: input $u_{ij}(t) \in \Re^u$, threshold $z_{ij}(t) \in \Re^z$, state $x_{ij}(t) \in \Re^x$, and output $y_{ij}(t) \in \Re^y$. A **CNN** cell is said to be isolated if it is not coupled to any other cell (Fig. 1).

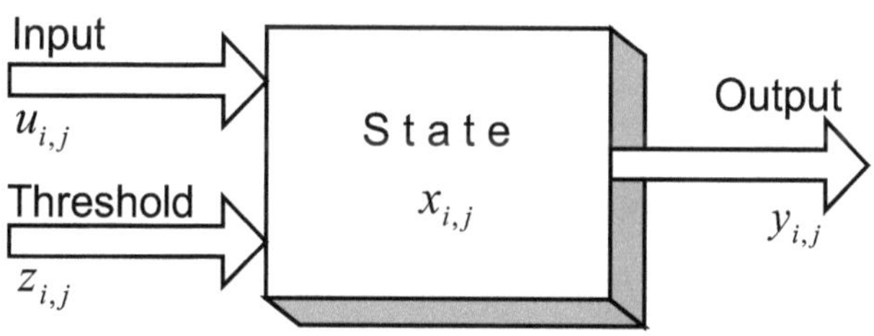

Fig. 1: Isolated cell: input u_{ij}, threshold z_{ij}, state $x_{ij}(t) \in \Re^x$, and output y_{ij} for a two-dimensional **CNN**.

In this work, we will assume that all isolated cells C_{ij} are identical, and that for simplicity we have that $z_{ij}(t)$ is a constant scalar. Besides, we assume that for any $x_{ij}(t_0)$ at $t = t_0$, any threshold $z_{ij}(t)$, and any input $u_{ij}(t)$, the state of each isolated cell C_{ij} is assumed to evolve for all $t > t_0$ as a nonautonomous set of ordinary differential equations

$$\dot{x}_{ij} = f(x_{ij}, z_{ij}, u_{ij}), \qquad i = 1, 2, ..., M; \quad j = 1, 2, ..., N$$
$$y_{ij} = g_{ij}(x_{ij})$$

where $g_{ij}(\cdot)$ is a nonlinear function of the state. However, in many cases the output of interest often coincides with the state, $y_{ij}(t) = x_{ij}(t)$.

The **standard CNN equations** used most widely in the literature, proposed in [2] for an M×N **CNN** array

$$\dot{x}_{ij} = -x_{ij} + z_{ij} + \sum_{kl \in S_{ij}(r)} a_{kl} y_{kl} + \sum_{kl \in S_{ij}(r)} b_{kl} u_{kl} \quad i = 1, 2, ..., M; \quad j = 1, 2, ..., N \tag{6}$$

$$y_{ij} = f(x_{ij}), \tag{7}$$

where $S_{ij}(r)$ is the sphere of influence of radius r; $\sum_{kl \in S_{ij}(r)} a_{kl} y_{kl}$ and $\sum_{kl \in S_{ij}(r)} b_{kl} u_{kl}$ are the local coupling, and

$$f(x_{ij}) = \frac{1}{2}\left(|x_{ij} + 1| - |x_{ij} - 1|\right) = \begin{cases} 1, & x_{ij} \geq 1 \\ x_{ij}, & |x_{ij}| < 1 \\ -1, & x_{ij} \leq -1 \end{cases}$$

For the particular case where $M = 3$ and $N = 4$, the Eqs. (6)-(7) assume the simpler form 3×4 **CNN** array

$$\begin{aligned} \dot{x}_1 &= -x_1 + a_{00} f(x_1) + a_{01} f(x_2) + b_{00} u_1(t), \\ \dot{x}_2 &= -x_2 + a_{0,-1} f(x_1) + a_{00} f(x_2) + b_{00} u_2(t), \\ y_1 &= f(x_1), \\ y_2 &= f(x_2). \end{aligned} \tag{8}$$

Example 1 [3] Consider the second-order nonautonomous **CNN**. If $a_{00} = 2, a_{0,-1} = -a_{0,1} = 1.2, b_{00} = 1, u_1(t) = 4.04 \sin\left(\frac{\pi}{2} t\right)$, and $u_2(t) = 0$; then Eq. (8) becomes

$$\begin{aligned} \dot{x}_1 &= -x_1 + 2 f(x_1) - 1.2 f(x_2) + 4.04 \sin\left(\frac{\pi}{2} t\right), \\ \dot{x}_2 &= -x_2 + 1.2 f(x_1) + 2 f(x_2), \end{aligned} \tag{9}$$

with nonlinear function

$$f(x) = \frac{1}{2}\left(|x + 1| - |x - 1|\right) = \begin{cases} 1, & x \geq 1 \\ x_{ij}, & |x| < 1 \\ -1, & x \leq -1 \end{cases} \tag{10}$$

Figure 2 shows a projection of the chaotic attractor of 3×4 **CNN** (9)-(10). The waveforms of $((x_1(t), x_2(t))$ corresponding to the $((x_1(t), x_2(t)) = (0.1, -0.1)$.

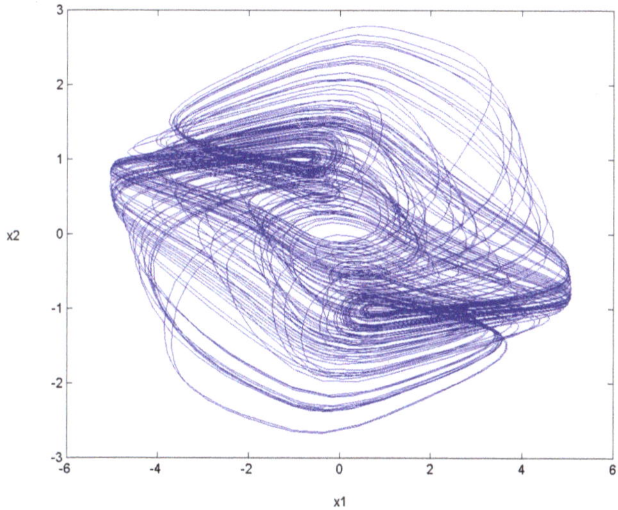

Fig. 2: Projection of the chaotic attractor of 3×4 **CNN** in the (x_1, x_2) plane.

The state equations describing the 3×4 **CNN** (9)-(10) in Hamiltonian canonical form with a destabilizing vector field (**master 3x4 CNN**) is given by

$$\begin{bmatrix} \dot{x}_1 \\ \dot{x}_2 \end{bmatrix} = \begin{bmatrix} 0 & -1.2 \\ 1.2 & 0 \end{bmatrix} \begin{bmatrix} f(x_1) \\ f(x_2) \end{bmatrix} + \begin{bmatrix} 2 & 0 \\ 0 & 2 \end{bmatrix} \begin{bmatrix} f(x_1) \\ f(x_2) \end{bmatrix} + \begin{bmatrix} 4.04 \sin\left(\frac{\pi}{2} t\right) - x_1 \\ -x_2 \end{bmatrix} \tag{11}$$

taking as the Hamiltonian energy function

$$H(x) = \int_0^{x_1} f(r_1) dr_1 + \int_0^{x_2} f(r_2) dr_2. \tag{12}$$

The destabilizing vector requires two signals for complete cancellation at the slave. Namely, the states $x_1(t)$ and $x_2(t)$. The output of the master (11) in this case, is then chosen as $y = (y_1, y_2)^T = (x_1, x_2)^T$. The matrices C, S, and I are found to be

$$C = \begin{bmatrix} 0 & 1 \end{bmatrix},\ I = \begin{bmatrix} 0 & -1.2 \\ 1.2 & 0 \end{bmatrix},\ S = \begin{bmatrix} 2 & 0 \\ 0 & 2 \end{bmatrix}.$$

The pair (C, S) is observable, and hence detectable. An injection of the synchronization error $e_2(t) = x_2(t) - \xi_2(t)$ suffices to have an asymptotically stable trajectory convergence. The **slave** (**3x4 CNN**) would then be designed as follows

$$\begin{bmatrix}\dot{\xi}_1\\ \dot{\xi}_2\end{bmatrix}=\begin{bmatrix}0 & -1.2\\ 1.2 & 0\end{bmatrix}\begin{bmatrix}f(\xi_1)\\ f(\xi_2)\end{bmatrix}+\begin{bmatrix}2 & 0\\ 0 & 2\end{bmatrix}\begin{bmatrix}f(\xi_1)\\ f(\xi_2)\end{bmatrix}+\begin{bmatrix}4.04\sin\left(\frac{\pi}{2}t\right)-\xi_1\\ -x_2\end{bmatrix}+\begin{bmatrix}k_1\\ k_2\end{bmatrix}e_2 \tag{13}$$

where $k=(k_1,k_2)^T$ is chosen in order to guarantee the asymptotic exponential stability to zero of the state reconstruction error trajectories (synchronization error). From (11) and (13) the synchronization error dynamics is governed by

$$\begin{bmatrix}\dot{e}_1\\ \dot{e}_2\end{bmatrix}=\begin{bmatrix}0 & -\frac{1}{2}k_1-1.2\\ -\frac{1}{2}k_1+1.2 & 0\end{bmatrix}\frac{\partial H(e)}{\partial e}+\begin{bmatrix}2 & -\frac{1}{2}k_1\\ -\frac{1}{2}k_1 & 2-\frac{1}{2}k_1\end{bmatrix}\frac{\partial H(e)}{\partial e}. \tag{14}$$

With initial states $(x_1(0),x_2(0))=(0.1,-0.1)$ and $(\xi_1(0),\xi_2(0))=(0.5,-0.5)$, and $k_1=k_2=2$ we obtain the following numerical results. Figure 3 shows synchronization between: a) $x_1(t)$ and $\xi_1(t)$, b) $x_2(t)$ and $\xi_2(t)$; solid line $x_i(t)$ and dashed line $\xi_i(t)$, $i=1,2$. c) and d) illustrate the time behaviors of the synchronization error trajectories $e_i(t)=x_i(t)-\xi_i(t)$, $i=1,2$. e) and f) x_i versus ξ_i in phase space.

Example 2: Time-delay oscillators represent examples of high-dimensional chaos generators. Now, the system considered is a cell equation in Cellular Neural Networks with delay [15]. Its model is given by

$$\dot{x}(t)=0.001x(t)-3.8\left(|x_\tau+1|-|x_\tau-1|\right)+2.85\left(\left|x_\tau+\frac{4}{3}\right|-\left|x_\tau-\frac{4}{3}\right|\right) \tag{15}$$

where $x_\tau=x(t-\tau)$. Its solution space is infinite-dimensional, with initial condition as any continuous function defined on the closed interval $[\tau,0]$. By considering $\tau=1$ and initial condition as a constant function equal to 0.5 on $[-1,0]$, and initial state $x(0)=-1$. Figure 4 shows a projection of the chaotic attractor of the cellular neural network with delay in the (x,x_τ) plane.

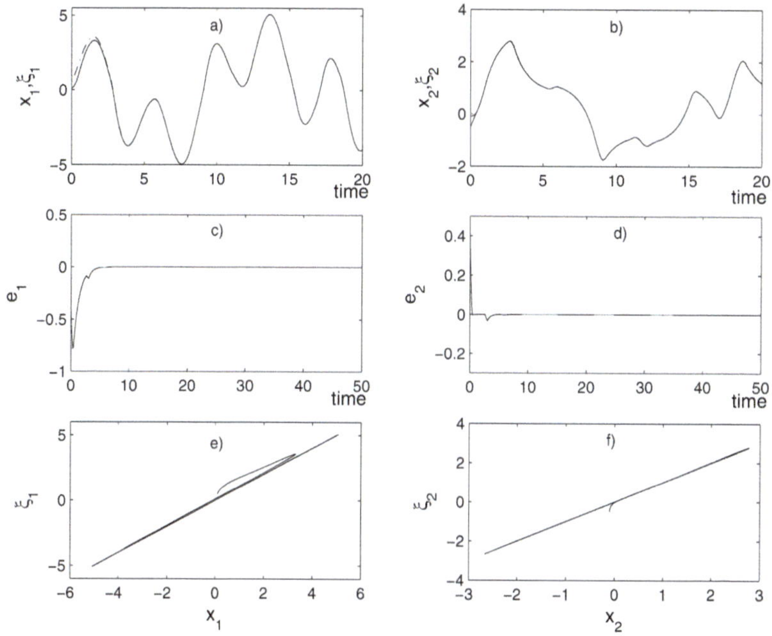

Fig. 3: a) and b) Synchronization between x_i and ξ_i; solid line x_i and dashed line $\xi_i(t)$, $i=1,2$. c) and d) the behaviors of the synchronization error trajectories $e_i(t)=x_i(t)-\xi_i(t)$, $i=1,2$. e) and f) x_i versus ξ_i in phase space.

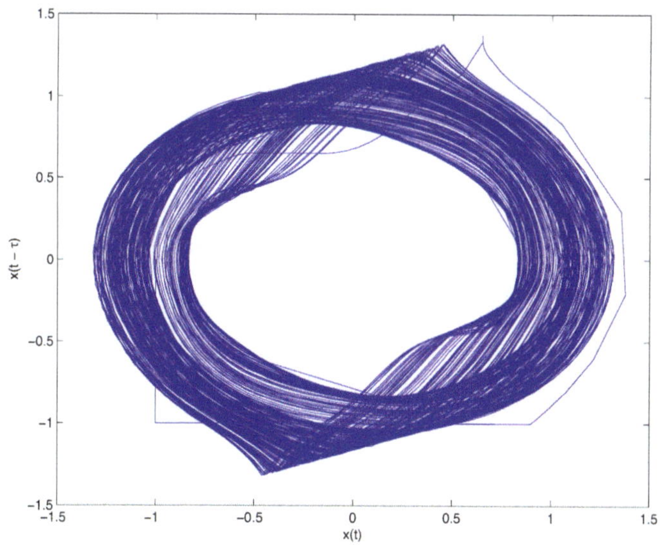

Fig. 4: Phase space dynamics for the Cellular Networks with delay projected onto the (x,x_τ) plane.

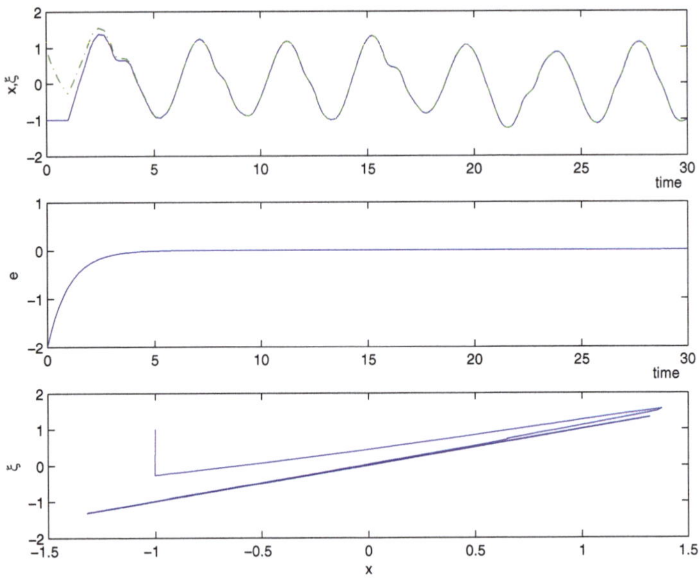

Fig. 5: Synchronization between the states $x(t)$ (solid line) and $\xi(t)$ (dashed line) (top of figure). Synchronization error (middle of figure). x versus ξ in phase space (bottom of figure).

The **CNN** with delay system (15) in Generalized canonical form (as master) is given by

$$\dot{x}(t) = 0.001\frac{\partial H(x)}{\partial x} - 3.8\left(|x_\tau + 1| - |x_\tau - 1|\right) + 2.85\left(\left|x_\tau + \frac{4}{3}\right| - \left|x_\tau - \frac{4}{3}\right|\right) \tag{16}$$

taking as Hamiltonian energy function

$$H(x) = \frac{1}{2}x^2 \tag{17}$$

with $\partial H(x)/\partial x = x$. It is clear that the system (16) is observable. The observer (as slave) for dynamics (16) is designed as

$$\dot{\xi}(t) = 0.001\xi(t) - 3.8\left(|\xi_\tau + 1| - |\xi_\tau - 1|\right) + 2.85\left(\left|\xi_\tau + \frac{4}{3}\right| - \left|\xi_\tau - \frac{4}{3}\right|\right) + k\,e(t), \tag{18}$$

where $e(t) = x(t) - \xi(t)$. From (16) and (18) the synchronization error dynamics is governed by

$$\dot{e}(t) = (0.001 - k)e(t). \tag{19}$$

Figure 5 depicts the synchronization between the state trajectories $x(t)$ (solid line) and $\xi(t)$ (dashed line) (top of figure), the time behavior of the synchronization error trajectory $e(t) = x(t) - \xi(t)$ (middle of figure), and x versus ξ (bottom of figure). When $x(0) = -1$ and $\xi(0) = 1$, and $k = 1$ are chosen.

4 Synchronization Stability Analysis

Now, we give conditions for asymptotic stability of the synchronization errors (14) and (19) between chaotic dynamics (11)-(13) and (16)-(18), respectively.

Theorem 1 [22] The state $x(t)$ of the nonlinear system (3) can be globally, exponentially, asymptotically estimated by the state $\xi(t)$ of an observer of the form (4), if the pair of matrices (C,W), or the pair (C,S), is either observable or, at least, detectable.

An observability condition on either of the pairs (C,W), or (C,S), is clearly a sufficient but not necessary condition for asymptotic state reconstruction. A necessary and sufficient condition for global asymptotic stability to zero of the estimation error is given by the following theorem.

Theorem 2 [22] The state $x(t)$ of the nonlinear system (3) can be globally, exponentially, asymptotically estimated, by the state $\xi(t)$ of the observer (4) if and only if there exists a constant matrix K such that the symmetric matrix

$$[W - KC] + [W - KC]^T = [S - KC] + [S - KC]^T = 2\left[S - \frac{1}{2}(KC + C^T K^T)\right]$$

is negative definite.

In particular, the matrix $2\left[S - \frac{1}{2}(KC + C^T K^T)\right]$ is negative definite (stability synchronization condition holds) for Example 1, if we choose k_1 and k_2 such that

$$k_1 \geq \sqrt{2k_2 - 4}\,,\ k_2 \geq 2,$$

i.e., if $k_1 = k_2 = k$, then $k \geq 1.6568$. And for Example 2, the synchronization error is stabilized at the origin for $k > 0.001$.

5 Confidential Communication

Finally, we apply the Hamiltonian synchronization of chaotic neural networks to transmit encrypted information. In particular, we use the modified chaos communication scheme (MCCS) for signal information masking with single transmission channel [16]; [14]. Figure 6 shows the MCCS (using previous Example 1) where: $m(t)$ is the confidential information to be hidden and transmitted, $x_2(t)$ is the chaotic signal of the network for masking purpose, $s(t) = x_2(t) + m(t)$ is the transmitted signal, and $m'(t) = s(t) - \xi_2(t)$ the recovered information. It was reported in [14] that due to $m(t)$ is also injected into the transmitter, the MCCS is able to recover faithfully the hidden information even if a noise level is present through the transmission channel.

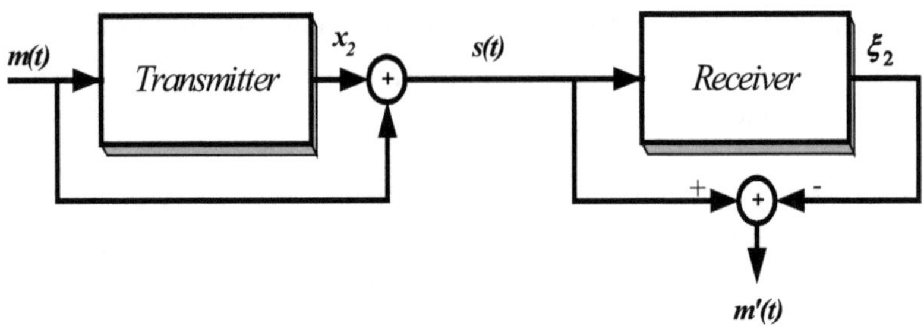

Fig. 6: Modified chaos-based communication scheme for signal masking using a single transmission channel.

Figure 7 illustrates the secret message communication of an audio message using the Example 1: the confidential message to be hidden and transmitted $m(t)$ (top of figure), the transmitted chaotic signal $s(t) = x_2(t) + m(t)$ (middle of figure), and the recovered audio message $m'(t)$ at the receiver (bottom of figure).

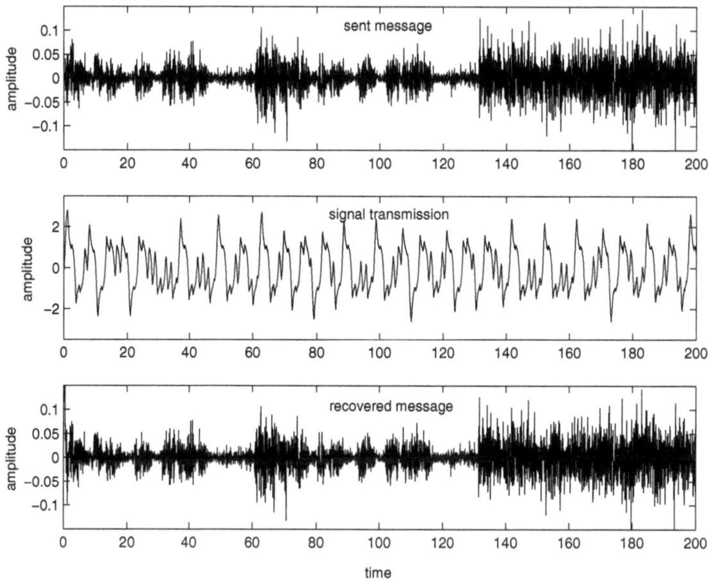

Fig. 7: Transmission and recovering of an audio message: Confidential message to be hidden and transmitted (top of figure). Transmitted chaotic signal $s(t) = x_2(t) + m(t)$ (middle of figure). Recovered audio message $m'(t)$ at the network receiver (bottom of figure).

6 Concluding Remarks

In this paper, we have presented the synchronization problem of chaotic neural networks from the perspective of Generalized Hamiltonian forms and observer design. The approach allows one to give a simple design procedure for the slave **CNN**. We have shown that synchronization of chaotic **CNNs** is possible from this viewpoint. The approach can be easily implemented on experimental setups. Moreover, we have shown based on chaotic **CNNs** synchronization the transmission of encrypted confidential information.

In a forthcoming work we will be concerned with a physical implementation of **CNN** with electronic circuits, and the synchronization of large chaotic neural networks and possible applications.

References

1. Aguilar, A. and Cruz-Hernández, C. (2002) "Synchronization of two hyperchaotic Rössler systems: Model matching approach," WSEAS Trans. Systems 1(2), 198-203.
2. Chua, L.O. and Yang, L. (1988) "Cellular neural networks: Theory and applications," IEEE Trans. Circuits Syst. I 35, 1257-1290.
3. Chua, L.O. (1998) CNN: A Paradigm for Complexity, World Scientific, Singapore.
4. Cruz-Hernández, C. and Nijmeijer, H. (1999) "Synchronization through extended Kalman filtering," New Trends in Nonlinear Observer Design eds. Nijmeijer & Fossen, Lecture Notes in Control and Information Sciences 244 Springer-Verlag, pp. 469-490.
5. Cruz-Hernández, C. and Nijmeijer, H. (2000) "Synchronization through filtering," Int. J. Bifurc. Chaos 10(4), 763-775.
6. Cruz-Hernández, C., Posadas-Castillo C., and Sira-Ramírez, H. (2002) "Synchronization of two hyperchaotic Chua circuits: A generalized Hamiltonian systems approach," Procs. of 15th IFAC World Congress, July 21-26, Barcelona, Spain.
7. Cruz-Hernández C. (2004) "Synchronization of time delay Chua's oscillator with application to secure communication," Nonlinear Dynamics and Systems Theory 4(1), 1-13
8. Cuomo, K.M., Oppenheim, A.V. and Strogratz, S.H. (1993) "Synchronization of Lorenz-based chaotic circuits with applications to communications," IEEE Trans. Circuits Syst. II 40(10), 626-633.
9. Dedieu, H., Kennedy, M.P., and Hasler, M. (1993) "Chaotic shift keying: Modulation and demodulation of a chaotic carrier using self-synchronizing Chua's circuits," IEEE Trans. Circuits Syst. II 40(10), 634-642.
10. Fradkov, A.L., and Pogromsky, A. Yu. (1998) Introduction to control of oscillations and chaos, World Scientific Publishing, Singapore.
11. López-Mancilla, D. and Cruz-Hernández, C. (2004) "An anlysis of robustness on the synchronization of chaotic systems under nonvanishing perturbations using sliding modes," WSEAS Trans. Math. 3(2), 364-269.
12. López-Mancilla, D. and Cruz-Hernández, C. (2005a) "Output synchronization of chaotic systems: model-matching approach with application to secure communication," Nonlinear Dynamics and Systems Theory. 5(2), 141-156.
13. López-Mancilla, D. and Cruz-Hernández, C. (2005b) "Output synchronization of chaotic oscillators and private communication," Procs. of 16th IFAC World Congress, July 3-8, Prague, Czech Republic.
14. López-Mancilla, D. and Cruz-Hernández, C. (2005c) "A note on chaos-based communication schemes," Revista Mexicana de Física 51(3), 265-269.
15. Lu, H., He, Y., and He, Z. (1998) "A chaos generator: Analysis of complex dynamics of a cell equation in delayed cellular neural networks," IEEE Trans. Circuits Syst. 45, 178-181.
16. Milanovic, V. and Zaghloul, M.E. (1996) "Improved masking algorithm for chaotic communication systems," Electronics Lett. 32(1), 11-12.

17. Nijmeijer, H. and Mareels, I.M.Y. (1997) "An observer looks at synchronization," IEEE Trans. Circuits Syst. I 44(10), 882-890.
18. Orzo, L., Vidnyanszky, Z., Hamori, J., and Roska, T. (1996) "CNN model of the feature-linked synchronized activities in the visual thalamo-cortical systems," Proc. 1996 Fourth IEEE Int. Workshop on Cellular Neural Networks and Their Applications (CNNA-96), Seville, Spain, 24-26 June, pp. 291-296.
19. Parlitz, U., Chua, L.O., Kocarev, Lj., Halle, K.S., and Shang, A. (1992) "Transmission of digital signals by chaotic synchronization," Int. J. Bifurc. Chaos 2(4), 973-977.
20. Pecora, L.M. and Carroll, T.L. (1990) "Synchronization in chaotic systems," Phys. Rev. Lett. 64, 821-824.
21. Pikovsky, A., Rosenblum, M., and Kurths, J. (2001) Synchronization: A Universal Concept in Nonlinear Sciences, Cambridge University Press.
22. Sira-Ramírez, H. and Cruz-Hernández, C. (2001) "Synchronization of Chaotic Systems: A Generalized Hamiltonian Systems Approach," Int. J. Bifurc. Chaos 11(5), 1381-1395. And in Procs. of American Control Conference (ACC'2000), Chicago, USA, pp. 769-773.
23. Special Issue (1992) on "Cellular neural networks," Int. J. Circuit Theor. Appl. 20.
24. Special Issue (1993) on "Cellular neural networks," IEEE Trans. Circuits Syst. I and II, 40 (in 2 separate issues).
25. Special Issue (1995) on "Nonlinear waves, patters, and spatio-temporal chaos," IEEE Trans. Circuits Syst. I, 42 (in 2 separate issues).
26. Special Issue (1997 Part I) on "Visions of nonlinear science in the 21st century," Int. J. Bifurc. Chaos, 7(9-10).
27. Special Issue (1997a) on "Chaos synchronization and control: Theory and applications," IEEE Trans. Circuits Syst. I, 44(10).
28. Special Issue (1997b) on "Control of chaos and synchronization," Syst. Contr. Lett. 31.
29. Special Issue (2000) on "Control and synchronization of chaos," Int. J. Bifurc. Chaos, 10(3-4).
30. Werblin, F., Roska, T., and Chua, L.O. (1994) "The analogic cellular neural network as a bionic eye," Int. J. Circuit Theor. Appl. 23, 541-569.
31. Yang, T. and Chua, L.O. (1996) "Secure communication via chaotic parameter modulation," IEEE Trans. Circuits Syst. I 43(9), 817-819.

Mediative Fuzzy Logic: A Novel Approach for Handling Contradictory Knowledge

Oscar Montiel, Oscar Castillo

[1]CITEDI-IPN, Av. del Parque #1310, Mesa de Otay Tijuana, B. C., México, oross@citedi.mx,

[2]Department of Computer Science, Tijuana Institute of Technology, P.O. Box 4207, Chula Vista CA 91909, USA., ocastillo@tectijuana.mx

Abstract. In this paper we are proposing a novel fuzzy method that can handle imperfect knowledge in a broader way than Intuitionistic fuzzy logic does (IFL). This fuzzy method can manage non-contradictory, doubtful, and contradictory information provided by experts, providing a mediated solution, so we called it Mediative Fuzzy Logic (MFL). We are comparing results of MFL, with IFL and traditional Fuzzy logic (FL).

1 Introduction

Uncertainty affects all decision making and appears in a number of different forms. The concept of information is fully connected with the concept of uncertainty; the most fundamental aspect of this connection is that uncertainty involved in any problem-solving situation is a result of some information deficiency, which may be incomplete, imprecise, fragmentary, not fully reliable, vague, *contradictory*, or deficient in some other way [1]. The general framework of fuzzy reasoning allows handling much of this uncertainty.

Nowadays, we can handle much of this uncertainty using Fuzzy logic type-1 or type-2 [2,3], also we are able to deal with hesitation using Intuitionistic fuzzy logic, but what happens when the information collected from different sources is somewhat or fully contradictory. What do we have to do if the knowledge base changes with time, and non-contradictory information becomes into doubtful or contradictory information, or any combination of these three situations? What should we infer from this kind of knowledge? The answer to these questions is to use a fuzzy logic system with logic rules for handling non-contradictory, contradictory or information with a hesitation margin. Mediative fuzzy logic is a novel

O. Montiel and O. Castillo: *Mediative Fuzzy Logic: A Novel Approach for Handling Contradictory Knowledge*, StudFuzz **208**, 75–91 (2007)
www.springerlink.com

approach presented for the first time in [4] which is able to deal with this kind of inconsistent information providing a common sense solution when contradiction exists, this is a mediated solution.

There are a lot of applications where information is inconsistent. In economics for estimating the Gross Domestic Product (GDP), it is possible to use different variables; some of them are distribution of income, personal consummation expenditures, personal ownership of goods, private investment, unit labor cost, exchange rate, inflation rates, and interest rates. In the same area for estimating the exportation rates it is necessary to use a combination of different variables, for example, the annual rate of inflation, the law of supply and demand, the dynamic of international market, etc. [5]. In medicine information from experiments can be somewhat inconsistent because living being might respond different to some experimental medication. Currently, randomized clinical trials have become the accepted scientific standard for evaluating therapeutic efficacy, and contradictory results from multiple randomized clinical trials on the same topic have been attributed either to methodological deficiencies in the design of one of the trials or to small sample sizes that did not provide assurance that a meaningful therapeutic difference would be detected [6]. In forecasting prediction, uncertainty is always a factor, because to obtain a reliable prediction it is necessary to have a number of decisions, each one based on a different group, in [7] says: Experts should be chosen "whose combined knowledge and expertise reflects the full scope of the problem domain. Heterogeneous experts are preferable to experts focused in a single specialty".

The aim of this paper is to present MFL as new fuzzy method for going around from traditional, intuitionistic, and now from meditative fuzzy logic. This is a transparent way from the point of view of the inference system. This paper is organized as follows. In section 2, we are giving some historical antecedent about different logic systems. In section 3, we are explaining Mediative Fuzzy Logic (MFL). In section 4, we are showing some experimental results, and finally we have the conclusions.

2 Historical Background

Throughout history, distinguish good from bad arguments has been of fundamental importance to ancient philosophy and modern science. The Greek philosopher Aristotle (384 BC – 322 BC) is considered a pioneer in the study of logic and, its creator in the traditional way. The Organon is his surviving collected works on logic [8]. Aristotelian logic is centered in the syllogism. In Traditional logic, a syllogism (deduction) is an inference that basically consists of three things: the major and minor premises, and the proposition (conclusion) which follows logically from the major and minor premises [9]. Aristotelian logic is “bivalent” or “two-valued”, that is, the semantics rules will assign to every sentence either the value "True" or the value "False". Two basic laws in this logic are the law of contradiction (*p cannot be both p and not p*), and the law of the excluded middle (*p must be either p or not p*).

In the Hellenistic period, the stoics work on logic was very wide, but in general, one can say that their logic is based on propositions rather than in logic of terms, like the Aristotelian logic. The Stoic treatment of certain problems about modality and bivalence are more significant for the shape of Stoicism as a whole. Chrysippus (280BC-206BC) in particular was convinced that bivalence and the law of excluded middle apply even to contingent statements about particular future events or states of affairs. The law of excluded middle says that for a proposition, p, and its contradictory, $\neg p$, it is necessarily true, while bivalence insists that the truth table that defines a connective like ‘or’ contains only two values, true and false [10].

In the mid-19th century, with the advent of symbolic logic, we had the next major step in the development of propositional logic with the work of the logicians Augustus DeMorgan (1806-1871) [11] and, George Boole (1815-1864). Boole was primarily interested in developing special mathematical to replace Aristotelian syllogistic logic. His work rapidly reaps benefits, he proposed "Boolean algebra” that was used to form the basis of the truth-functional propositional logics utilized in computer design and programming [12,13]. In the late 19th century, Gottlob Frege (1848-1925) claimed that all mathematics could be derived from purely logical principles and definitions and he considered verbal concepts to be expressible as symbolic functions with one or more variables [14].

L. E. J. Brouwer (1881-1966) published in 1907 in his doctoral dissertation the fundamentals of intuitionism [15], his student Arend Heyting (1898-1980) did much to put intuitionism in mathematical logic, he created the Heyting algebra for constructing models of intuitionistic logic [16]. Gerhard Gentzen (1909-1945), in (1934) introduces systems of natural deduction for intuitionist and classical pure predicate calculus [17], his cornerstone was cut-elimination theorem which implies that we can put every proof into a (not necessarily unique) normal form. He introduces two formal systems (sequent calculi) LK and LJ. The LJ system is obtained with small changes into the LK system and it is suffice for turning it into a proof system for intuitionistic logic.

Nowadays, Intuitionistic logic is a branch of logic which emphasizes that any mathematical object is considered to be a product of a mind, and therefore, the existence of an object is equivalent to the possibility of its construction. This contrasts with the classical approach, which states that the existence of an entity can be proved by refuting its non-existence. For the intuitionist, this is invalid; the refutation of the non-existence does not mean that it is possible to find a *constructive* proof of existence. Intuitionists reject the *Law of the Excluded Middle* which allows proof by contradiction. Intuitionistic logic has come to be of great interest to computer scientists, as it is a constructive logic, and is hence a logic of what computers can do.

Bivalent logic was the prevailing view in the development of logic up to XX century. In 1917, Jan Łukasiewicz (1878-1956) developed the three-value propositional calculus, inventing ternary logic [18]. His major mathematical work centered on mathematical logic. He thought innovatively about traditional propositional logic, the principle of non-contradiction and the law of excluded middle. Łukasiewicz worked on multi-valued logics, including his own three-valued propositional calculus, the first non-classical logical calculus. He is responsible for one of the most elegant axiomatizations of classical propositional logic; it has just three axioms and is one of the most used axiomatizations today [19].

Paraconsistent logic is a logic rejecting the principle of non-contradiction, a logic is said to be *paraconsistent* if its relation of logical consequence is not explosive. The first paraconsistent calculi was independently proposed by Newton C. da Costa (1929-) [20] and Ja kowski, and are also related to D. Nelson's ideas [21]. Paraconsistent logic was proposed in 1976 by the Peruvian philosopher Miró Quesada, it is a non-trivial logic which allows inconsistencies. The modern history of paraconsistent logic is relatively short. The expression "paraconsistent logic" is at present time well-established and it will make no sense to change it. It can be interpreted in many different ways which correspond to the many different views on a logic which permits to reason in presence of contradictions. There are many different paraconsistent logics, for example, *non-adjunctive*, *non-truth-functional*, *many-valued*, and *relevant.*

Fuzzy sets, and the notions of inclusion, union, intersection, relation, etc, were introduced in 1965 by Dr. Lofti Zadeh [2], as an extension of Boolean logic. Fuzzy logic deals with the concept of partial truth, in other words, the truth values used in Boolean logic are replaced with degrees of truth. Zadeh is the creator of the concept Fuzzy logic type-1 and type-2. Type-2 fuzzy sets are fuzzy sets whose membership functions are themselves type-1 fuzzy sets; they are very useful in circumstances where it is difficult to determine an exact membership function for a fuzzy set [22].

K. Atanassov in 1983 proposed the concept of Intuitionistic fuzzy sets (IFS) [23], as an extension of the well-known Fuzzy sets defined by Zadeh. IFS introduces a new component, degree of nonmembership with the requirement that the sum of membership and nonmbership functions must be less than or equal to 1. The complement of the two degrees to 1 is called the hesitation margin. George Gargov proposed the name of intuitionistic fuzzy sets with the motivation that their fuzzification denies the law of excluded middle, wish is one of the main ideas of intuitionism [24].

3 Mediative Fuzzy Logic

Since knowledge provided by experts can have big variations and sometimes can be contradictory, we are proposing to use a Contradiction fuzzy set to calculate a mediation value for solving the conflict. Mediative Fuzzy Logic is proposed as an extension of Intuistionistic fuzzy Logic [23,25]. Mediative fuzzy logic (MFL) is based in traditional fuzzy logic with the ability of handling contradictory and doubtful information, so we can say that also it is an intutitionistic and paraconsistent fuzzy system.

A traditional fuzzy set in X [25], given by

$$A = \{(x, \mu_A(x)) | x \in X\} \tag{1}$$

where $\mu_A : X \rightarrow [0, 1]$ is the membership function of the fuzzy set A.

An intuitionistic fuzzy set B is given by

$$B = \{(x, \mu_B(x), \nu_B(x)) | x \in X\} \tag{2}$$

where $\mu_B : X \rightarrow [0, 1]$ and $\nu_B : X \rightarrow [0, 1]$ are such that

$$0 \leq \mu_B(x) + \nu_B(x) \leq 1 \tag{3}$$

and $\mu_B(x)$; $\nu_B(x) \in [0, 1]$ denote a degree of membership and a degree of non-membership of $x \in A$, respectively.

For each intuitionistic fuzzy set in X we have a "hesitation margin" $\pi_B(x)$, this is an intuitionistic fuzzy index of $x \in B$, it expresses a hesitation degree of whether x belongs to A or not. It is obvious that $0 \leq \pi_B(x) \leq 1$, for each $x \in X$.

$$\pi_B(x) = 1 - \mu_B(x) - \nu_B(x) \tag{4}$$

Therefore if we want to fully describe an intuitionistic fuzzy set, we must use any two functions from the triplet [10].

1. Membership function
2. Non-membership function
3. Hesitation margin

The application of intuitionistic fuzzy sets instead of fuzzy sets, means the introduction of another degree of freedom into a set description, in other words, in addition to μ_B we also have ν_B or π_B. Fuzzy inference in intuitionistic has to consider the fact that we have the membership functions μ as well as the non-membership functions ν. Hence, the output of an intuitionistic fuzzy system can be calculated as follows:

$$IFS = (1-\pi)FS_{\mu} + \pi FS_{\nu} \tag{5}$$

where FS_{μ} is the traditional output of a fuzzy system using the membership function μ, and FS_{ν} is the output of a fuzzy system using the non-membership function ν. Note in equation (6), when $\pi=0$ the *IFS* is reduced to the output of a traditional fuzzy system, but if we take into account the hesitation margin of π the resulting IFS will be different.

In similar way, a contradiction fuzzy set *C* in *X* is given by:

$$\zeta_C(x) = \min(\mu_C(x), \nu_C(x)) \tag{6}$$

where $\mu_C(x)$ represents the agreement membership function, and for the variable $\nu_C(x)$ we have the non-agreement membership function.

We are using the agreement and non-agreement instead membership and non-membership, because we think these names are more adequate when we have contradictory fuzzy sets.

We are proposing three expressions for calculating the inference at the system's output, these are

$$MFS = \left(1-\pi-\frac{\zeta}{2}\right)FS_{\mu} + \left(\pi+\frac{\zeta}{2}\right)FS_{\nu} \tag{7}$$

$$MFS = \min\left(\left((1-\pi)*FS_{\mu} + \pi * FS_{\nu}\right), \left(1-\frac{\zeta}{2}\right)\right) \tag{8}$$

$$MFS = \left((1-\pi)*FS_{\mu} + \pi * FS_{\nu}\right)*\left(1-\frac{\zeta}{2}\right) \tag{9}$$

In this case, when the contradictory index ζ is equal to zero, the system's output can be reduced to an intituionistic fuzzy output or, in case that $\pi=0$, it can be reduced to a traditional fuzzy output.

4 Experimental Results

For testing the system we dealt with the problem of population control. This is an interesting problem that can be adapted to different areas. We focused in controlling the population size of an evolutionary algorithm by preserving, killing or creating individuals in the population. Dynamic population size algorithms attempt to optimize the balance between efficiency and the quality of solutions discovered by varying the number of individuals being investigated over the course of the evolutionary algorithm's run. We used Sugeno Inference system to calculate FS_μ and FS_ν, so the system is divided in two main parts: the inference system of the agreement function side, and the inference system of the non-agreement function side.

At the FS_μ side, we defined the variable percentage of cycling (pcCycling) with three terms, Small, Medium and Large. The universe of discourse is in the range [0,100]. We used a Sugeno Inference System, which in turn have three variables for the outputs: MFSCreate, MFSKill and MFSPreserve. They correspond to the amount of individuals that we have to create, kill and preserve in the population. Each output variable has three constant terms, so we have:

1. For the MFSCreate variable, the terms are: Nothing=0, Little=0.5, and Many=1.
2. For FSKill we have: Nothing=0, Little=0.5, All=1.
3. For MFSPreserve we have: Nothing=0, More or Less=0.5, All=1.

The rules for the FS_μ side are:

if (pcCycled is small) then (create is nothing)(kill is nothing)(preserve is all)
if (pcCycled is medium) then (create is little)(kill is little)(preserve is moreOrLess)
if (pcCycled is large) then (create is many)(kill is all)(preserve is nothing)

At the side FS_ν, we defined the input variable NMFpcCycled with three terms: NoSmall, NoMedium, and NoLarge, they are shown in Fig. 2. They are applied to a Sugeno Inference System with three output variables: nCreate, nKill, and nPreserve. In similar way, they are contributing to the calculation of the amount of individuals to create, kill and preserve, respectively. Each output variable has three constant terms, they are:

1. For nCreate we have the output terms: Nothing=0, Little=0.5, Many=1.
2. For nKill we have: Nothing=0, Little=0.5, and All=1.
3. For nPreserve we have: Nothing=0, More or Less=0.5, and All =1.

The corresponding rules are:

if (NMFpcCycled is Nsmall) then (create is nothing)(kill is nothing)(preserve is all)

if (NMFpcCycled is Nmedium) then (create is little)(kill is nothing)(preserve is moreOrLess)

if (NMFpcCycled is Nlarge) then (create is many)(kill is all)(preserve is nothing)

Using the agreement function (FS_{μ}) and the non-agreement functions (FS_{ν}) we obtained the hesitation fuzzy set and the contradictory fuzzy set.

We performed experiments for the aboventioned problem obtaining results for traditional and intuitionistic fuzzy systems. Figures 5, 6, and 7 show results of a traditional fuzzy system (*FS*), and in figures 8, 9, and 10 we have the intuitionistic fuzzy outputs (*IFS*). Moreover, we did experiments for calculating the meditative fuzzy output using equations (7), (8), and (9). Next we are commenting about them.

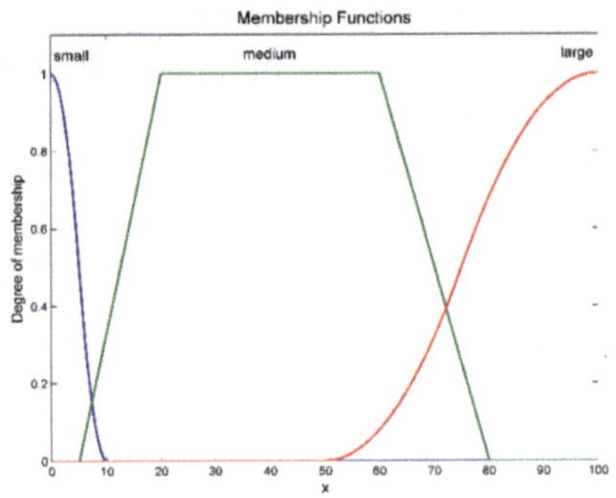

Fig. 1. Membership functions in a traditional fuzzy system (FS).

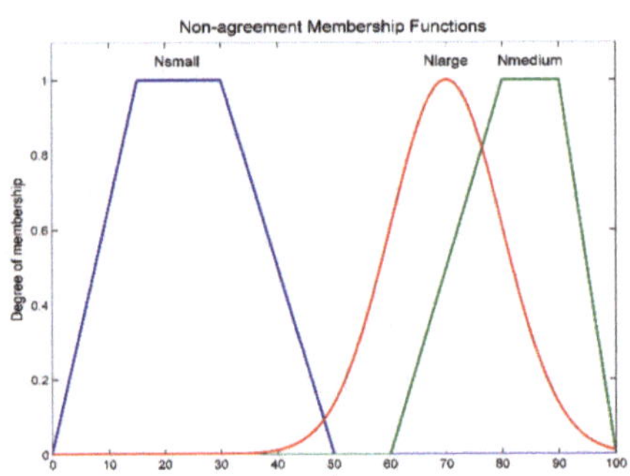

Fig. 2. Non-agreement membership functions for Mediative Fuzzy Inference System (MFS).

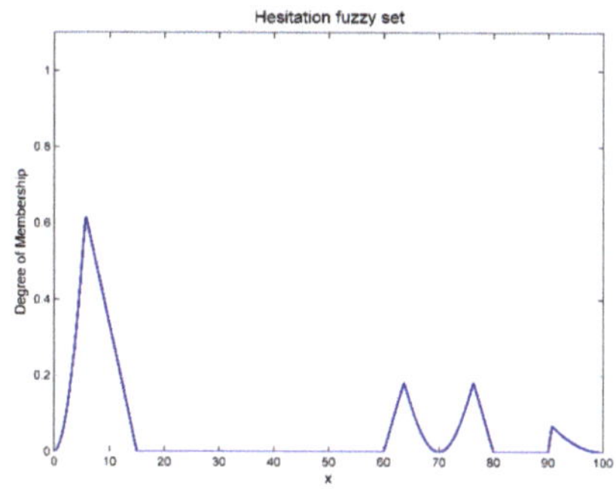

Fig. 3. Hesitation fuzzy set. We applied equation (4) to each complementary subset of membership and non-membership functions, in this case agreement and non-agreement membership functions.

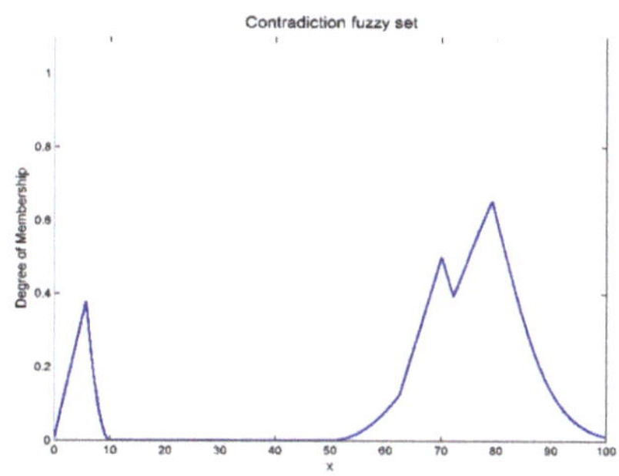

Fig. 4. Contradiction fuzzy set. We obtained this set applying equation (6) to each subset of agreement and non-agreement membership functions.

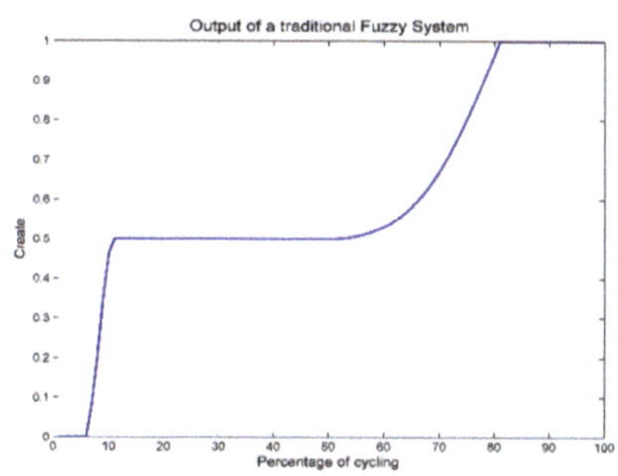

Fig. 5. Traditional FS for the output Create. We can observe that the system is inferring that we have to create 50% more individual in the actual population size when we have a percentage of cycling between 12 and 55.

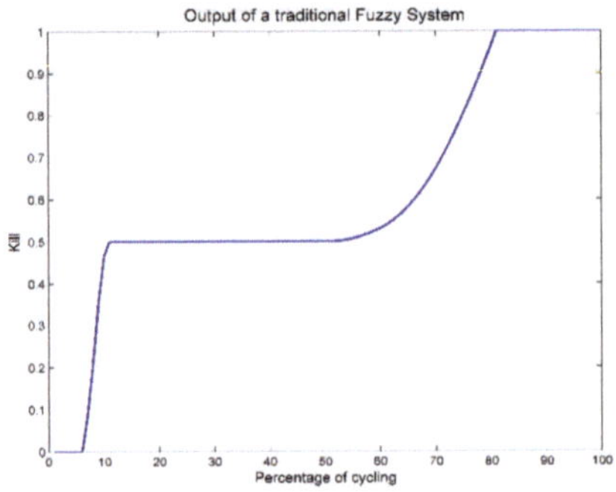

Fig. 6. The output Kill of FS, says that we have to remove 50% of the less fit individuals when we have more or less a percentage of cycling between 12 and 55.

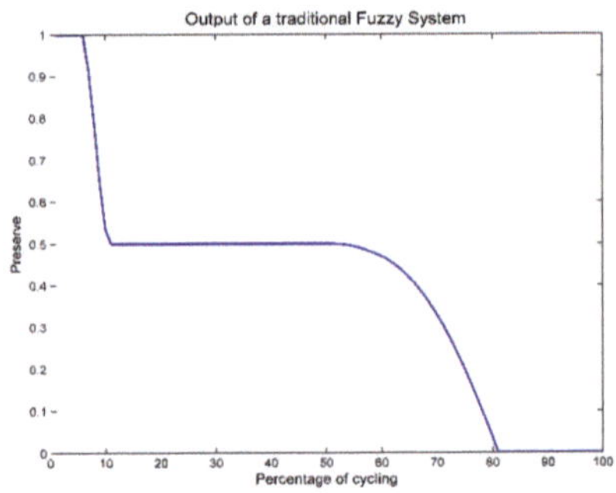

Fig. 7. The output Preserve of traditional FS says how many individuals we have to preserve, this is depending on the degree of cycling. Note that this result is in accordance with Figs. 6 and 7.

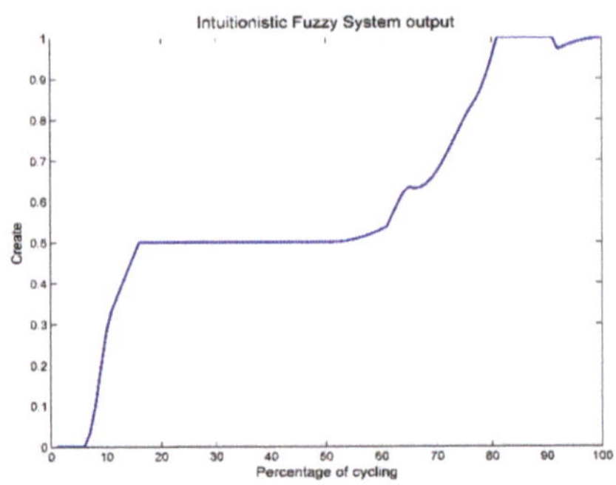

Fig. 8. We can observe that although there is contradictory knowledge, we have a softener transition in the lower part of Percentage of cycling, but when contradiction increases we cannot say the same. We used eq. (5), with FS_{Create} y FSn_{Create}.

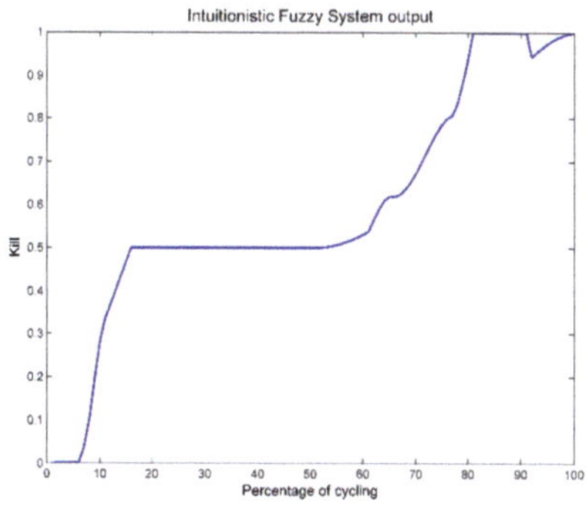

Fig. 9. In fact, a comparison using contradictory knowledge in IFS is not fear since the idea of this logic is not to use this kind of knowledge, but it is interesting to plot the inference output to compare results with MFS.

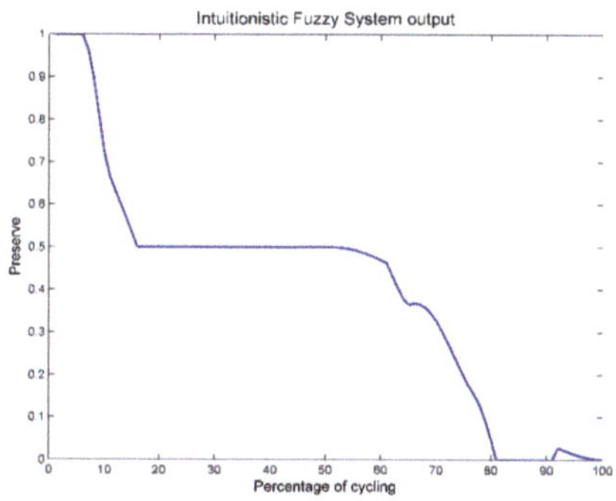

Fig. 10. IFS do not reflect contradictory knowledge at the systems' output.

Experiment #1. Using equation (7).

Equation (7) is transformed in equations (10), (11), and (12) for the three different outputs *MFSCreate*, *MFSKill*, and *MFSPreserve*. Figures 11, 12 and 13 correspond to these outputs.

$$MFSCreate = \left(1 - \pi - \frac{\zeta}{2}\right)FS_{Create} + \left(\pi + \frac{\zeta}{2}\right)FS_{nCreate} \tag{1}$$

$$MFSKill = \left(1 - \pi - \frac{\zeta}{2}\right)FS_{Kill} + \left(\pi + \frac{\zeta}{2}\right)FS_{nKill} \tag{2}$$

$$MFSPreserve = \left(1 - \pi - \frac{\zeta}{2}\right)FS_{\Pr eserve} + \left(\pi + \frac{\zeta}{2}\right)FS_{n\Pr eserve} \tag{3}$$

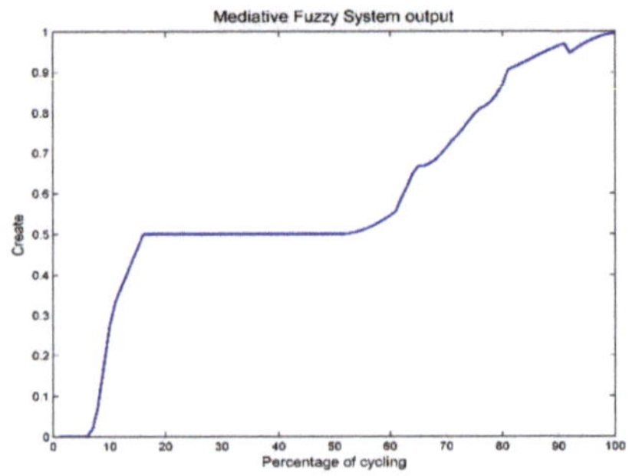

Fig. 11. MFS reflects contradictory knowledge at the systems' output. Note that here we have a softener transition values in the range between 50 and 100. We used equation (7) for plotting Figs. 11, 12 and 13. Experiment #1.

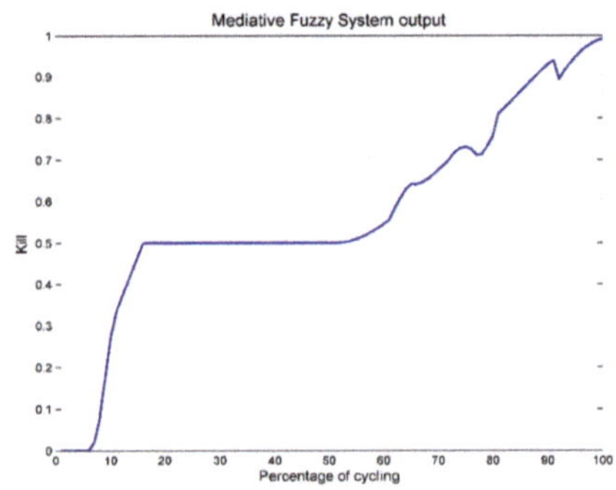

Fig. 12. Although, we have the highest degree of contradiction around the value 80, inference gives, for this region, reasonably good output values. Experiment #1.

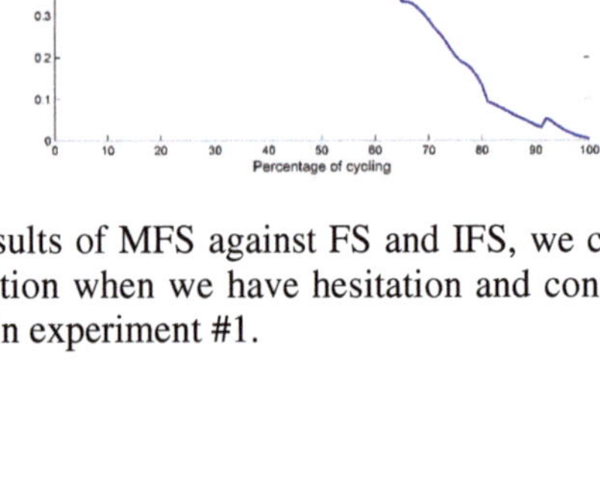

Fig. 13 Comparing results of MFS against FS and IFS, we can see that MFS can gives a softener transition when we have hesitation and contradiction fuzzy sets. We used equation (7) in experiment #1.

Experiment #2. Using equation (8).

Similar than experiment #1, we can calculate the three corresponding outputs for the system using equation (8). Figures 14, 15 and, 16 correspond to the calculated output for the variables: *MFSCreate*, *MFSKill*, and *MFSPreserve*.

Experiment #3. Using equation (9).

In the same way than experiment #1, we used equation (9) to calculate the three meditative fuzzy outputs of the system. Figures 17, 18, and 19 corresponds to this experiment.

In Fig. 3 we are showing the hesitiation fuzzy set for the Membership functions of Figs. 1 and 2, they are the agreement and non-agreement membership function respectively. Figure 3 shows the hesitation fuzzy set obtained using equation (4). Figure 4 shows the contradiction fuzzy set obtained using equation (6). Figures 5, 6, and 7 correspond to the outputs *MFSCreate*, *MFSKill*, and *MFSPreserve*, we can see in these figures that the hesitation and contradiction fuzzy set did not impact the output. Figs. 8, 9 and 10 show that the corresponding outputs were impacted by the hesitation fuzzy set and they were calculated using the IFS given in (5). The outputs in Figs. 11 to 19 were impacted by the hesitation and contradiction fuzzy set, they were calculated using MFL. We used equation (7) to calculate the output in Figs. 11, 12, and 13. Equation (8) was used to obtain Figs. 14, 15 and 16. Finally, Figs. 17, 18, and 19 were obtained using equation (9). In general, we observed that the best results were obtained using equation (7), with this equation we obtained a softener inference output in all the test that we made, this can be observed comparing Fig. 13 against Figs. 16 and 19.

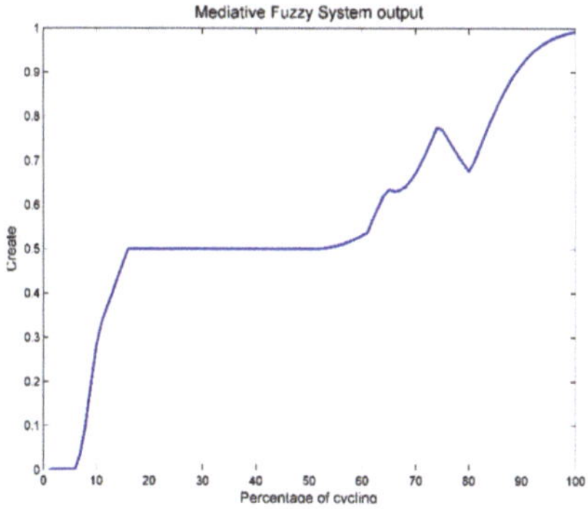

Fig. 14. MFS for the output *MFSCreate*. Figs. 14, 15 and 16 were plotted using equation (8) as base. Experiment #2.

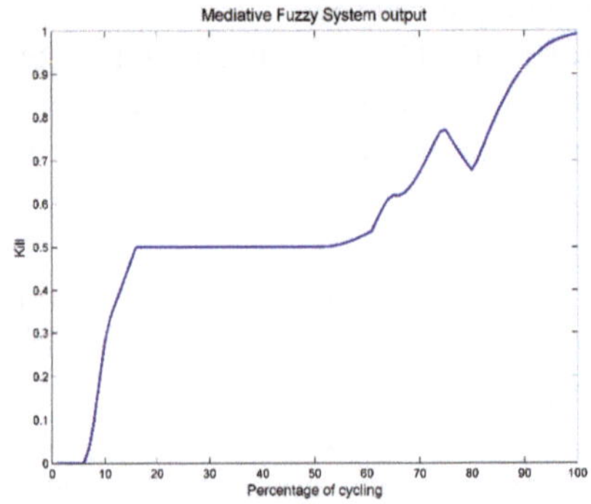

Fig. 15. MFS for the output *MFSKill.* Experiment #2.

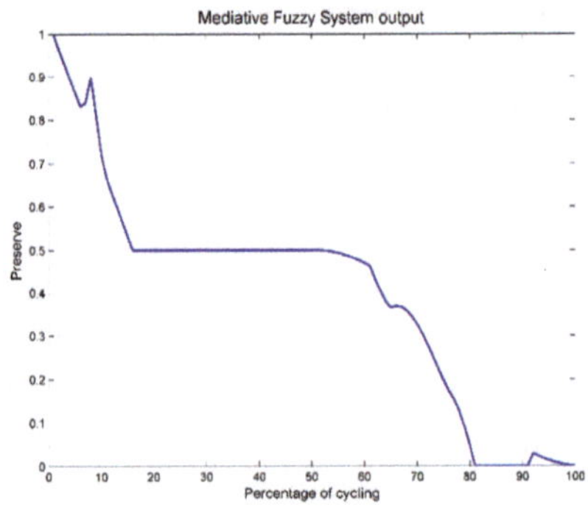

Fig. 16. Output for the variable *MFSPreserve.* Experiment #2.

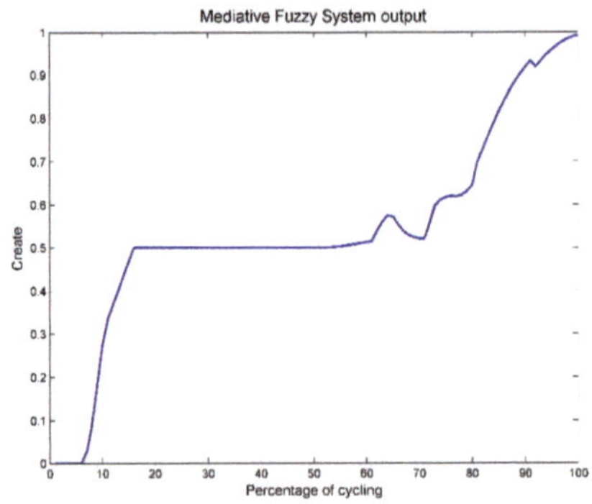

Fig. 17. Output for the variable *MFSCreate*. Figs. 17, 18, and 19 were plotted using equation (9) as base. Experiment #3.

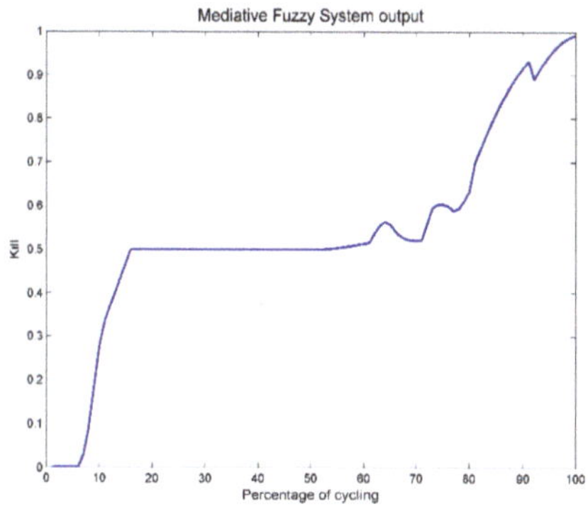

Fig. 18. Output for the variable *MFSKill*. Experiment #3.

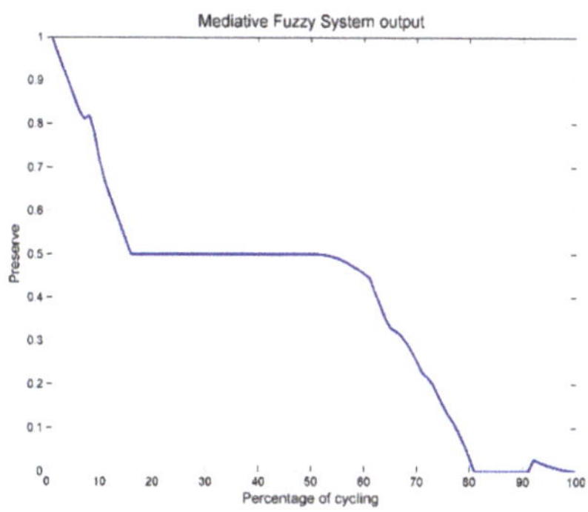

Fig. 19 Output for the variable Preserve in Experiment #3.

5 Conclusions

Through time fuzzy logic type-1 and type-2 have demonstrated their usefulness for handling uncertainty in uncountable applications. Intuitionistic fuzzy logic is relatively a new concept which introduces the degree of nonmbership as a new component, this technology also have found several application niches. Mediative fuzzy logic is a novel approach that enables us to handle imperfect knowledge in a broader way than traditional and intuitionistic fuzzy logic do. MFL is a sort of paraconsistent fuzzy logic because it can handle contradictory knowledge using fuzzy operators. MFL provides a mediated solution in case of a contradiction, moreover it can be reduced automatically to intuitionistic and traditional fuzzy logic in an automatized way, this is depending on how the membership functions (agreement and non-agreement functions) are established. We introduced three equations to perform the meditative inference. In this experiment we found the best results using equation (7) that is an extension of equation (5), i.e. it is an extension of the formula to calculate the intuitionistic fuzzy output. MFL is a good option when we have knowledge from different human experts, because it is common that experts do not fully agree all the time, so we can obtain contradiction fuzzy sets to represent the amount of disagree with the purpose of impacting the inference result. Traditional FL, and IFL will not impact the output when we have contradictory knowledge.

References

1. George J. Klir, Bo Yuan, *Fuzzy Sets and Fuzzy Logic: Theory and Applications*, Ed. Prentice Hall, USA, 1995.
2. L. A. Zadeh, "Fuzzy Sets", Information and Control, Vol. 8, p.p. 338-353, 1965.
3. Jerry M. Mendel, *Uncertain Rule-Based Fuzzy Logic Systems, Introduction and new directions*, Ed. Prentice Hall, USA, 2000.
4. O. Montiel, O. Castillo, P. Melin, A. Rodríguez Días, R. Sepúlveda, ICAI-2005.
5. Donald A. Bal, Wendell H. McCulloch, JR., "International Business. Introduction and Essentials", Fifth Edition, pp. 138-140, 225, USA, 1993.
6. Horwitz R. I.: *Complexity and contradiction in clinical trial research.* Am. J. Med., 82: 498-510,1987.
7. J. Scott Armstrong, "Principles of Forecasting. A Handbook for researchers and Practitioners", Edited by J. Scott Armstrong, University of University of Pennsylvania, Wharton School, Philadelphia, PA., USA, 2001.
8. Aristotle, The Basic Works of Aristotle, Modern Library Classics, Richard McKeon Ed., 2001.
9. Robin Smith, Aristotle's logic, Stanford Encyclopedia of Philosophy, 2004, http://plato.stanford.edu/entries/aristotle-logic/.
10. Dirk Baltzly, Stanford Encyclopedia of Philosophy, 2004, http://plato.stanford.edu/entries/stoicism/
11. [J J O'Connor and E F Robertson, Augustus DeMorgan, MacTutor History of Mathematics: Indexes of Biographies *(University of St. Andrews)*, 2004, *http://www-groups.dcs.st-andrews.ac.uk/~history/Mathematicians/De_Morgan.html*
12. George Boole, The Calculus of Logic, *Cambridge and Dublin Mathematical Journal,* Vol. III (1848), 1848, pp. 183-98
13. J J O'Connor and E F Robertson, George Boole, MacTutor History of Mathematics: Indexes of Biographies (University of St. Andrews), 2004, http://www-groups.dcs.st-andrews.ac.uk/~history/Mathematicians/Boole.html

14. J J O'Connor and E F Robertson,Friedrich Ludwig Gottlob Frege, MacTutor History of Mathematics: Indexes of Biographies (University of St. Andrews), http://www-groups.dcs.st-andrews.ac.uk/~history/Mathematicians/Frege.html
15. J J O'Connor and E F Robertson, Luitzen Egbertus Jan Brouwer, MacTutor History of Mathematics: Indexes of Biographies (University of St. Andrews), http://www-groups.dcs.st-andrews.ac.uk/~history/Mathematicians/Brouwer.html
16. J J O'Connor and E F Robertson , Arend Heyting, *MacTutor History of Mathematics: Indexes of Biographies* (University of St. Andrews), 2004, http://www-history.mcs.st-andrews.ac.uk/Mathematicians/Heyting.html
17. J J O'Connor and E F Robertson , Gerhard Gentzen, MacTutor History of Mathematics: Indexes of Biographies (University of St. Andrews), 2004, http://www-history.mcs.st-andrews.ac.uk/Mathematicians/Gentzen.html
18. J J O'Connor and E F Robertson , Jan Lukasiewicz, MacTutor History of Mathematics: Indexes of Biographies (University of St. Andrews), 2004, http://www-history.mcs.st-andrews.ac.uk/Mathematicians/Lukasiewicz.html
19. Wikipedia the free encyclopedia, vailable from the web page: http://en.wikipedia.org/wiki/Jan_Lukasiewicz
20. Wikipedia the free encyclopedia, available from the webpage: http://en.wikipedia.org/wiki/Newton_da_Costa
21. How to build your own paraconsistent logic: an introduction to the Logics of Formal (In)Consistency. W. A. Carnielli. In: J. Marcos, D. Batens, and W. A. Carnielli, organizers, *Proceedings of the Workshop on Paraconsistent Logic* (WoPaLo), held in Trento, Italy, 5–9 August 2002, as part of the 14th European Summer School on Logic, Language and Information (ESSLLI 2002), pp. 58–72.
22. Jerry M. Mendel and Robert I. Bob John, Type-2 Fuzzy Sets Made Simple, IEEE Transactions on Fuzzy Systems, Vol. 10, No. 2, April 2002.
23. Atanassov, K., "Intuitionistic Fuzzy Sets: Theory and Applications", Springer-Verlag, Heidelberg, Germany, 1999.
24. Mariana Nikilova, Nikolai Nikolov, Chris Cornelis, Grad Deschrijver, “Survey of the Research on Intuitionistic Fuzzy Sets”, In: Advanced Studies in Contemporary Mathematics **4**(2), 2002, p. 127-157.
25. O. Castillo, P. Melin, *A New Method for Fuzzy Inference in Intuitionistic Fuzzy Systems*, Proceedings of the International Conference NAFIPS 2003, IEEE Press, Chicago, Illinois, USA, Julio 2003, pp. 20-25.

Part II Intelligent Control Applications

Direct and Indirect Adaptive Neural Control of Nonlinear Systems

Ieroham Baruch

CINVESTAV-IPN, Department of Automatic Control, Ave. IPN No 2508, col. Zacatenco A.P. 14-740 07380 Mexico D.F., Mexico,
baruch@ctrl.cinvestav.mx

Abstract. A comparative study of various control systems using neural networks is done. The paper proposes to use a Recurrent Trainable Neural Network (RTNN) identifier with backpropagation method of learning. Two methods of adaptive neural control with integral plus state action are applied – an indirect and a direct trajectory tracking control. The first one is the indirect Sliding Mode Control (SMC) with I-term where the SMC is resolved using states and parameters identified by RTNN. The second one is the direct adaptive control with I-term where the adaptive control is resolved by a RTNN controller. The good tracking abilities of both methods are confirmed by simulation results obtained using a MIMO mechanical plant and a 1-DOF mechanical system with friction plant model. The results show that both control schemes could compensate constant offsets and that - without I- term did not.

1 Introduction

Recent advances in understanding of the working principles of artificial neural networks has given a tremendous boost to identification and control tools of nonlinear systems, [1], [2], [3]. Most of the current applications rely on the classical NARMA approach, where a feed-forward neural network is used to synthesize the nonlinear map, [4], [5]. This approach has some disadvantages, [2], like that: the network inputs are a number of past system inputs and outputs, so to find out the optimum number of past values, a trial and error must be carried on; the model is naturally formulated in discrete time with fixed sampling period, so if the sampling period is changed the network, must be trained again; problems associated with stability, convergence and rate of convergence of this networks are not

I. Baruch: *Direct and Indirect Adaptive Neural Control of Non-Linear Systems*, StudFuzz **208**, 95–114 (2007)
www.springerlink.com

clearly understood and there is not a framework available for its analysis in vector-matricial form, [6]; it is a necessary condition, that the plant order has to be known. Besides to avoid these difficulties, a new Recurrent Trainable Neural Networks (RTNN) topology, and a Backpropagation (BP) like learning algorithm, [7], has been proposed, but they have still what to do. So, the objective of this paper is to derive a normalized BP vector-matricial learning algorithm, to prove its stability, to derive two adaptive control algorithms and to give some simulation results, illustrating its capabilities.

2 Description of the RTNN Topology and Learning

The given in [7], [8] RTNN topology is expressed by the following vector-matricial equations:

$$X(k+1) = J(k)\, X(k) + B(k)\, U(k) \tag{1}$$

$$J(k) = \text{block-diag}\,[\, Ji(k)\,];\ |\, Ji(k)\,| < 1;\ i=1,\dots,n \tag{2}$$

$$Z(k) = \Gamma\,[\, X(k)\,] \tag{3}$$

$$Y(k) = \Phi\,[\, C(k)\, Z(k)\,] \tag{4}$$

Where: Y(.), X(.), U(.) are output, state and input variables with dimensions l, n, m, respectively; J(.) is a (nxn) block-diagonal weight matrix of the hidden layer feedback; B(.), C(.) are input and output weight matrices of dimensions (nxm) and (lxm), respectively; Γ(.), Φ(.) are vector-valued activation functions of respective dimensions, with functional elements like: saturation, sigmoid or hyperbolic tangent. The eigenvalues of the RTNN model must be placed in the unit circle, so some restrictions on the weight elements of the matrix J(.), are imposed during the learning (see equation (2)).

The RTNN topology has a linear time varying structure properties like: controllability, observability, and identifiability (trainability), which are proved in [9], [10]. These properties of the RTNN structure signify that starting from the block- diagonal matrix structure of J(.) , we can find a correspondence in the block structure of the matrices B(.) and C(.), that's show us how to find out the ability of learning of this RTNN. The main advantage of this discrete RTNN (which is really a Jordan Canonical RNN model), is of being an universal hybrid neural network model with one or two feedforward layers, and one recurrent hidden layer, where the weight matrix J(.) is a block-diagonal one. So, the RTNN posses a minimal number of learning weights and the performance of the RTNN is fully parallel. The described RTNN architecture could be used as one step ahead state predictor/estimator and systems identifier. Another property of the RTNN model is that it is globally nonlinear, but locally linear. That is why the matrices J(.), B(.), C(.), generated by learning, could be used to design a controller law. Furthermore, the RTNN model is robust, due to the dynamic weight adaptation law, based on the sensitivity model of the RTNN, and the performance index, which is as follows:

$$\xi(k) = (1/2)\, \Sigma_j\, [\, E_j(k)\,]^2, \quad j \in C \tag{5}$$

Here the performance index $\xi(.)$ is nonlinear function of the weight matrices of the output and the hidden RTNN layers, respectively. The general RTNN - BP learning algorithm, written in a vector-matricial form, is given by the following equation:

$$W(k+1) = W(k) + \eta\, \Delta W(k) + \alpha\, \Delta W(k-1) \tag{6}$$

Where: W(.) is the weight matrix, being modified {J(.), B(.), C(.)}; ΔW is the weight matrix correction { ΔJ(.), ΔB(.), ΔC(.)} which is defined as $\Delta W(k) = -\frac{\partial \xi}{\partial W}$; is a learning rate normalized parameter's diagonal matrix, and is a momentum term normalized learning parameter's diagonal matrix. The momentum term of this learning algorithm is used when some error oscillations occur. The structure of the normalized learning and momentum rate terms with respect to the error is shown in [8]. The weight matrix elements update for the discrete time model of the RTNN has been derived and applied in [7], but here it will be expressed in vector-matricial form, [8], using the diagrammatic method, proposed in [11]. The weight update algorithm is:

$$\Delta C(k) = E_1(k)\, Z^T(k);\; E_1(k) = \Phi'[\, Y(k)\,]\, E(k);\; E(k) = Y_d(k) - Y(k) \tag{7}$$

$$\Delta J(k) = E_3(k)\, X^T(k);\; E_3(k) = \Gamma'[\, Z(k)\,]\, E_2(k);\; E_2(k) = C^T(k)\, E_1(k) \tag{8}$$

$$\Delta vJ(k) = E_3(k) \otimes X(k) \tag{9}$$

$$\Delta B(k) = E_3(k)\, U^T(k) \tag{10}$$

Where: $\Delta J(.)$, $\Delta B(.)$, $\Delta C(.)$ are weight corrections of the of the learned matrices J(.), B(.), C(.), respectively; $E(k) = Y_d(k) - Y(k)$ is an l-error vector of the output RTNN layer, where Y_d (.) is a desired target vector (plant output) and Y(.) is a RTNN output vector, both with dimensions l; X(.) is a n-state vector, and E_j (.) is a j-th error vector with respective dimension; $\Gamma'(.)$, $\Phi'(.)$ are diagonal Jacobean matrices with appropriate dimensions, which elements are derivatives of the activation functions. The equation (8) represents the learning of the feedback weight matrix of the hidden layer, which is supposed as a full (nxn) matrix. The equation (9) gives the learning solution when this matrix is diagonal, which is our case, where vJ(.) is the diagonal of the matrix J with dimension n.

2.1 Stability Proof of the Learning Algorithm

The stability and the properties of the BP - RTNN learning algorithm, given by the equation (6), are proved by one theorem and two lemmas.

Theorem of stability. *Let the RTNN with Jordan Canonical Structure, [7], is given by equations (1), (2), (3), (4) and the nonlinear plant model, [8], is as follows:*

$$X_{d.}(k+1) = F[\, X_d(k), U(k)\,] \tag{11}$$

$$Y_d(k) = G[\, X_d(k)\,] \tag{12}$$

Where: $\{Y_d(.), X_d(.), U(.)\}$ are output, state and input variables with dimensions l, n_d, m, respectively; F(.), G(.) are vector valued nonlinear functions with respective dimensions. Under the assumption of RTNN identifiability made, the application of the BP learning algorithm for J(.), B(.), C(.), in general matricial form, described by equation (6), and the learning rates η (k), α (k) (here they are considered as time-dependent and normalized with respect to the error) are derived using the following Lyapunov function, [8], [12]:

$$L(k) = \|\, J(k)\, \|^2 + \|\, B(k)\, \|^2 + \|\, C(k)\, \|^2 \tag{13}$$

Then the identification error is bounded, i.e.:

$$\Delta L(k) \leq - \eta(k) \, |E(k)|^2 - \alpha(k) \, |E(k\text{-}1)|^2 + d; \; \Delta L(k) = L(k) - L(k\text{-}1) \qquad (14)$$

Where all: the unmodelled dynamics, the approximation errors and the perturbations, are represented by the d-term, and the complete proof of that theorem and two lemmas are given in [8], [12].

3 Description of the Designed Adaptive Neural Control Methods

Two different adaptive neural control methods will be derived: an indirect adaptive neural control with I-term and a direct adaptive neural control with I-term.

3.1 Indirect Adaptive Neural Control with I-term

The block diagram of the indirect adaptive neural control is shown on Fig. 1. Here the indirect control is realized as a sliding mode one. The scheme contains identification and state estimation RTNN, an adaptive sliding mode controller, and an I-term.

Let us first design the Sliding Mode Control (SMC). The stable nonlinear plant is identified by a RTNN with topology, given by equations (1), (2), (3), (4) which is learned by the stable BP-learning algorithm, given by equation (6), completed by equations (7), (8), (9), (10), (11), where the identification error $E_i(k) = Y_d(k) - Y(k)$ tends to zero ($E_i \to 0$, $k \to \infty$). The linearization of the activation functions of the learned identification RTNN model, which approximates the nonlinear plant, given by (11), (12), leads to the following linear local plant model:

$$X(k+1) = JX(k) + BU(k) \qquad (15)$$

$$Y(k) = CX(k) \qquad (16)$$

Where l=m, is supposed. The new point in the proposed sliding mode control here, with respect to the original works of Utkin, [13], [14], [15], [16], is that the sliding surface here is defined with respect to the plant output and not to the state. Let us define the sliding surface with respect to the output error of reference tracking, [17]:

$$S(k+1)=E(k+1)+\sum_{i=1}^{p} \gamma_i E(k-i+1); \quad |\gamma_i| < 1 \tag{17}$$

Where: S(.) is the sliding surface error function; E(.) is the systems output tracking error; γ_i are parameters of the desired error function; p is the order of the error function. The additional inequality in (17) is a stability condition, required for the sliding surface error function. The tracking error is defined as:

$$E(k) = R(k) - Y(k) \tag{18}$$

Where R(.) is a l-dimensional reference vector and Y(.) is an output vector with the same dimension.

The objective of the SMC system design is to find a control action which maintains the systems error on the sliding surface which assure that

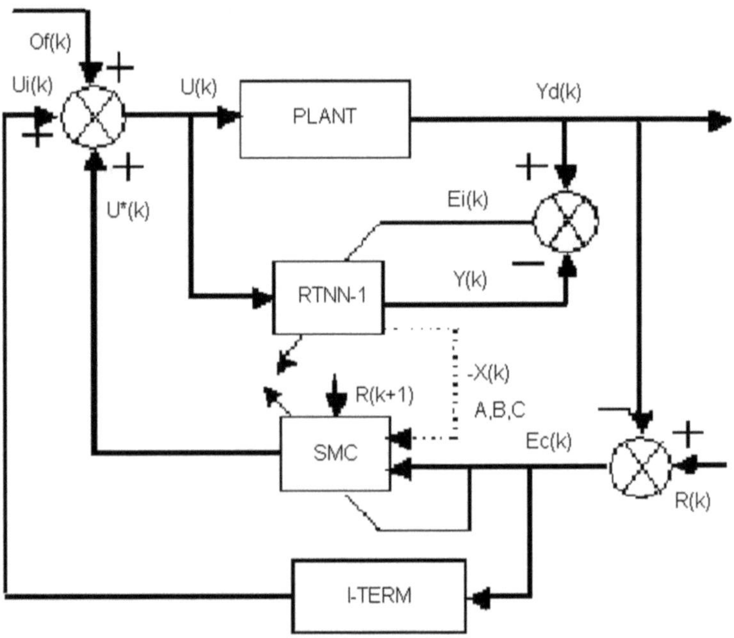

Fig. 1. Block-diagram of an indirect adaptive neural control system with I-term

the output tracking error reaches zero in p steps, where $p<n$. So, the control objective is fulfilled if:

$$S(k+1) = 0 \tag{19}$$

The iteration of the error, defined by (18) gives:

$$E(k+1) = R(k+1) - Y(k+1) \tag{20}$$

Now, let us iterate (16) and substitute (15) in it so to obtain the input/output local plant model, which yields:

$$Y(k+1) = CX(k+1) = C[JX(k) + BU(k)] \tag{21}$$

From (17), (19), and (20), it is easy to obtain:

$$R(k+1) - Y(k+1) + \sum_{i=1}^{p} \gamma_i E(k-i+1) = 0 \tag{22}$$

The substitution of (21) in (22) gives:

$$R(k+1) - CJX(k) - CBU(k) + \sum_{i=1}^{p} \gamma_i E(k-i+1) = 0 \tag{23}$$

As the local approximation plant model, given by (15), (16), is controllable, observable and stable, the matrix J is diagonal, and l=m, the matrix product (CB) is nonsingular, and the plant states X(.) are smooth non-increasing functions. Now, from (23) it is possible to obtain the equivalent control U_{eq} (.) capable to lead the system to the sliding surface which yields:

$$U_{eq}(k) = (CB)^{-1}[-CJX(k) + R(k+1) + \sum_{i=1}^{p} \gamma_i E(k-i+1)] \tag{24}$$

Following [15], [16], the SMC avoiding chattering is taken using a saturation function inside a bounded control level Uo, taking into account plant uncertainties. Furthermore, as the sliding surface contains the output plant trajectory, than the chattering is completely avoided. So the SMC U*(.) takes the form:

$$U^*(k) = \begin{cases} U_{eq}(k), & \text{if } \|U_{eq}(k)\| < Uo \\ -Uo\, U_{eq}(k)/\|U_{eq}(k)\|, & \text{if } \|U_{eq}(k)\| \geq Uo. \end{cases} \quad (25)$$

The proposed SMC copes with the characteristics of the wide class of plant model reduction neural control with reference model, defined by Narendra and Parthasarathy, [3], and represents an indirect adaptive neural control, given by Baruch, [18].

Let us suppose that the plant is a second order nonlinear mechanical plant, so we could accept p=1. In order to study the stability of the closed loop control system, let us accept Uo=1, and linearize the saturation function, given by (25), supposing its gain to be equal to one. Then the SMC yields:

$$U^*(k)=(CB)^{-1}\,[-CJX(k)+R(k+1)+\gamma\, E_c(k)] \quad (26)$$

Where: γ is a (lxl) diagonal control gain matrix.

Now, let us add an I-term to the SMC and an offset perturbation term to the plant input, so to obtain an indirect control with integral action. Following the block-diagram, given on Fig.1, we could express the input of the plant as it is:

$$U(k) = U^*(k) + Of(k) + U_i(k) \quad (27)$$

Where: U*(.) is the dynamic compensation control part, based on SMC; Of(.) is a constant offset perturbation term, taking to account all imperfections of the plant model; U_i (.) is the I-term control part, which is:

$$U_i(k+1) = U_i(k) + T_0\, K_i\, E_c(k) \quad (28)$$

Where: T_0 is a period of discretization; K_i is a diagonal (lxl) I-term gain matrix. The substitution of the control component U*(.), given by (26), in (27), and then – the obtained control signal U(.) - in the linear model (21), taking into account (19), give us after some mathematical manipulations an expression for the error dynamics, i.e.:

$$E_c(k+1) = -\gamma\, E_c(k) - (CB)U_i(k) - (CB)Of(k) \qquad (29)$$

The equations (27), (28), could be rewritten in z-operators form and the closed-loop systems error dynamics could be derived as:

$$U_i(z) = (z\text{-}1)^{-1} T_0\, K_i\, E_c(z) \qquad (30)$$

$$(zI + \gamma)\, Ec = -\, (CB)\, Ui(z) - (CB)\, Of(z) \qquad (31)$$

$$[(z\text{-}1)(zI + \gamma) + T_0\, (CB)\, K_i]\, Ec(z) = -\, (z\text{-}1)\, (CB)\, Of(z) \qquad (32)$$

As it could be seen from the equation (32), the closed-loop systems stability could be assured by an appropriate choice of the diagonal gain matrices γ and $K_{i,}$ respectively. It could be seen also that the effect of the I-term on the control error resulted in the introduction of a difference on the offset which reduces substantially that error, especially for constant offset, and accelerates the RTNN learning.

3.2 Direct Adaptive Neural Control with I-term

The block diagram of the direct adaptive neural control is shown on Fig. 2. The scheme contains identification and state estimation RTNN-1, an adaptive recurrent neural controller RTNN-2, and an I-term. In order to derive the closed-loop system dynamics, let us to linearize the plant equations (11), (12) and represent them in z-operation form, as it is:

$$Y_d(z) = W_d(z)\, U(z) \qquad (33)$$

Where $W_d(z)$ id a plant transfer function. Following the block-diagram of Fig. 2, the linearized controller equation is obtained in state space form like this:

$$X_c(k+1) = J_c X_c(k) - B_{c,1} V(k) - B_{c,2} X(k) + B_{c,3} R(k) \quad (34)$$

$$U^*(k) = C_c X_c(k) \quad (35)$$

Where: X^c (.) is a n_c-dimensional state vector ($n_c \le l+m+n$ is supposed); $U^*(.)$ is a m-dimensional controller output vector; J_c , C_c, $B_{c,1}$, $B_{c,2}$, $B_{c,3}$, are (n_c x n_c), (m x n_c), (n_c x l), (n_c x n), (n_c x l), weight matrices, respectively (l=m is supposed). The l-dimensional I-term vector V(.) and the m-dimensional plant control variable are:

$$V(k+1) = V(k) + To\, Y_d(k) \quad (36)$$

$$U(k) = U^*(k) + Of(k) \quad (37)$$

Where Of(.) is a m-dimensional offset vector variable. To derive the dynamics of the closed-loop system we need to define also the following statements and z-transfer functions, derived from its corresponding state space representations:

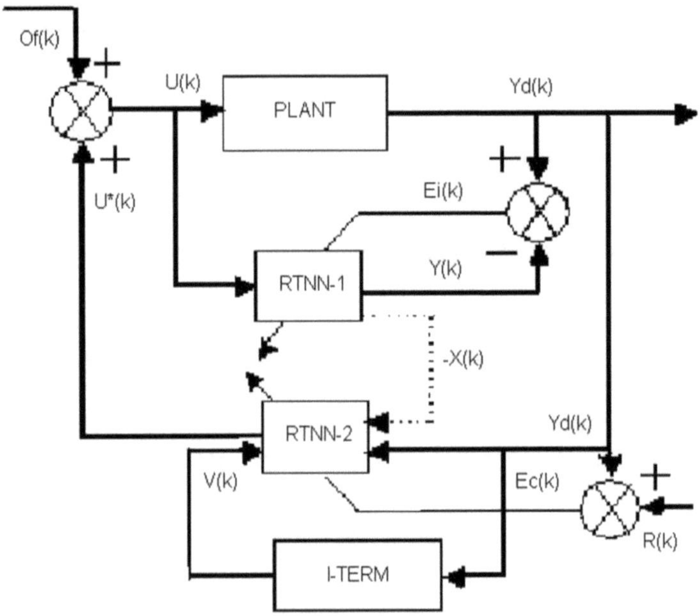

Fig. 2. Block-diagram of an direct adaptive neural control system with I-term

$$Y_d(z) = W_d(z)\, U^*(z) + W_d(z)\, Of(z) \tag{38}$$

$$V(z) = I(z)\, Y_d(z);\ I(z) = (zI - I)^{-1}\, To \tag{39}$$

$$X(z) = P(z)\, U(z);\ P(z) = (zI - J)^{-1}\, B \tag{40}$$

$$U^*(z) = - Q_1(z)\, V(z) - Q_2(z)\, P(z)\, U(z) + Q_3(z)\, R(z) \tag{41}$$

$$Q_1(z) = C_c\, (zI - J_c)^{-1}\, B_{c,1}\ ;\ Q_2(z) = C_c\, (zI - J_c)^{-1}\, B_{c,2}\ ;\ Q_3(z) = C_c\, (zI - J_c)^{-1}\, B_{c,3} \tag{42}$$

The RTNN learning BP algorithm, given by the equations (6) to (10) is proved to be convergent, (see the Theorem of Stability), and the RTNN model is proved to be stable, controllable and observable, [8], [12], so the identification and control errors $E^i(k) = Y_p(k) - Y^i(k)$, and $E^c(k)=R(k)-Y_p(k)$ tends to zero. The plant is supposed to be BIBO stable. So the transfer functions (38) to (42) are stable with minimum phase. Using (37) to (42) and performing some manipulations, finally yields:

$$U^*(z) = [I+ Q_2(z)\, P(z)]^{-1}\, [-Q_1(z)\, V(z) - Q_2(z)\, P(z)\, Of(z) + Q_3(z)\, R(z)] \tag{43}$$

$$\begin{aligned} &\{(z-1)\, I+ W_d(z)\, [I+ Q_2(z)\, P(z)]^{-1}\, Q_1(z) To\}\, Y_d(z) \\ &= W_d(z)\, [I+ Q_2(z)\, P(z)]^{-1}\, Q_3(z)\, (z-1)\, R(z) \\ &+ W_d(z)\, \{I - [I+ Q_2(z)\, P(z)]^{-1}\, Q_2(z) P(z)\}\, (z-1)\, Of(z) \end{aligned} \tag{44}$$

The equation (44) shows that the closed loop system remains also stable, and it is obvious that the I-term reduces the steady-state reference and off-set parts of the systems error, which tend to zero when k tends to infinity $(z \to 1)$.

4 Simulation Results

Two examples illustrating both the indirect and the direct neural control schemes with I-term will be presented in the following paragraphs.

4.1 Example 1. MIMO Mechanical Plant Controlled by a SMC

The plant, considered here, is a third-order nonlinear MIMO kinematics system whit two inputs, two outputs, three states, described by the following equations, [19]:

$$x_1(k+1) = 0.9x_1(k)\sin[x_2(k)] + \left[2 + 1.5\frac{x_1(k)u_1(k)}{1 + x_1^2(k)u_1^2(k)}\right]u_1 + \left[x_1(k) + \frac{2x_1(k)}{1 + x_1^2(k)}\right]u_2 \tag{45}$$

$$x_2(k+1) = x_3(k)\{1 + \sin[4x_3(k)]\} + \frac{x_3(k)}{1 + x_3^2} \tag{46}$$

$$x_3(k+1) = \{3 + \sin[2x_1(k)]\}u_2(k) \tag{47}$$

$$y_1(k) = x_1(k); \tag{48}$$

$$y_2(k) = x_2(k); \tag{49}$$

Where: $x(k)=[x_1(k), x_2(k), x_3(k)]^T$ represents the 3-state vector, $u(k) =[u_1(k), u_2(k)]^T$ the 2-dimensional input vector, and $y(k) =[y_1(k), y_2(k))]^T$ the 2-dimensional output vector, at the instant k. The reference signal is sum of two sinusoids with different amplitudes and frequencies. Results of simulation experiments with duration of 50 seconds are shown in 20- second-graphics. The results, obtained using the control scheme, given on Fig.1 and 40% constant offset (load disturbance), corrupting the plant input, are shown on Fig. 2 a to i. Results, obtained by control scheme without integral term and 40% constant offset, are given on Fig. 3 a to i.

The graphics, given on Fig. 3 a, b, compare reference signals with plant outputs in the last 20 seconds of the control simulation. The next two graphics compare plant outputs with corresponding outputs of the identification RTNN during the last 20 seconds of the plant identification (Fig. 3, c, d). The following two graphics, (Fig. 3, e, f), represents control signals in the same period of time. The next two graphics represent the Mean Squared Error (MSE %) of control and the MSE% of identification (Fig. 3, g, h), both decreasing rapidly to small values. The last graphics (Fig. 3, i) represents the 5 states of the system, issued by the identification RTNN (architecture 2, 5, 2, $\eta = 0.01$, $\alpha = 0.001$) which are entry to the SMC. Results, obtained with a control system without I-term and a constant offset of 40%, are shown on Fig. 4, a – I, which could be compared with that of Fig. 3, a - i. As it could be seen, the system without integral action is sensitive to load disturbances, and the MSE% (see Fig. 3 g) remains great, which means that the SMC without I-term could not compensate the offset at all.

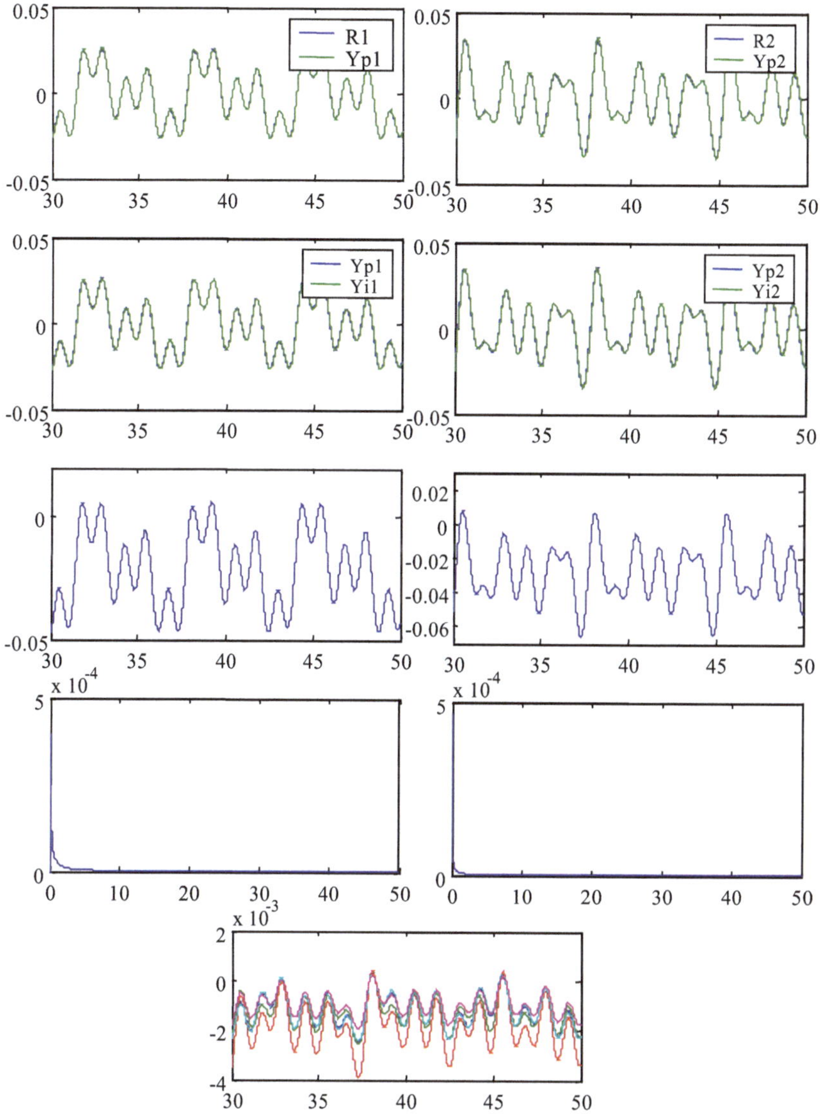

Fig. 3. Simulation results of a SMC with I-term and a 40% constant offset; a) comparison between the first plant output Y1 and R1 in the last 20 seconds of the simulation; b) comparison between the second plant output Y2 and R2 in the last 20 seconds of the simulation; c) comparison between Y1 and Yi1 of the identification RTNN in the last 20 seconds of the simulation; d) comparison between Y2 and Yi2 of the identification RTNN in the last 20 seconds of the simulation; e) first control signal U1; f) second control signal U2; g) mean squared error of control (MSE%); h) mean squared error of identification (MSE%); i) systems state variables, estimated by RTNN.

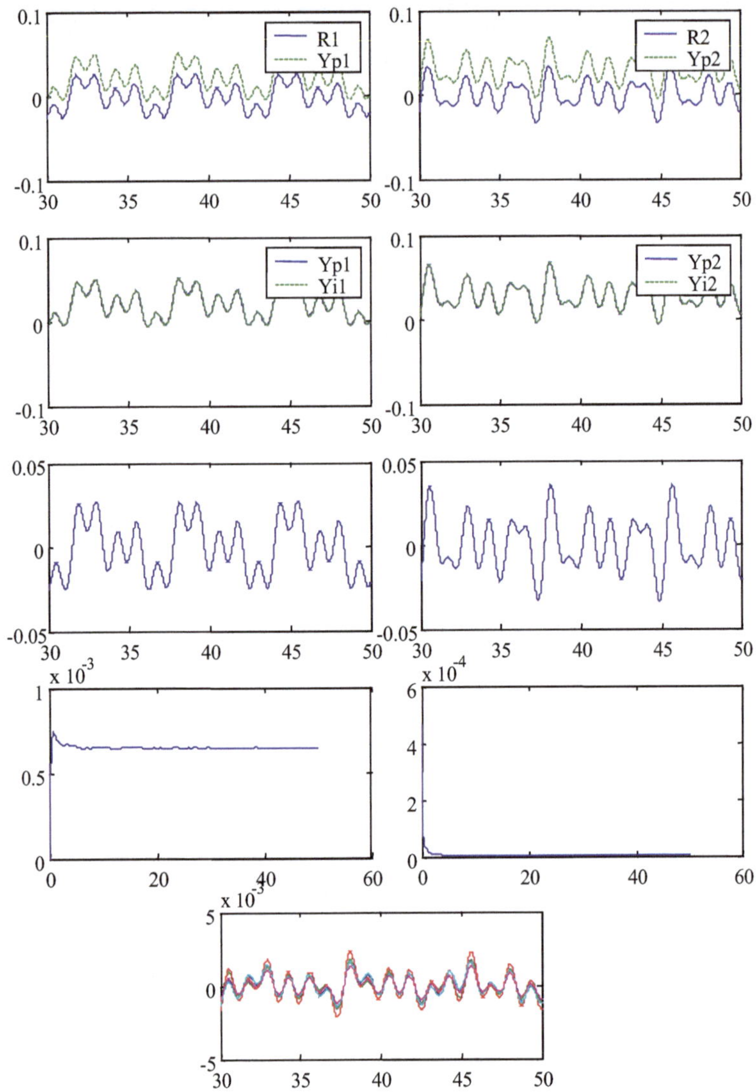

Fig. 4. Simulation results of a SMC without I-term and a 40% constant offset; a) comparison of Y1 and R in the last 20 seconds of the simulation; b) comparison of Y2 and R in the last 20 seconds of the simulation; c) comparison of Y1 and the first output Yi1 of the identification RTNN in the last 20 seconds of the simulation; d) comparison of Y2 and the second output Yi2 of the identification RTNN in the last 20 seconds of the simulation; e) first control signal U1; f) second control signal U2; g) mean squared error of control (MSE%); h) mean squared error of identification (MSE%); i) systems state variables, estimated by RTNN.

4.2 Example 2. 1-DOF Mechanical Plant with Friction Controlled by a Direct Adaptive Neural Controller

Let us consider a 1-DOF mechanical system with friction, which general model, taken from [20], is given by the equation:

$$m\ddot{q} + fr(\dot{q},t) + v(t) = k_o u(t) \tag{50}$$

Where: m is the mass, q(t) is the relative displacement; ω(t)=dq(t)/dt is the velocity, fr(ω,t) is the friction force, u(t) is the control force, k_o is the system gain, and v(t) is a bounded external load disturbance, with unknown upper bound d:

$$|v| \leq d \;;\; t > 0 \tag{51}$$

The equations, describing the behavior of the friction force, taken from [20], are given as:

$$fr(\omega,t) = F_{slip}(\omega)\lambda(\omega) + F_{stick}(u)[1-\lambda(\omega)] \tag{52}$$

$$\lambda(\omega) = \begin{cases} 1 & |\omega| > \alpha \\ 0 & |\omega| \leq \alpha \end{cases} \tag{53}$$

$$F_{stick}(u) = \begin{cases} F_s^+ & u \geq F_s^+ > 0 \\ u & F_s^- < u < F_s^- \\ F_s^- & u \leq F_s^- < 0 \end{cases} \tag{54}$$

$$F_{slip}(\omega) = F_d^+\mu(\omega) + F_d^-\mu(-\omega) \tag{55}$$

$$\mu(\omega) = \begin{cases} 1 & \omega > 0 \\ 0 & \omega \leq 0 \end{cases} \tag{56}$$

$$F_d^+(\omega) = F_s^+ - \Delta F^+\left[1 - e^{-(\omega/\omega_{cr}^+)}\right] + \beta^+\omega \tag{57}$$

$$F_d^-(\omega) = F_s^- - \Delta F^-\left[1 - e^{-(\omega/\omega_{cr}^-)}\right] + \beta^-\omega \tag{58}$$

The friction model have the following friction parameters, [19]: α = 0.001 m/s; $F_s^+ = 4.2$ N; $F_s^- = -4.0$ N; $\Delta F^+ = 1.8$ N ; $\Delta F^- = -1.7$ N ; $\omega_{cr} = 0.1$ m/s; β = 0.5 Ns/m. Let us also consider that the time of discretization is $T_o = 0.1$ s, the system gain is $k_o = 8$, the mass is m = 1 kg, and the load disturbance depends on the position and the velocity ($v(t) = d_1 q(t) + d_2\omega(t)$; $d_1 = 0.25$; $d_2 = -0.7$). So the discrete-time model of the 1-DOF mass mechanical system with friction is obtained in the form:

$$X_1(k+1) = X_2(k) \tag{59}$$

$$X_2(k+1) = 0.025\, X_1(k) - 0.3\, X_2(k) + 0.8\, U(k) - 0.1\, Fr(k) \tag{60}$$

$$\Omega(k) = X_2(k) - X_1(k) \tag{61}$$

$$Y(k) = 0.1\, X_1(k) \tag{62}$$

Where: $X_1(k)$, $X_2(k)$ are system states; $\Omega(k)$ is system velocity and $Y(k)$ is system position; k is a discrete time variable, and the friction force $Fr(k)$ is governed by the equations (52) to (58) with given values of friction parameters. The reference signal is a saturated sinusoid, given by:

$$r(k) = \text{sat}\,[\,\sin(\pi k/10)\,] \tag{63}$$

The RTNN characteristics for this control scheme are: the RTNN-1 topology is chosen as (1, 5, 1); the RTNN-2 topology is (7, 5, 1); the learning rate parameters for both RTNNs are - $\eta = 0.01$, $\alpha = 0$. Results of simulation experiments are shown in 20-seconds graphics. The results obtained using the control scheme given on Fig. 2 and 40% constant offset, are shown on Fig. 5 a-h. Results, obtained by control scheme without integral term and 40% constant offset, for the same case, are given on Fig. 6 a-h. The graphic, given on Fig. 5 a, compares the reference signal with the plant output in the last 20 seconds of the mechanical plant control simulation. The next graphics (Fig. 5 b) compares the plant output with the output of the RTNN-1 during the last 20 seconds of the plant identification. The following graphics, (Fig. 5 c), represents the control signal in the same period of time. The graphics, given on (Fig. 5 d), represents the five states of the system, issued by the RTNN-1, which are entry to the control RTNN-2. The next two graphics represent the instantaneous error of control and the instantaneous error of identification (Fig. 5 e, f). The last two graphics represent the Mean Squared Error (MSE%) of control and the MSE% of identification (Fig. 5 g, h), both decreasing rapidly to small values. Results, obtained with a control system without integral block, and a 40% constant offset, are shown on Fig. 6 a-h, given in the same order, so to be compared with that of Fig. 5 a-h.

As it could be seen, the system without integral action is sensitive to constant load disturbances, and as it is seen in Fig. 6 g, the MSE% is greater, than that of Fig. 5 g., which signifies that the control compensate the constant offset badly.

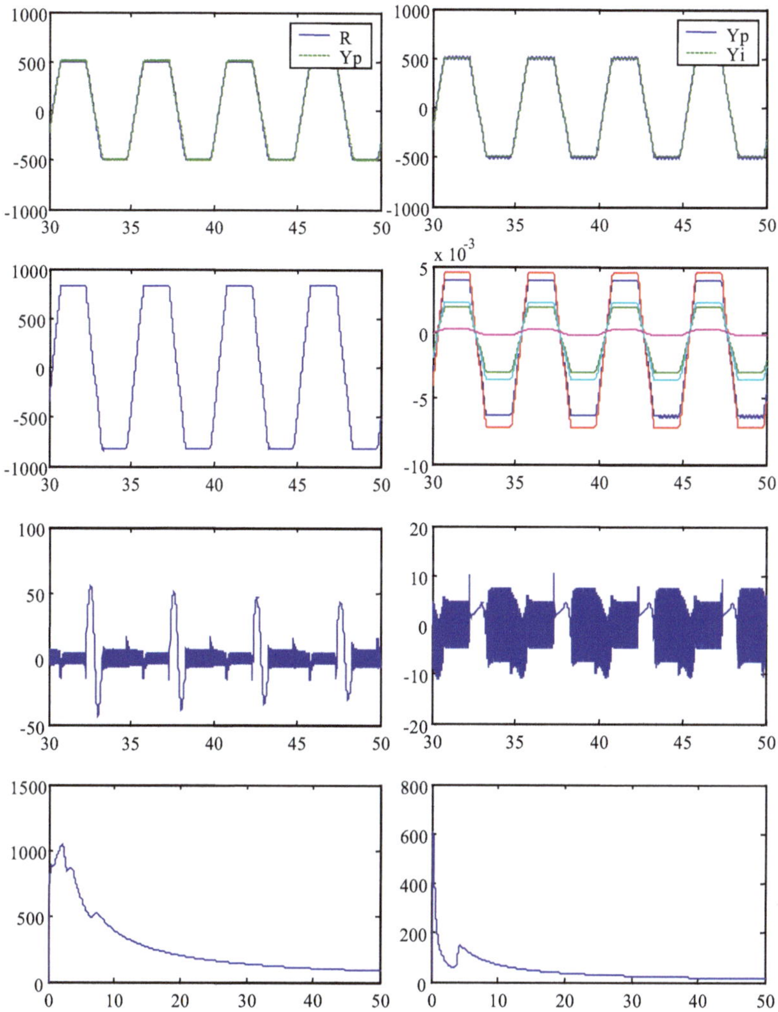

Fig. 5. A direct adaptive trajectory tracking control with I-term and 40% constant offset; a) comparison between the plant output and the reference signal; b) comparison between the plant output and the output of the identification RTNN-1; c) control signal; d) the five systems states, issued by the RTNN-1; e) instantaneous error of control; f) instantaneous error of identification; g) mean squared error of control (MSE%); h) mean squared error of identification (MSE%).

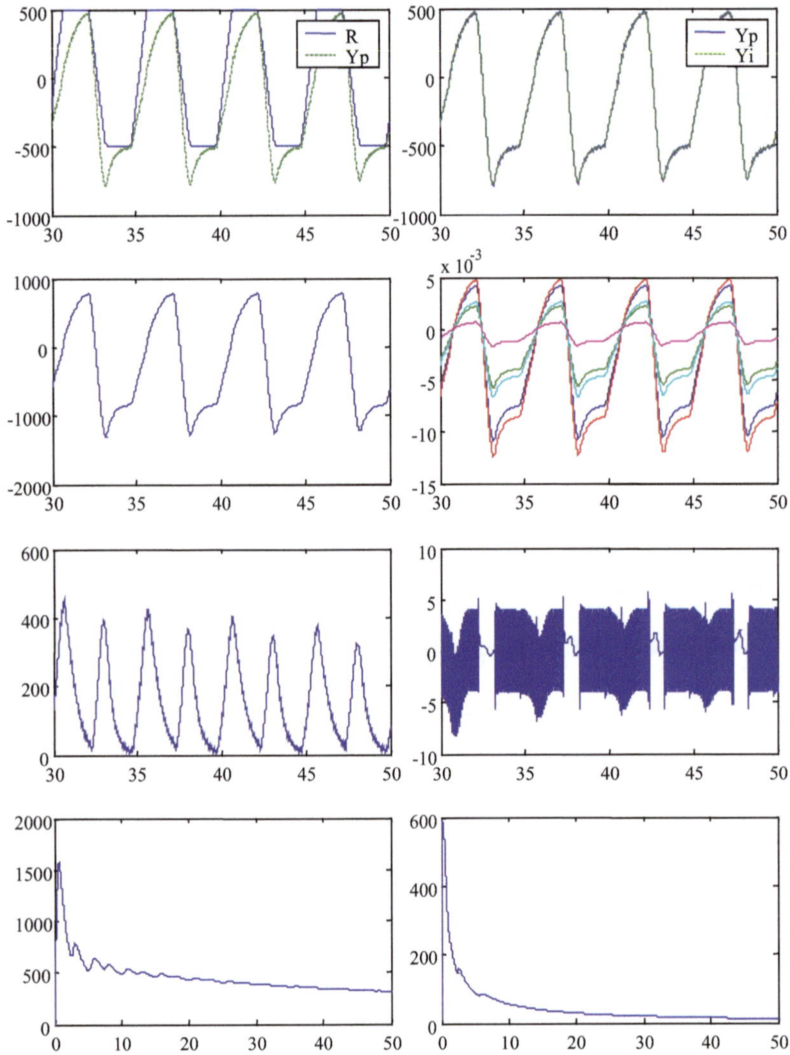

Fig. 6. A direct adaptive trajectory tracking control without I-term and 40% constant offset; a) comparison between the plant output and the reference signal; b) comparison between the plant output and the output of the identification RTNN-1; c) control signal; d) the five systems states, issued by the RTNN-1; e) instantaneous error of control; f) instantaneous error of identification; g) mean squared error of control; h) mean squared error of identification

5 Conclusions

A comparative study of various control systems using neural networks is done. The paper proposes to use a Recurrent Trainable Neural Network (RTNN) with Backpropagation method of learning as a plant parameters identifier and state estimator. The paper applied two methods of adaptive neural control with integral plus state action – an indirect and a direct trajectory tracking control. The first one is a Sliding Mode Control (SMC) with I-term where the SMC is resolved using states and parameters identified by RTNN. The second one is the direct adaptive neural control with I-term where the control is resolved by a RTNN controller. The good tracking abilities of both methods are illustrated by simulation results obtained using a MIMO mechanical plant and a 1-DOF mechanical system with friction plant model. The results show that both control schemes with I-term could compensate constant offsets and that - without I- term did not.

References

1. Miller III, W.T., Sutton, R.S., Werbos, P.J.: Neural Networks for Control, MIT Press, London (1992).
2. Hunt, K.J., Sbarbaro, D., Zbikowski, R., Gawthrop, P.J.: Neural Networks for Control Systems, A Survey. Automatica, 28 (1992) 1083-1112.
3. Narendra, K.S., Parthasarathy, K.: Identification and Control of Dynamic Systems Using Neural Networks. IEEE Trans. Neural Networks, 1 (1990) 4-27.
4. Chen, S., Billings, S. A.: Neural Networks for Nonlinear Dynamics System Modeling and Identification. International Journal of Control, 56 (1992)319-346.
5. Pao, S.A., Phillips, S.M., Sobajic, D.J.: Neural-Net Computing and Intelligent Control Systems. International Journal of Control, 56 (1992) 263-289.
6. Jin, L., Gupta, M.: Stable Dynamic Backpropagation Learning in Recurrent Neural Networks. IEEE Transactions on Neural Networks, 10 (1999) 1321-1334.
7. Baruch, I.S., Stoyanov, I.P., Gortcheva, E.: Topology and Learning of a Class RNN. ELEKTRIK, Supplement, 4 (1996) 35-42.
8. Nava, F., Baruch, I.S., Poznyak, A.S., Nenkova, B.: Stability Proofs of Advanced Recurrent Neural Networks Topology and Learning. Comptes Rendus (Proceedings of the Bulgarian Academy of Sciences), ISSN 0861-1459, 57 (2004) 27-32.
9. Sontag, E. Sussmann, H.: Complete Controllability of Continuous Time Recurrent Neural Network. System and Control Letters, 30 (1997) 177-183.
10. Albertini, F., Sontag, E.: State Observability in Recurrent Neural Networks. System and Control Letters, 22 (1994) 235-244.
11. Wan, E., Beaufays, F.: Diagrammatic Method for Deriving and Relating Temporal Neural Networks Algorithms. Neural Computations, 8 (1996) 182-201.
12. Baruch, I.S., Flores, J.M., Nava, F., Ramirez, I.R., Nenkova, B.: An Advanced Neural Network Topology and Learning, Applied for Identification and Control of a D.C. Motor. In: Proc. 1-st. Int. IEEE Symp. Intelligent Systems, Varna, Bulgaria, Sept. (2002) 289-295.

13. Utkin, V.I.: Sliding Mode in Control and Optimization. Springer-Verlag, Berlin (1992).
14. Utkin, V.I.: Sliding Mode Control in Dynamic Systems. In: Proc. of the 32-nd Conference on Decision and Control, San Antonio, Texas, Dec. (1993) 2446-2451.
15. Utkin, V.I.: Adaptive Discrete-Time Sliding Mode Control of Infinite-Dimensional Systems. In: Proc. of the 37-th Conference on Decision and Control, Tampa, Florida, Dec. (1998) 4033-4038.
16. Young, K.D., Utkin, V.I., Ozguner, U.: A Control Engineer's Guide to Sliding Mode Control. IEEE Trans. on Control Systems Technology, 7 (1999) 328-342.
17. Baruch, I.S., Hernandez, L.A., Barrera-Cortes, J.: Adaptive Discrete-Time Sliding Mode Control Using Recurrent Neural Networks. In: Proc. of the 2-nd IFAC Symposium on System, Structure and Control, Oaxaca, Mexico, Dec. 8-10, (2004) file [SSSC112.pdf].
18. Baruch, I.S., Martinez, A.C., Garrido, R.: An Adaptive Neural Dynamics Compensator with Integral-Plus-State Action. In: Proc. of the 23-th Int. Conf. on Artificial Neural Networks and 10-th Int. Conf. on Neural Information Processing, ICANN/ICONIP, Istambul, Turkey, June 26-29, (2003) 320-323.
19. Narendra, K., Mukhopadhyay, S.: Adaptive Control of Nonlinear Multivariable Systems Using Neural Networks. Neural Networks, 7 (1994) 737-752.
20. Lee, S.W., Kim, J.H.: Robust Adaptive Stick-Slip Friction Compensation. IEEE Trans. Ind. Electronics, 42 (1995) 474-479.

Simple Tuning of Fuzzy Controllers

Eduardo Gómez-Ramírez

Laboratorio de Investigación y Desarrollo de Tecnología Avanzada, LIDETEA, Dirección de Posgrado e Investigación, Universidad La Salle, Benjamín Franklin 47 Col. Condesa, 06140, México, D.F. México,
E-mail: egr@ci.ulsa.mx

Abstract. The number of applications in the industry using the PID controllers is bigger than fuzzy controllers. One reason is the problem of the tuning, because it implies the handling of a great quantity of variables like: the shape, number and ranges of the membership functions, the percentage of overlap among them and the design of the rule base. The problem is more complicated when it is necessary to control multivariable systems due that the number of parameters. The importance of the tuning problem implies to obtain fuzzy system that decrease the settling time of the processes in which it is applied, or in some cases, the settling time must be fixed to some specific value. In this work a very simple algorithm is presented for the tuning of a fuzzy controller using only one variable to adjust the performance of the system. The results are based on the relation that exists between the shape of the membership functions and the settling time. Some simulations are presented to exemplified the algorithm proposed.

1 Introduction

The implementation of a fuzzy controller is not a hard work. The knowledge of the dynamic of the system can be used to find a very good estimation of the rules and the sets of membership functions. But to find an optimal and well tuned fuzzy controller is other history. Normally the methodology for tuning fuzzy controller is a heuristic work and for every problem it is necessary to consider some elements like: the bandwidth, the error in steady state, or the settling time. In some cases it is possible to use this information to find the optimal parameters of the controller. In the case of a PID controller it is necessary to find three parameters (proportional gain, derivative time, and integral parameters). In the case of fuzzy

E. Gómez-Ramírez: *Simple Tuning of Fuzzy Controllers*, StudFuzz **208**, 115–133 (2007)
www.springerlink.com

controllers, there are many parameters to compute like, number of membership functions used, the ranges of every function, the rules, the shape, the percentage of overlap, etc. [1, 2 3]. Many people prefer to use a very well known PID controller that a fuzzy controller due the tuning complexity. Many times this is a very important reason to avoid the use in the industry of this type of "intelligent" controller.

The tuning of any controller's type implies the adjustment of the parameters to obtain a desired behavior or a good approach with a minimal error of the desire response. The different methods published in the area for the problem of fuzzy controllers' tuning use methodologies like evolutionary computation [4, 5, 6, 7] and artificial neural networks [8, 9]. These methods search the solution according to objective functions, parameter estimation, gradient error, etc., but in many cases these alternatives have serious convergence problems and a very complex mathematical representation. The computation time is big, or it is possible that the solution computed is only a local minimum of the general solution.

In this paper a very simple method for tuning fuzzy controllers is presented using only two parameters. In this case, the paper is based in the relation between the stabilization time and the range of the membership functions. This papers uses the results of a previous work [10] using only one parameter for tuning all the membership functions. The paper is structured in the following way: In section 2 the relationship that exists between the positions of the membership functions with the transfer characteristic is presented. In section 3 the non-linear system used is described with the controller's description for the tuning. The section 4 outlines an algorithm of parametric tuning that modifies the operation points that define the group of membership functions. In the section 5 the results of the simulations are shown for different values of the tuning factor and different graphics that show the behavior of the settling time in function of the tuning factors and finally the conclusions of the work.

2 Fuzzy Controller Characterization

The fuzzy controller's behavior, considering its answer speed, sensitive and reaction under disturbances can be described using the transfer characteristic and the position of the operation points. This is related with the election of the fuzzy controller's gain dy/dx (where y is the output and x is the input of the system), in different regions of the domain x.

Case 1. For a flat slope in the middle of the domain x and increasing slopes toward increasing $|x|$ values, choose larger distances between operations points in the middle of domain (see figure 1). This means:

$$\text{For} \quad |x_2| > |x_1| \Rightarrow |dy/dx|_{x_2} > |dy/dx|_{x_1} \tag{1}$$

Case 2. For a steep slope in the middle of the domain x and decreasing slopes toward increasing $|x|$ values, choose smaller distances between operations points in the middle of domain (see figure 2). This means:

$$\text{For} \quad |x_2| > |x_1| \Rightarrow |dy/dx|_{x_2} < |dy/dx|_{x_1} \tag{2}$$

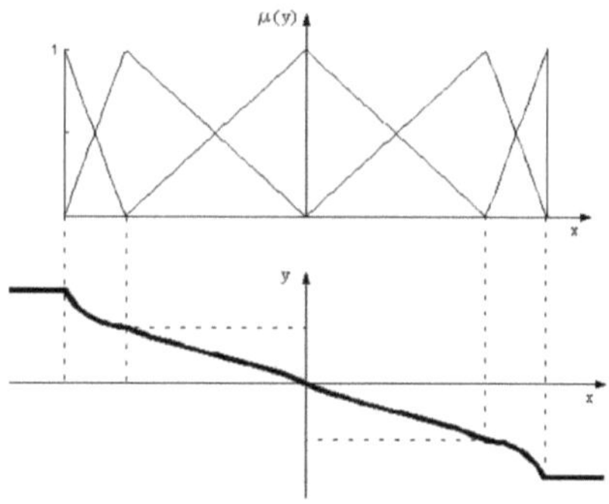

Fig. 1. Relationship between the location of the membership functions and the transfer characteristic for the case 1.

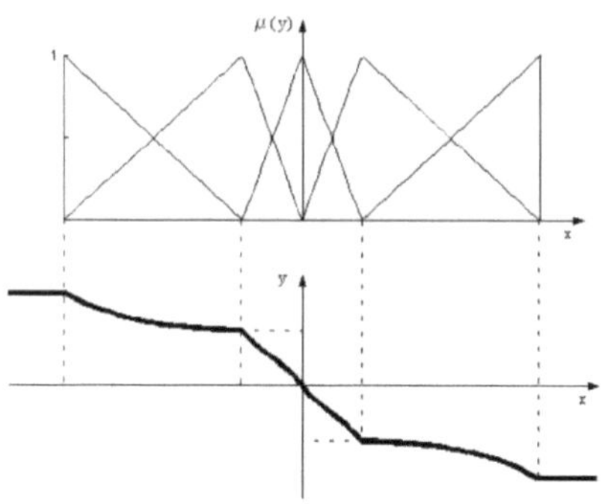

Fig. 2. Relationship between the location of the membership functions and the transfer characteristic for the case 2.

Option 1 should be chosen if for small errors a slow reaction to disturbances of the system under control is required. Option 2 should be chosen if for small errors the system is supposed to be sensitive with respect to disturbances.

In the previous figures the values for the intervals of the membership function are important for the slopes and the speed of the controller response. If the membership functions "expand" (figure 1) then the response is slower than a compress group of membership functions (figure 2).

3 Dynamic System and Fuzzy Controller

For the analysis and simulations with the tuning algorithm a second order system has been considered:

$$\frac{1}{0.45s^2+2s+1} \tag{3}$$

overdamped with a damping ratio $\xi = 1.4907$ and a natural frequency $\omega_n = 1.4907\ rad/s$.

The fuzzy controller designed for the control of the plant described previously is a system TISO (two inputs-one output) where the inputs are the error and the change of error while the output is the control action. Each one of the controller's variables has been divided in 5 fuzzy regions. The fuzzy associative memory, integrated by 25 rules, it is shown in the figure 4.

Table 1 Controller Fuzzy variables

Input variables		Output variable
error	*change of error*	*control action*
GN: Big negative	GN: Big negative	DG: Big diminution
MN: Medium negative	MN: Medium negative	DP: Small diminution
Z: Zero	Z: Zero	M: Hold
MP: Medium positive	MP: Medium positive	AP: Small increase
GP: Big positive	GP: Big positive	AG: Big increase

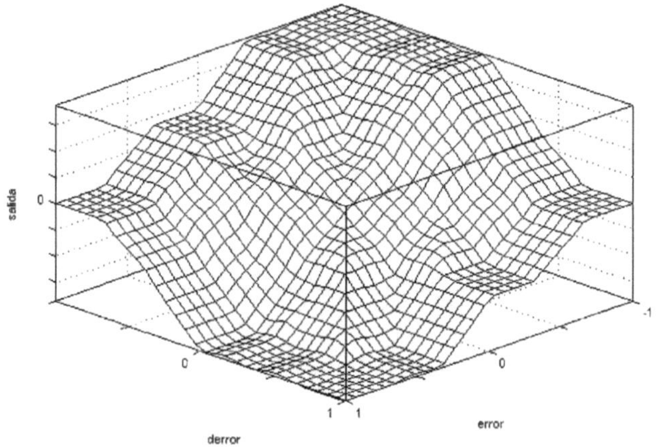

Fig. 4. Example of control surface of a fuzzy controller.

e / $\dot{e}$	GN	MN	Z	MP	GP
GN	AG	AG	AP	DP	DG
MN	AG	AP	M	M	DG
Z	AG	AP	M	DP	DG
MP	AG	M	M	DP	DG
GP	AG	AP	DP	DG	DG

Fig. 5. Fuzzy associative memory for the control system.

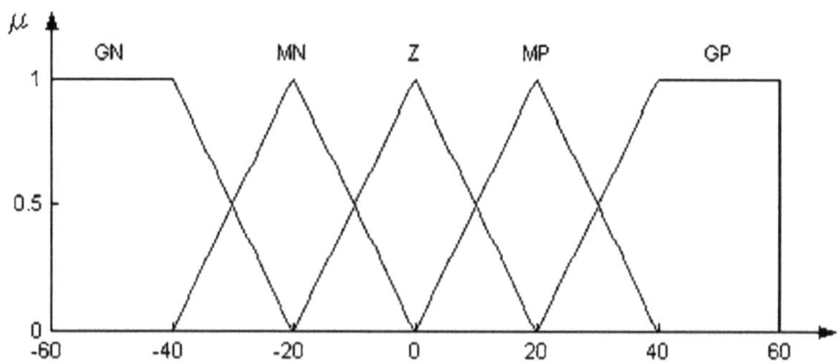

Fig. 6. Membership functions for the input variable *error*.

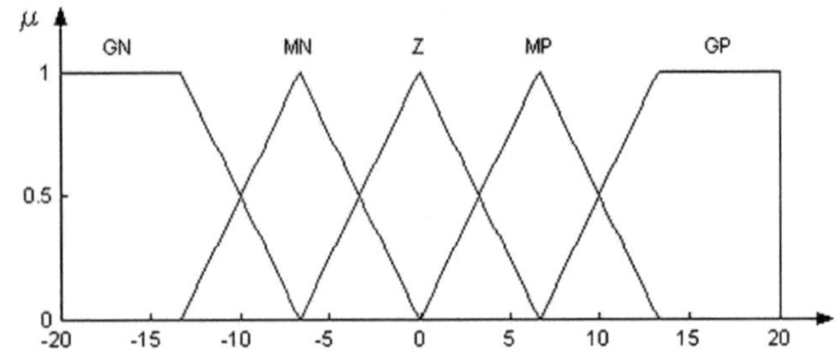

Fig. 7. Membership functions for the input variable *change of error.*

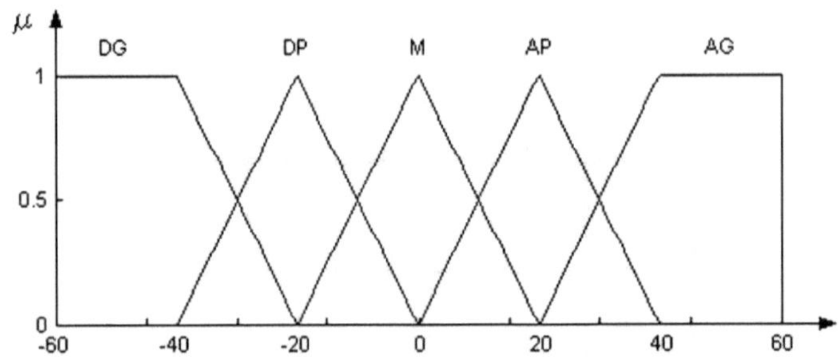

Fig. 8. Membership functions for the output variable *control action.*

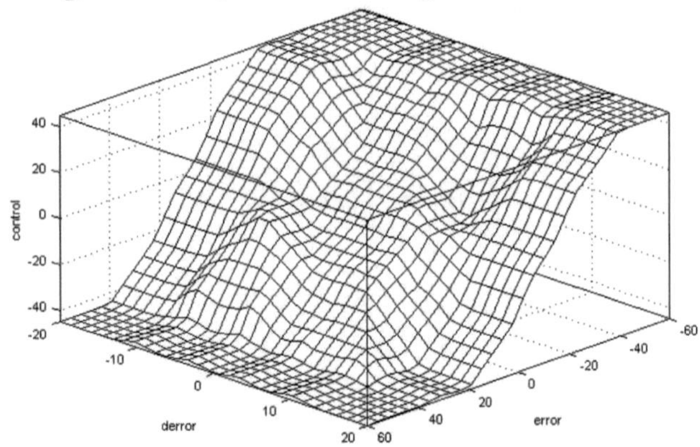

Fig. 9. Control surface for the fuzzy controller with the membership functions under their initial conditions.

The membership functions were defined in triangular shape for the middle and in a trapezoidal shape in the extremes; such that always have it overlap in the grade of membership $\mu(x) = 0.5$ (figures 6, 7 and 8). These membership functions will be considered later as the initial conditions for the proposed algorithm. The control surface for the fuzzy controller under its initial conditions is shown in the figure 9.

The answer of the system, with a step input of amplitude 40, is shown in figure 10.

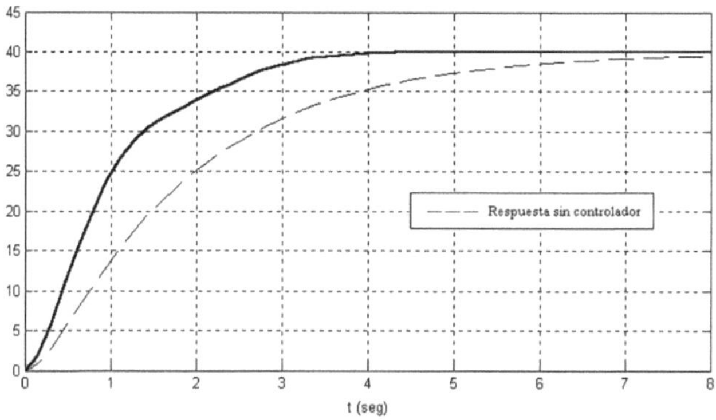

Fig. 10. Answer of the system with the membership functions under their initial conditions.

3.1 Tuning Algorithm

The objective of the tuning algorithm is to be able to manipulate, by means of a single variable and in a simple way, the settling time of the system, from the answer without controller until the response equivalent to 1/5 of the settling time of the answer without controller. The response must be fulfilled too with the constraints of small overshoots and without persistent oscillations, which means, a very smooth response.

This algorithm is based on the properties of the transfer characteristic or, in this case, of the control surface that it allows to modify the controller's behavior by means of modifications in the position and support of the membership functions maintaining fixed the fuzzy controller's structure. Obtaining a slower answer for configurations with wide or expanded membership functions in the center (see fig. 1) and reduced in the ends, and the other way, a faster answer for configurations with reduced or compressed membership functions in the center and wide in the ends (fig. 2).

The tuning algorithm only modifies the membership functions of the input variables and the membership functions of the output fuzzy variable remains constant since this disposition is only in function of a proportion of the range of the control action, in other words, they always remain uniformly spaced.

3.2 Tuning Factor Selection

The tuning factor is a number $k \in [0, 1]$ that determines the grade of tuning adjustment obtaining for $k = 0$ the biggest settling time (fig. 1) and for $k = 1$ the smallest settling time (fig. 2).

3.3 Normalization of Ranges of the Fuzzy Controller's Variables

In this step the range of each input fuzzy variable is modified so that their upper and lower limits are equal to +1 and -1, respectively.

3.4 Tuning Factor Processing

When the range is normalized in the range between 0 and 1 the values can be computing using a power function, because to expand it is only necessary to use an exponent less than 1 and to compress an exponent more than 1 (see fig. 12). Recall the previous step that all the values belong to the interval [-1,1]. In an experimental way different values of exponents were tested (table 2) such that the new vector of operation points will be given by:

$$Vop_{final} = (Vop_{initial})^{r(k)} \tag{4}$$

Where $Vop_{initial}$ are the values normalized of the membership function in the x-axis and *r(k)* is a polynomial.

Table 2. Important values of *r(k)*.

k	r
0	1/40
0.5	1
1	3

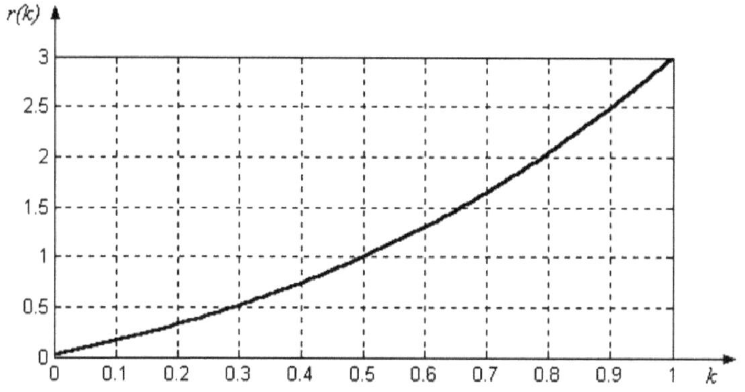

Fig. 11. Plot of function *r(k)*.

The initial coefficients of the polynomial were obtained using mean square method. The values of *k* were defined in this way to be able to make an estimate over all their range $k \in [0, 1]$. The values of *r*, since it is an exponent, they were defined considering the increasing or decreasing of a number that is powered to the exponent *r*. Remember that the goal is to expand (slow response) for *k=0* and to compress (fast response) k=1. For values below *r* = 1/40 the answer of the system was not satisfactory and in the same way, for values more than *r* = 3. With all these elements the polynomial obtained was:

$$r(k) = \frac{30k^3 + 37k^2 + 52k + 1}{40} \tag{5}$$

This *r(k)* was found testing the optimal response for different dynamical systems (linear and non-linear) and finding the optimal parameters of the polynomial that fix the function for different values of *k* (*k* = 0, 0.5, 1) (figure 11).

To visualize the effect of this processing it is useful the graph of curves of adjustment for the vectors of operation points (figure 12). The values that can take an operation point (positive section) are in the horizontal axis and in the vertical axis, the values that takes this operation point once it has been powered to the exponent *r(k)* where $k \in [0, 1]$.

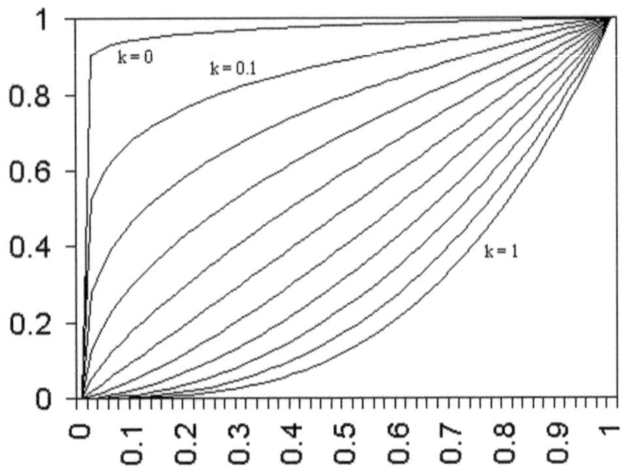

Fig. 12. Curves of adjustment for the operation points.

3.5 Denormalization of the Ranges of the Fuzzy Variables

In this step it is necessary to convert the normalized range to the previous range of the system. This can be computed only multiplying the *Vop* vector by a constant factor.

4 Results of the Simulation. The Same Gain for Both Inputs

The cases will be analyzed for 3 different values of *k*, *k* = 0, 0.5, 1, showing the effect in the membership functions of the fuzzy variables, the control surface and the graph result of the simulation. For all the analyzed cases it will be used as input a step function with amplitude 40, the parameters that allow evaluating the quality of the tuning are the settling time (considered to 98% of the value of the answer in stationary state), the overshoots and the oscillations. Also for all the analyzed cases the controller's structure is fixed, that means, the fuzzy associative memory is the same in all the examples. In the simulations it is included (on-line dotted) the answer of the system without controller to compare with the response using different tunings of the controller. The controller's fuzzy variables and the membership functions for the initial conditions are shown in the figures 6, 7 and 8: *error*, *change of error* and *control action* respectively.

4.1 Case 1: Adjusting the Membership Functions with a Tuning Factor $k = 0$

The function $r(k)$ takes the value $r(0) = 1/40$. With the tuning process the vectors of operation points for the fuzzy input variables are the following ones:

$$Vop\ error_{final} = [-59.3947, -58.3743, 0, 58.3473, 59.3947]$$
$$Vop\ d/dt(error)_{final} = [-19.7982, -19.4581, 0, 19.4581, 19.7982]$$

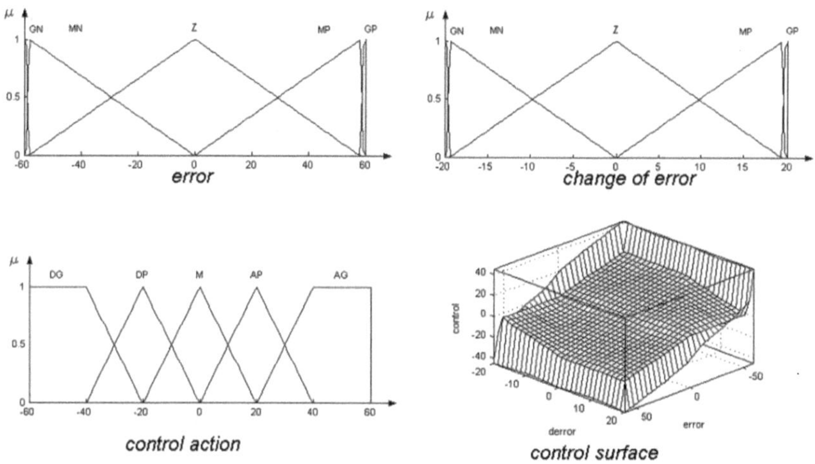

Fig. 13. Membership functions of the fuzzy variables and control surface for $k = 0$.

Making the simulation with the controller's characteristics shown in the figure 13 the following answer was obtained:

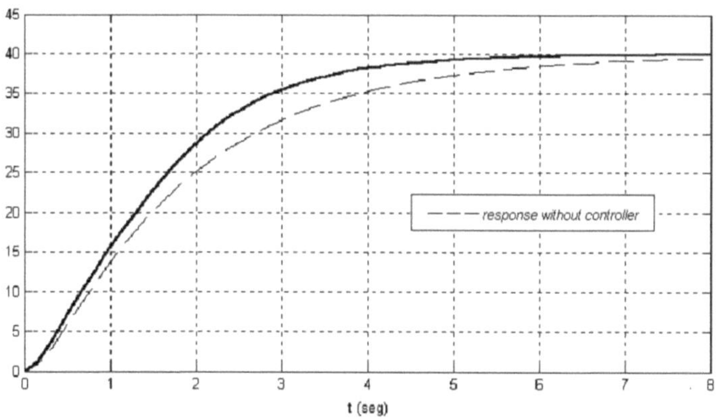

Fig. 14 Answer of the system for the case 1 with $k = 0$.

In the figure 14 it is shown that with the tuning factor $k = 0$ the controller's effect on the answer of the system, due to the tuning, is small, approaching to the answer without controller. In this case the settling time is the biggest that can be obtained, $t_s = 4.96\ s$.

4.2 Case 2: Adjusting the Membership Functions with a Tuning Factor $k = 0.5$.

The function $r(k)$ takes the value $r(0.5) = 1$. With the tuning process the vectors of operation points for the fuzzy input variables are the following ones:

$$Vop\ error_{final} = [-40, -20, 0, 20, 40]$$
$$Vop\ d/dt(error)_{final} = [-13.332, -6.666, 0, 6.666, 13.332]$$

Computing the simulation with the controller's characteristics shown in the figure 15 the following answer was obtained:

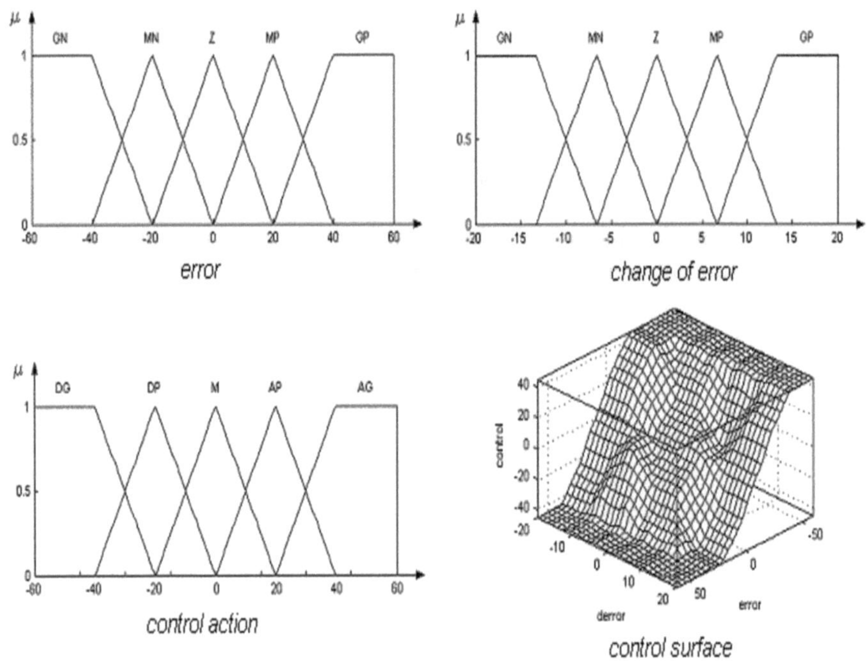

Fig. 15. Membership functions of the fuzzy variables and control surface for $k = 0.5$.

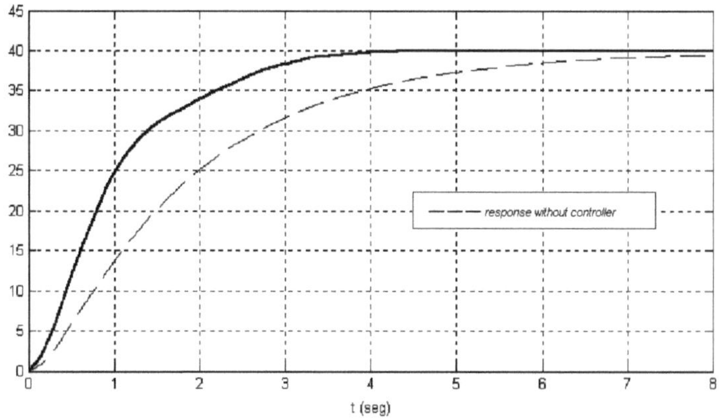

Fig. 16. Answer of the system for the case 2 with $k = 0.5$.

This case, with the tuning factor $k = 0.5$, is equal to operate with the initial conditions of the membership functions. The settling time is $t_s = 3.36$ *s*.

4.3 Case 3: Adjusting the Membership Functions with a Tuning Factor k = 1

The function $r(k)$ takes the value $r(1) = 3$. With the tuning process the vectors of operation points for the fuzzy input variables are the following ones:

$$Vop\ error_{final} = [-17.7724, -2.2215, 0, 2.2215, 17.7724]$$

$$Vop\ d/dt(error)_{final} = [-5.9241, -0.7405, 0, 0.7405, 5.9241]$$

Computing the simulation with the controller's characteristics shown in the figure 17 the answer was obtained in figure 18, where the settling time is $t_s = 1.6$ *s* and it is the less value that can be obtained.

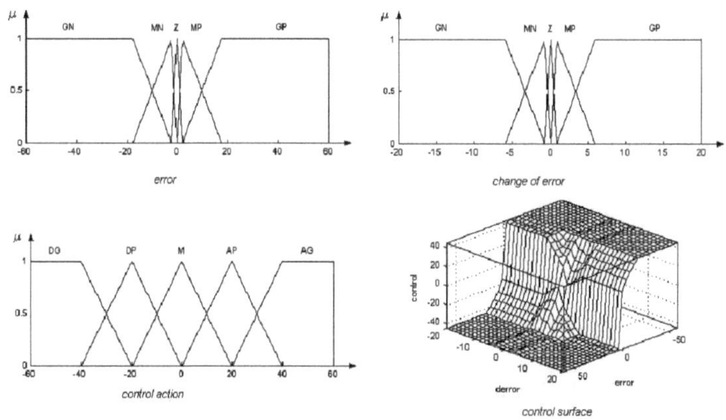

Fig. 17. Membership functions of the fuzzy variables and control surface for $K = 1$.

The controller's effect on the answer of the system has begun to cause a small overshoot, due the bigger compression of the membership functions. If the value of *r(k)* is increased, the settling time is not reduced and it only causes bigger overshoots and oscillations around the reference.

To visualize the effect of different values of the tuning factor *k* over the settling time of the answer of the system simulations with increments $\Delta k = 0.05$ in the interval $[0, 1]$ were computed. The result is shown in figure 19.

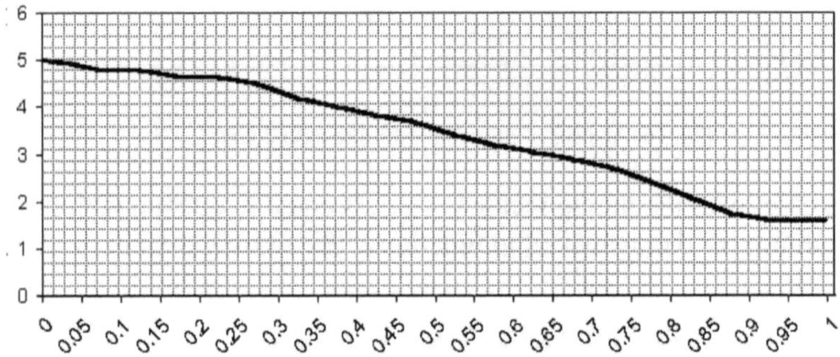

Fig. 19. Settling time versus tuning factor *k*.

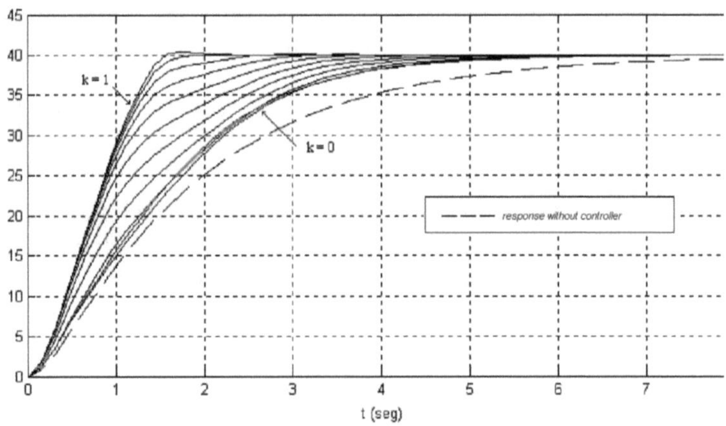

Fig. 20. Comparative graph of the system answers for different values of *k*.

Figure 20 shows the behavior of the answer of the system for increments $\Delta k = 0.1$. The next curve to the left is the corresponding for a tuning factor $k = 1$, with a settling time $t_s = 1.6$ *s*. For each increment, beginning in $k = 0$, a curve was plotted showing in the figure a smaller settling time.

Making use of the simulations, and the graphs in figures 19 and 20, it is possible to see that the optimal value of *k* for the tuning is $k = 0.9$. This value generates a settling time t_s = 1.6 *s* without a great overshoot and without oscillations (fig. 21).

Additionally, the fuzzy controller's performance was compared with a controller PID (Proportional-integral-derivative) whose parameters are the following ones $K_p = 25$, $T_i = 1.35$ and $T_d = 5$, and being that the differences are minimum as for time of establishment and general behavior (figures 22).

The disadvantage found in the controller PID is its inefficiency in comparison with the fuzzy controller since the control action generated by the PID can take very big values that are impossible to consider in a real implementation. On the other hand, the fuzzy controller uses real range of values.

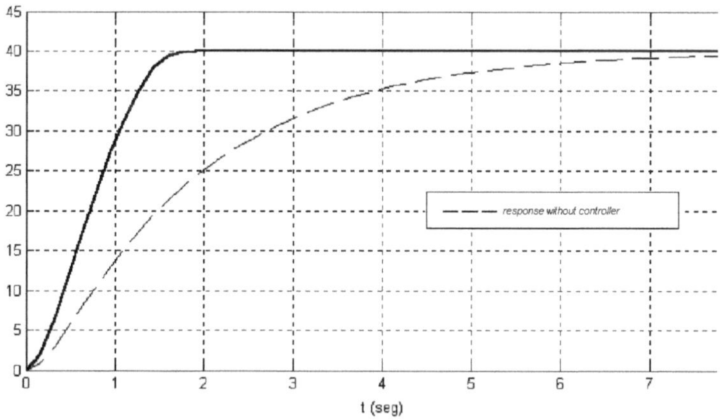

Fig. 21. Response of the system with $k = 0.9$.

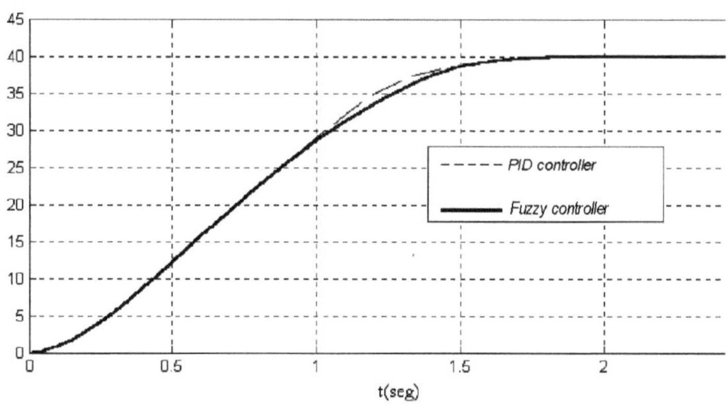

Fig. 22. Comparative graph of fuzzy controller answer versus PID.

Considering that this is the fastest answer that can be gotten with the controller PID, limited to the nature of the system, that is to say, limiting the range of the values that can take the control action to same values that those considered in the fuzzy controller's definition, it can be said that the tuning made on the fuzzy controller is satisfactory since it allows to vary the time of answer with very good behavior in the whole range of the tuning factor. Note that it is not evident to find three parameters of the PID for the optimal tuning and in the case of the fuzzy controllers it is necessary to increase or decrease the parameter k depending the settling time desired.

4.4 Simulation Results. Different Gains for all Inputs

In this case, a different gain k was considered for all inputs. The methodology used is exactly the same of the previous sections. The next figures show the stabilization time, the gain for the error k_1 and the gain used for the derivative of the error k_2 using a Δk of 0.01. For the axes k_1 and k_2 the value of 21 is equivalent to 1 in the figures 23, 25 and 26. Fig. 19 can be obtained for the case of $k_1=k_2$ in this surface (see fig. 24).

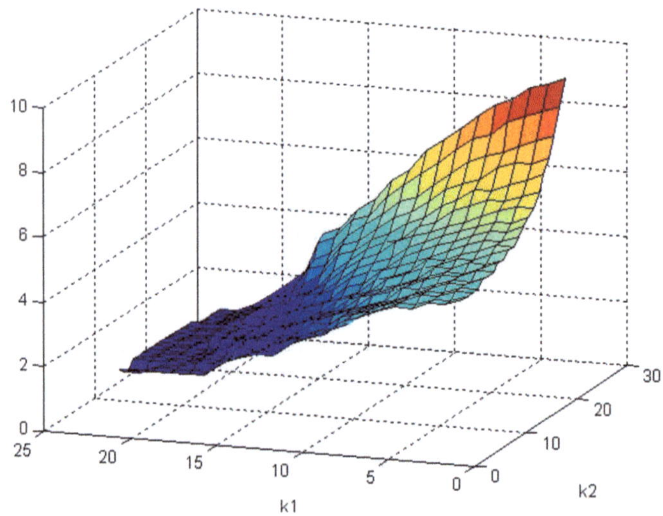

Fig. 23. Settling time versus tuning factor k_1 *and* k_2.

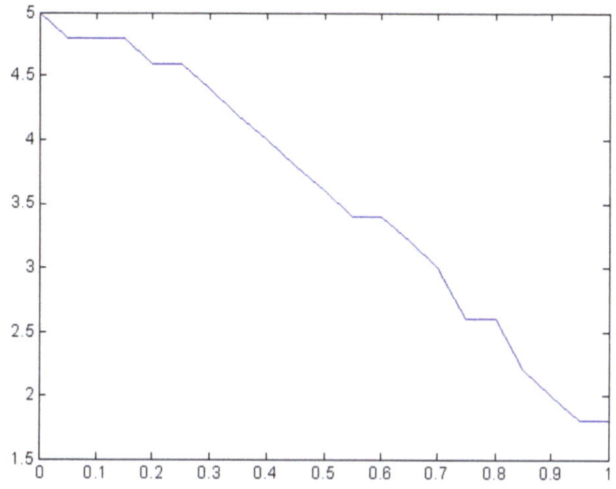

Fig. 24. Settling time versus tuning factor $k_1 = k_2$.

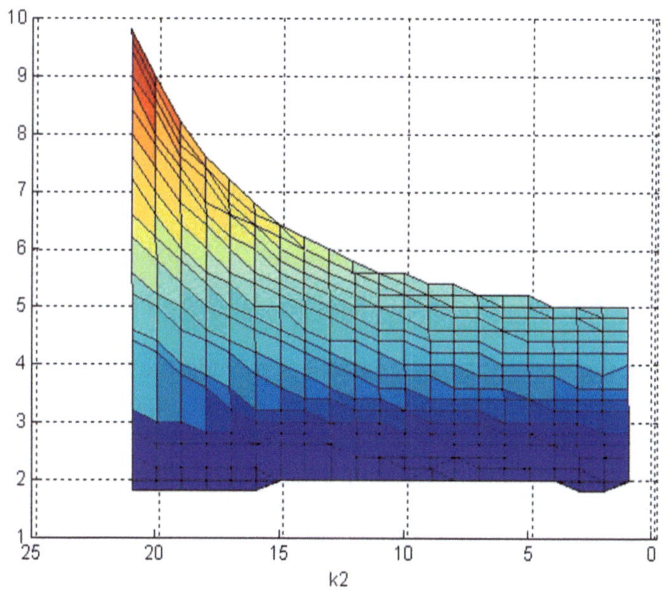

Fig. 25. View for tuning factor k_2.

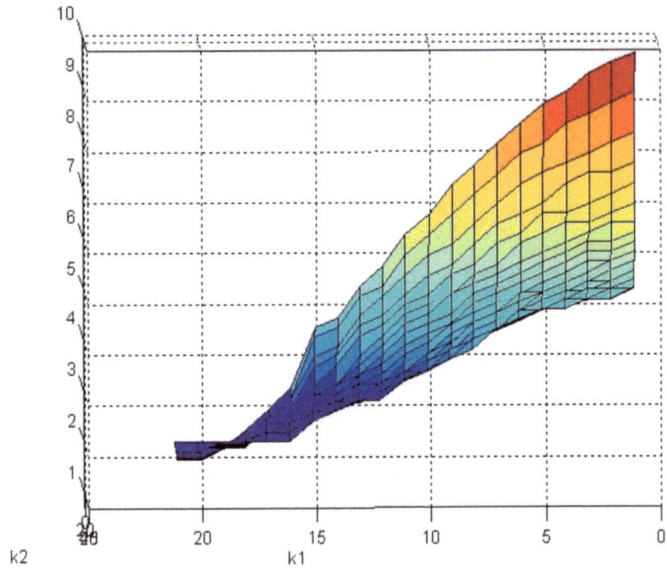

Fig. 26. View for tuning factor k_1 ·

With this option the controller has other degree of freedom to control the response of the dynamic system. Fig. 25 shows how the effect of tuning factor k_2, is less important than factor k_1. Keeping the factor k_1=1, the response of the system is in the interval 1.8-2.0 secs.

5 Conclusions

The tuning methods of fuzzy controllers include the handling of a great quantity of variables that makes very difficult, and many times non-satisfactory the search of structures and good parameters. The method proposed uses only one variable and operates considering the transfer characteristic, or in this case the control surface that is the fuzzy controller's property that defines their behavior allowing that the system can response with bigger or smaller speed and precision. The function $r(k)$ can be generalized to any system that uses a fuzzy controller varying the values $r(0)$ and $r(1)$ as well as the coefficients of the function $r(k)$ depending on the desired behavior. Another perspective is to create a self-tuning algorithm that modifies by itself the factors k_1 and k_2 to find the desired response. In this point, the use of fuzzy controllers presents attractive aspects for its implementation in real systems.

References

1. Cox E.; "Fuzzy Fundamentals", IEEE Spectrum, USA, October (1992)
2. Palm R., "Fuzzy control approaches, General design schemes, Structure of a fuzzy controller", Handbook of Fuzzy Computation. Institute of Physics, Bristol. USA, (1998).
3. Palm R., "Fuzzy control approaches, Sliding mode fuzzy control", Handbook of Fuzzy Computation. Inst. of Physics, Bristol. USA. (1998).
4. Casillas, J., Cordón O., del Jesús M.J. and Herrera F., Genetic Tuning of fuzzy rule deeps structures preserving interpretability and its interaction with fuzzy rule set reduction. IEEE Trans on Fuzzy systems, vol. 13, num 1, (2005) 27-45.
5. Kinzel J., Klawoon F. y Kruse R., "Modifications of genetic algorithms for designing and optimizing fuzzy controllers". 1st IEEE Conf. Evol. Computations, ICEC'94. Orlando, FL, USA. (1994)
6. Lee M. A. & Tagaki H., "Integrating design stages of fuzzy systems using genetic algorithms", 2nd IEEE Int. Conf. On Fuzzy Systems, Fuzz-IEEE'93 San Francisco, CA, USA. , (1993) 612-617.
7. Bonissone P., Kedhkar P. y Chen Y., "Genetic algorithms for automated tuning of fuzzy controllers: a transportation application". 5th IEEE Conf. On Fuzzy Systems, Fuzz-IEEE'96, New Orleans, LA, USA, (1996).
8. Jang J.. "ANFIS: adaptive-network-based-fuzzy-inference-system". IEEE Trans. Syst. Man Cybernet, SMC-23, USA. (1993) 665-685.
9. Kawamura A., Watanabe N., Okada H.& Asakawa K.. "A prototype of neuro-fuzzy cooperation systems". 1st IEEE Int. Conf. On Fuzzy Systems, Fuzz-IEEE'92, San Diego, CA, USA, (1992).
10. Gómez-Ramírez E., & Chávez-Plascencia A., How to tune Fuzzy Controllers, FUZZ-IEEE 2004. Budapest, Hungary, July 25-29, (2004).

From Type-1 to Type-2 Fuzzy Logic Control: A Stability and Robustness Study

Nohé Cázarez, Oscar Castillo, Luís Aguilar, Selene Cárdenas

[1] Tijuana Institute of Technology, Calzada del Tecnológico S/N, Tijuana, BC, México, ocastillo@tectijuana.mx,

[2] CITEDI-IPN, Av. Del Parque #1310, Mesa de Otay, Tijuana, BC, México, laguilar@citedi.mx

Abstract. Stability is one of the more important aspects in the traditional knowledge of Automatic Control. Type-2 Fuzzy Logic is an emerging and promising area for achieving Intelligent Control (in this case, Fuzzy Control). In this work we use the Fuzzy Lyapunov Synthesis as proposed by Margaliot [11] to build a Lyapunov Stable Type-1 Fuzzy Logic Control System, and then we make an extension from a Type-1 to a Type-2 Fuzzy Logic Control System, ensuring the stability on the control system and proving the robustness of the correponding fuzzy controller.

1 Introduction

Fuzzy logic controllers (FLC's) are one of the most useful control schemes for plants in which we have difficulties in deriving mathematical models or having performance limitations with conventional linear control schemes. Error e and change of error $\dot{e}$ are the most used fuzzy input variables in most fuzzy control works, regardless of the complexity of controlled plants. Also, either control input u (PD-type) or incremental control input Δu (PI-type) is typically used as a fuzzy output variable representing the rule consequent ("then" part of a rule) [6].

Stability has been one of the central issues concerning fuzzy control since Mamdani's pioneer work [9], [10]. Most of the critical comments to fuzzy control are due to the lack of a general method for its stability analysis.

But as Zadeh often points out, fuzzy control has been accepted by the fact that it is task-oriented control, while conventional control is characterized as setpoint-oriented control, and hence do not need a mathematical analysis of stability. And as Sugeno says, in general, in most industrial ap-

N. Cázarez et al.: *From Type-1 to Type-2 Fuzzy Logic Control: A Stability and Robustness Study*, StudFuzz **208**, 135–149 (2007)
www.springerlink.com

plications, the stability of control is not fully guaranteed and the reliability of a control hardware system is considered to be more important than the stability [15].

The success of fuzzy control, however, does not imply that we do not need a stability theory for it. Perhaps the main drawback of the lack of stability analysis would be that we cannot take a model-based approach to fuzzy control design. In conventional control theory, a feedback controller can be primarily designed so that a close-loop system becomes stable [13], [14]. This approach of course restricts us to set-point-oriented control, but stability theory will certainly give us a wider view on the future development of fuzzy control.

Therefore, many researchers have worked to improve the performance of the FLC's and ensure their stability. Li and Gatland in [7] and [8] proposed a more systematic design method for PD and PI-type FLC's. Choi, Kwak and Kim [4] presents a single-input FLC ensuring stability. Ying [18] presents a practical design method for nonlinear fuzzy controllers, and many other researchers have results on the matter of the stability of FLC's, in [1] Castillo et al., and Cázarez et al. [2] presents an extension of the Margaliot work [11] to build stable type-2 fuzzy logic controllers in Lyapunov sense.

This work is based on Margaliot et al. [11] work and in Castillo et al. [1] and Cázarez et al. [2] results, we use the Fuzzy Lyapunov Synthesis [11] to built an Stable Type-2 Fuzzy Logic Controller for a 1DOF manipulator robot, first without gravity effect to probe stability, and then with gravity effect to probe the robustness of the controller. The same criterion can be used for any number of DOF manipulator robots, linear or nonlinear, and any kind of plants.

This work if organized as follows: In Section II we presents an introductory explanation of type-1 and type-2 FLC's. In Section III we extend the Margaliot result to built a general rule base for any type (1 or 2) of FLC's. Experimental results are presented in Section IV and the concluding remarks are collected in Section V.

2 Fuzzy Logic Controllers

2.1 Type-1 Fuzzy Logic Control

Type-1 FLCs are both intuitive and numerical systems that map crisp inputs to a crisp output. Every FLC is associated with a set of rules with meaningful linguistic interpretations, such as

R^l : If x_1 is F_1^l and x_2 is F_2^l and … and x_n is F_n^l Then w is G^l

which can be obtained either from numerical data, or experts familiar with the problem at hand. Based on this kind of statement, actions are combined with rules in an antecedent/consequent format, and then aggregated according to approximate reasoning theory, to produce a nonlinear mapping from input space $U = U_1 x U_2 x ... U_n$ to the output space W, where $F_k^l \subset U_k, k = 1,2,...,n$, are the antecedent type-1 membership functions, and $G^l \subset W$ is the consequent type-1 membership function. The input linguistic variables are denoted by $u_k, k = 1,2,...,n$, and the output linguistic variable is denoted by w.

A Fuzzy Logic System (FLS), as the kernel of a FLC, consist of four basic elements (Fig. 1): the type-1 fuzzyfier, the fuzzy rule-base, the inference engine, and the type-1 defuzzifier. The fuzzy rule-base is a collection of rules in the form of R^l, which are combined in the inference engine, to produce a fuzzy output. The type-1 fuzzyfier maps the crisp input into type-1 fuzzy sets, which are subsequently used as inputs to the inference engine, whereas the type-1 defuzzyfier maps the type-1 fuzzy sets produced by the inference engine into crisp numbers.

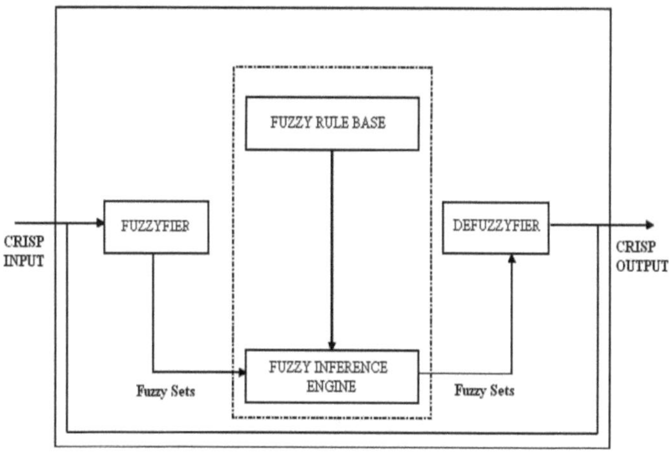

Fig. 1. Structure of type-1 fuzzy logic system

Fuzzy sets can be interpreted as membership functions u_X that associate with each element x of the universe of discourse, U, a number $u_X(x)$ in the interval [0,1]:

$$u_X : U \to [0,1] \tag{1}$$

For more detail of Type-1 FLS see [17], [5], [3].

2.2 Type-2 Fuzzy Logic Control

As the type-1 fuzzy set, the concept of type-2 fuzzy set was introduced by Zadeh [19] as an extension of the concept of an ordinary fuzzy set.

A FLS described using at least one type-2 fuzzy set is called a type-2 FLS. Type-1 FLSs are unable to directly handle rule uncertainties, because they use type-1 fuzzy sets that are certain. On the other hand, type-2 FLSs, are very useful in circumstances where it is difficult to determine an exact, and measurement uncertainties [12].

It is known that type-2 fuzzy set let us to model and to minimize the effects of uncertainties in rule-based FLS. Unfortunately, type-2 fuzzy sets are more difficult to use and understand that type-1 fuzzy sets; hence, their use is not widespread yet.

Similar to a type-1 FLS, a type-2 FLS includes type-2 fuzzyfier, rule-base, inference engine and substitutes the defuzzifier by the output processor. The output processor includes a type-reducer and a type-2 defuzzyfier; it generates a type-1 fuzzy set output (from the type reducer) or a crisp number (from the defuzzyfier). A type-2 FLS is again characterized by IF-THEN rules, but its antecedent of consequent sets are now type-2. Type-2 FLSs, can be used when the circumstances are too uncertain to determine exact membership grades. A model of a type-2 FLS is shown in Fig. 2.

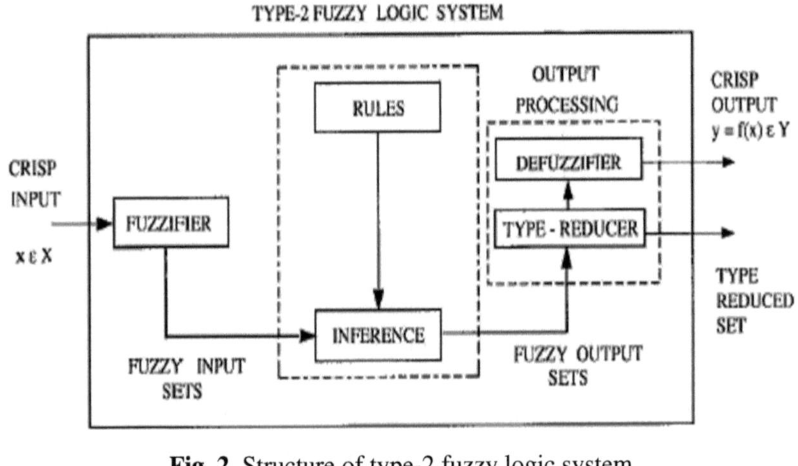

Fig. 2. Structure of type-2 fuzzy logic system

In the case of the implementation of the type-2 FLCs, we have the same characteristics as in type-1 FLC, but we used type-2 fuzzy sets as membership functions for the inputs and for the outputs. Figure. 3 shows the structure of a control loop with a FLC.

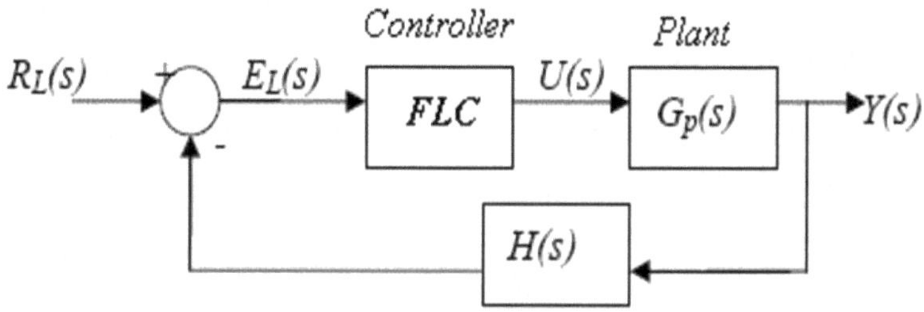

Fig. 3. Fuzzy control loop

3 Systematic Design of the Stable Fuzzy Controller

For our description we consider the problem of designing a stabilizing controller for a 1DOF manipulator robot system depicted in Fig.4. The state-variables are $x_1 = \theta$ - the robot arm angle, and $x_2 = \dot{\theta}$ - its angular velocity. The system's actual dynamical equation, which we will assume unknown, is as the shows in (2)[14]:

$$M(q)\ddot{q} + C(q,\dot{q})\dot{q} + g(q) = \tau \tag{2}$$

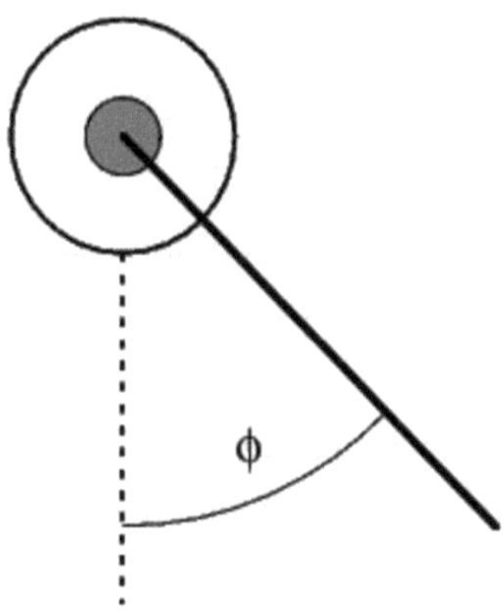

Fig. 4. 1DOF Manipulator robot

To apply the fuzzy Lyapunov synthesis method, we assume that the exact equations are unknown and that we have only the following partial knowledge about the plant (see Fig. 4):

1) The system may have really two degrees of freedom θ and $\dot{\theta}$, referred to as x_1 and x_2, respectively. Hence, $\dot{x}_1 = x_2$.
2) $\dot{x}_2$ is proportional to u, that is, when u increases (decreases) $\dot{x}_2$ increases (decreases).

To facilitate our control design we are going to suppose no gravity effect in our model, see (3).

$$ml^2\ddot{q} = \tau \tag{3}$$

Our objective is to design the rule-base of a fuzzy controller that will carry the robot arm to a desired position $x_1 = \theta d$. We choose (4) as our Lyapunov function candidate. Clearly, V is positive-definite.

$$V(x_1, x_2) = \frac{1}{2}(x_1^2 + x_2^2) \tag{4}$$

Differentiating V, we have (5),

$$\dot{V} = x_1\dot{x}_1 + x_2\dot{x}_2 = x_1x_2 + x_2\dot{x}_2 \tag{5}$$

Hence, we require:

$$x_1x_2 + x_2\dot{x}_2 < 0 \tag{6}$$

We can now derive sufficient conditions so that condition (6) holds: If x_1 and x_2 have opposite signs, then $x_1x_2 < 0$ and (6) will hold if $\dot{x}_2 = 0$; if x_1 and x_2 are both positive, then (6) will hold if $\dot{x}_2 < -x_1$; and if x_1 and x_2 are both negative, then (6) will hold if $\dot{x}_2 > -x_1$.

We can translate these conditions into the following fuzzy rules:

- If x_1 is *positive* and x_2 is *positive* Then $\dot{x}_2$ must be *negative big*
- If x_1 is *negative* and x_2 is *negative* Then $\dot{x}_2$ must be *positive big*

- If x_1 is *positive* and x_2 is *negative* Then $\dot{x}_2$ must be *zero*
- If x_1 is *negative* and x_2 is *positive* Then $\dot{x}_2$ must be *zero*

However, using our knowledge that $\dot{x}_2$ is proportional to u, we can replace each $\dot{x}_2$ with u to obtain the fuzzy rule-base for the stabilizing controller:

- If x_1 is *positive* and x_2 is *positive* Then u must be *negative big*
- If x_1 is *negative* and x_2 is *negative* Then u must be *positive big*
- If x_1 is *positive* and x_2 is *negative* Then u must be *zero*
- If x_1 is *negative* and x_2 is *positive* Then u must be *zero*

It is interesting to note that the fuzzy partitions for x_1, x_2, and u follow elegantly from expression (5). Because $\dot{V} = x_2(x_1 + \dot{x}_2)$, and since we require that $\dot{V}$ be negative, it is natural to examine the signs of x_1 and x_2; hence, the obvious fuzzy partition is *positive, negative*. The partition for $\dot{x}_2$, namely *negative big, zero, positive big* is obtained similarly when we plug the linguistic values *positive, negative* for x_1 and x_2 in (5). To ensure that $\dot{x}_2 < -x_1$ $(\dot{x}_2 > -x_1)$ is satisfied even though we do not know x_1's exact magnitude, only that it is *positive* (*negative*), we must set $\dot{x}_2$ to *negative big* (*positive big*). Obviously, it is also possible to start with a given, pre-defined, partition for the variables and then plug each value in the expression for $\dot{V}$ to find the rules. Nevertheless, regardless of what comes first, we see that fuzzy Lyapunov synthesis transforms classical Lyapunov synthesis from the world of exact mathematical quantities to the world of computing with words [20].

To complete the controllers design, we must model the linguistic terms in the rule-base using fuzzy membership functions and determine an inference method. Following [16], we characterize the linguistic terms *positive, negative, negative big, zero* and *positive big* by the type-1 membership functions shows in Fig. 5 for a Type-1 Fuzzy Logic Controller, and by the type-2 membership functions shows in Fig. 6 for a Type-2 Fuzzy Logic Controller. Note that the type-2 membership functions are extended type-1 membership functions.

To this end, we had systematically developed a FLC rule-base that follows the Lyapunov Stability criterion. At Section IV we present some experimental results using our fuzzy rule-base to build a Type-2 Fuzzy Logic Controller.

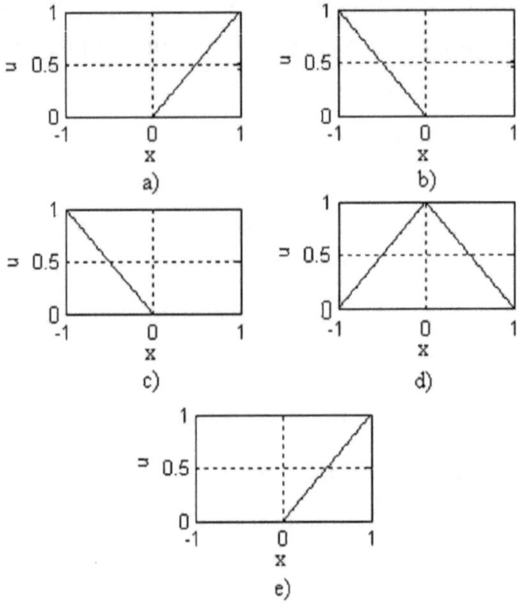

Fig. 5. Kind of type-1 membership functions: a) positive, b) negative, c) negative big, d) zero and e)positive big

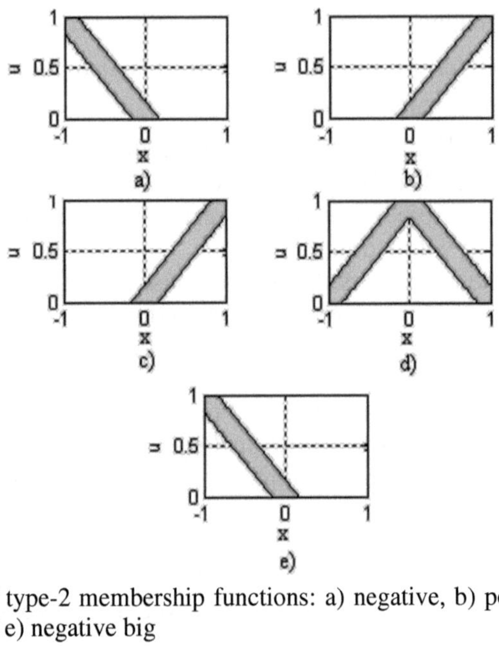

Fig. 6. Kind of type-2 membership functions: a) negative, b) positive, c) positive big, d) zero and e) negative big

4 Experimental Results

In Section III we had systematically develop a stable FLC rule-base, now we are going to show some experimental results using our stable rule-base to built Type-2 FLC. The plant description used in the experiments is the same shown in Section III.

Our experiments were done with Type-1 Fuzzy Sets and Interval Type-2 Fuzzy Sets. In the Type-2 Fuzzy Sets the membership grade of every domain point is a crisp set whose domain is some interval contained in [0,1] [12]. On Fig. 6 we show some Interval Type-2 Fuzzy Sets, for each fuzzy set, the grey area is known as the Footprint Of Uncertainty (FOU) [12], and this one is bounded by an upper and a lower membership function as shown in Fig. 7.

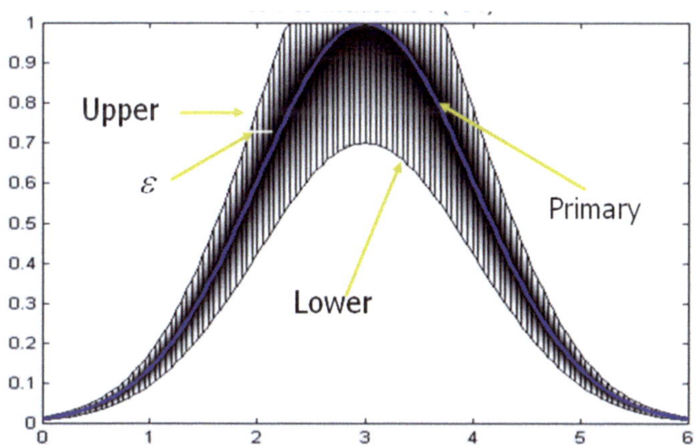

Fig. 7. Type-2 Fuzzy Set

In our experiments we increase and decrease the value of ε to the left and to the right side having a εL and a εR values respectively to determine how much can be extended or perturbed the FOU with out loss of stability in the FLC.

We did make simulations with initial conditions θ having values in the whole circumference [0, 2π], and the desired angle θd having values in the same range. The initial conditions considered in the experiments shown in this paper are an angle $\theta = 0rad$ and $\theta_d = 0.1rad$.

In Fig. 8 we show a simulation of the plant made with a Type-1 FLC, as can be seen, the plant has been regulated in around 8seg, and in Fig. 9 we show the graph of equation (5) which is always negative defined and consequently the system is stable.

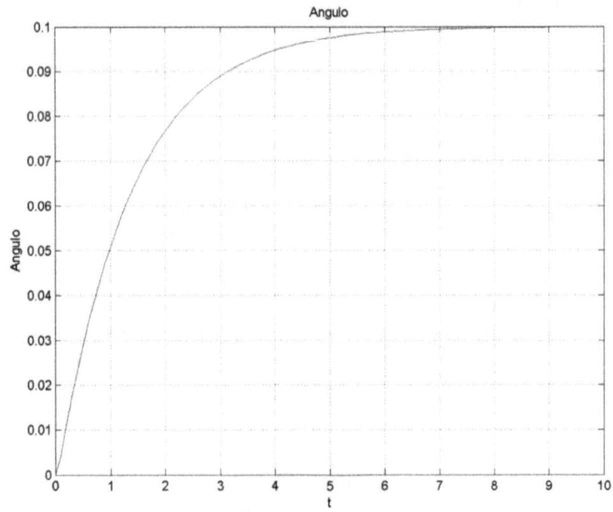

Fig. 8. Response for the Type-1 FLC

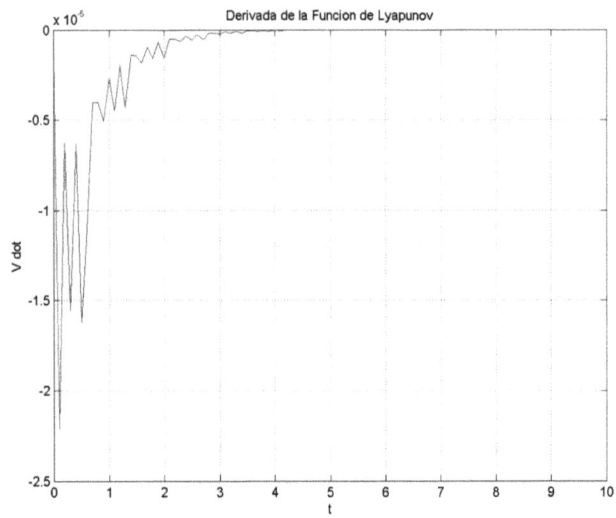

Fig. 9. $\dot{V}$ for the Type-1 FLC

Figure 10 shows the simulation results of the plant made with the Type-2 FLC increasing and decreasing ε in the range of [0,1], as can be seen the plant has been regulated in the around of 10 seg, and the graph of (5) depicted at Fig. 11 is always negative defined and consequently the system is stable. As we can see, the time response is increasing about de value of ε is increasing.

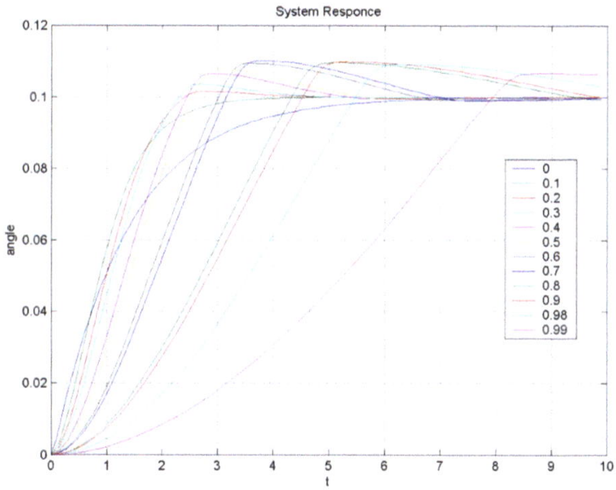

Fig. 10. Response for the Type-2 FLC ($\varepsilon \rightarrow [0,1)$)

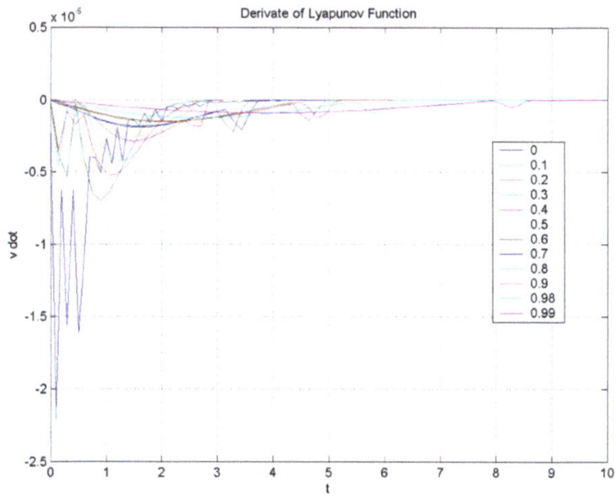

Fig. 11. $\dot{V}$ for the Type-2 FLC ($\varepsilon \rightarrow [0,1]$)

With the variation of ε in the definition of the FOU, the control surface changes proportional to the change of ε, for that reason, the values of u for $\varepsilon \geq 1$ is practically zero, and the plant do not have physical response.

To test the robustness of the built Fuzzy Controller, now we are going to use the same controller designed in Section III, but at this time, we are going to use it to control (2) considering the gravity effect as shows in (7).

$$ml^2\ddot{q} + gml\cos q = \tau \tag{7}$$

At Fig. 12 we can see a simulation of the plant made with a Type-1 FLC, as can be seen, the plant has been regulated in around 8 seg, and Fig.13 shows the graph of (5) which is always negative defined and consequently the system is stable.

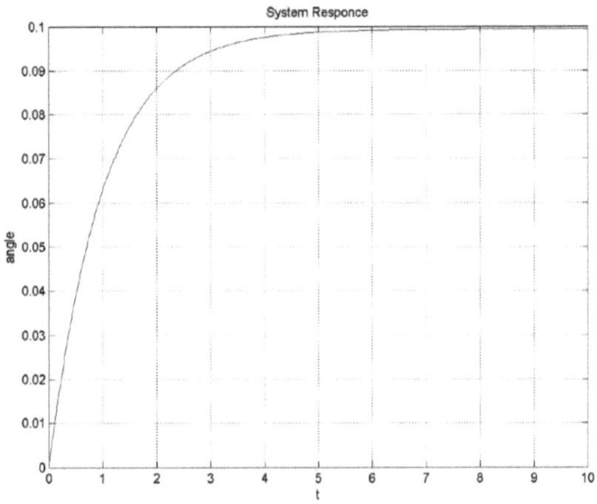

Fig. 12. Response for the Type-1 FLC

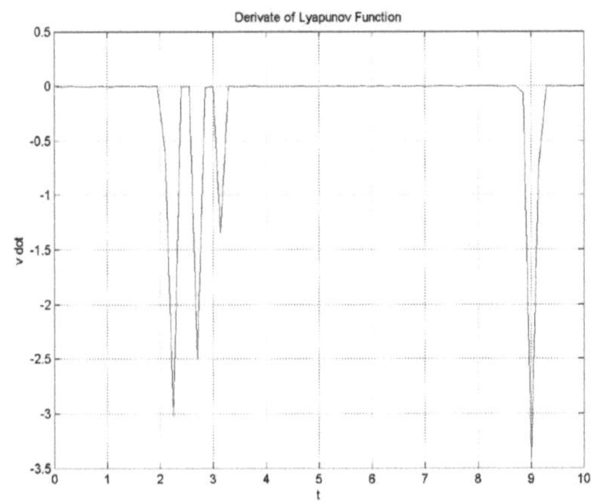

Fig. 13. $\dot{V}$ for the Type-1 FLC

Figure 14 shows the simulation results of the plant made with the Type-2 FLC increasing and decreasing ε in the range of [0,1], and the graph of (5) depicted at Fig. 15 is always negative defined and consequently the system is stable. As we can see, that if we use an adaptive gain like in [1] all the cases of ε can be regulated around 8 seg.

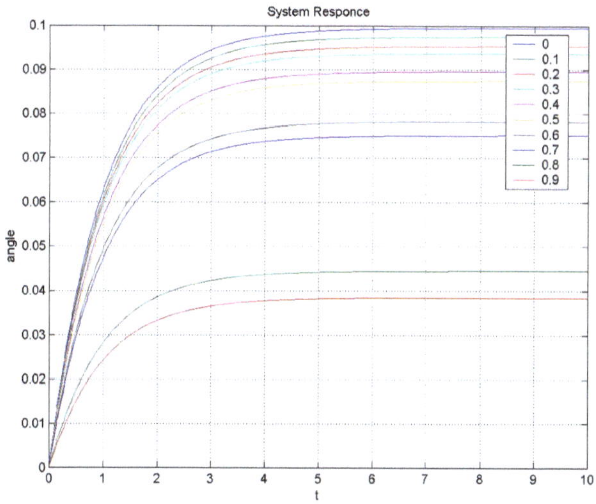

Fig. 14. Response for the Type-2 FLC ($\varepsilon \rightarrow [0,1)$)

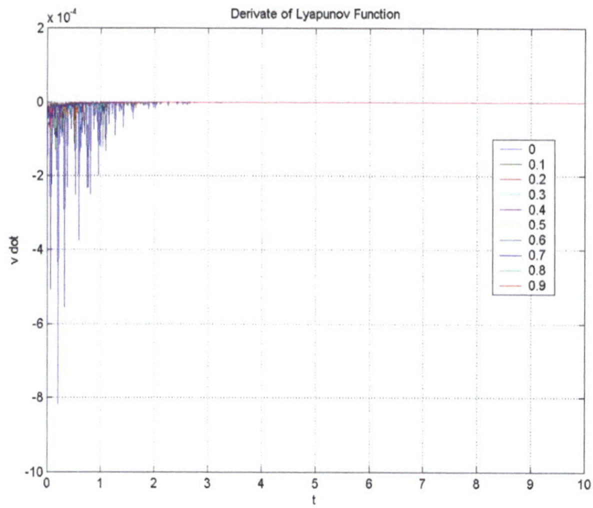

Fig. 15. $\dot{V}$ for the Type-2 FLC ($\varepsilon \rightarrow [0,1]$)

5 Conclusions

As in [1] and [2], the Margaliot approach for the design of FLC is now proved to be valid for both, Type-1 and Type-2 Fuzzy Logic Controllers.

On Type-2 FLC's membership functions, we can perturb or change the definition domain of the FOU without losing stability of the controller; in the case seen at this paper, like in [1] we have to use an adaptive gain to regulate the plant in a desired time.

For our example of the 1DOF manipulator robot, the stability holds extending the FOU on the domain [0,1), this same was happened in [1] and [2]; we proved that a FLC designed following the Fuzzy Lyapunov Synthesis is stable and robust.

Acknowledgment

The authors thank Tijuana Institute of Technology, CITEDI-IPN, CONACYT and COSNET for supporting our research activities.

References

1. O. Castillo, L. Aguilar, N. Cázarez and D. Rico, "Intelligent Control of Dynamical Systems with Type-2 Fuzzy and Stability Study", Proceedings of the International Conference on Artificial Intelligence (IC-AI'2005, IMCCS&CE'2005), pp. ??, Las Vegas, USA, June 27-30, 2005 (to appear)
2. N. R. Cázarez, S. Cárdenas, L. Aguilar and O. Castillo, "Lyapunov Stability on Type-2 Fuzzy Logic Control", IEEE-CIS International Seminar on Computational Intelligence, México Distrito Federal, México, October 17-18 2005 (submitted)
3. G. Chen and T.T. Pham, "Introduction to fuzzy sets, fuzzy logic, and fuzzy control systems", CRC Press, USA, 2000.
4. B.J. Choi, S.W. Kwak, and B. K. Kim, "Design and Stability Analysis of Single-Input Fuzzy Logic Controller", IEEE Trans. Fuzzy Systems, vol. 30, pp. 303-309, 2000
5. J.-S.R. Jang, C.-T. Sun and E. Mizutani, "Neuro-Fuzzy and Soft Computing: a computational approach to learning and machine intelligence", Prentice Hall, USA, 1997.
6. J. Lee, "On methods for improving performance of PI-type fuzzy logic controllers", IEEE Trans. Fuzzy Systems, vol. 1, pp. 298-301, 1993.
7. H.-X. Li and H.B. Gatland, "A new methodology for designing a fuzzy logic controller", IEEE Trans. Systems Man and Cybernetics, vol. 25, pp. 505-512, 1995.

8. H.-X. Li and H.B. Gatland, "Conventional fuzzy control and its enhancement", IEEE Trans. Systems Man and Cybernetics, vol. 25, pp. 791-797, 1996.
9. E. H. Mamdani and S. Assilian, "An experiment in linguistic systhesis with a fuzzy logic controller", Int. J. Man-Machine Studies, vol. 7, pp. 1-13, 1975.
10. E.H. Mamdani, "Advances in the linguistic synthesis of fuzzy controllers", Int. J. Man-Machine Studies, vol. 8, pp. 669-679, 1976.
11. M. Margaliot and G. Langholz, "New Approaches to Fuzzy Modeling and Control: Design and Analysis", World Scientific, Singapore, 2000.
12. J.M. Mendel, "Uncertain Rule-Based Fuzzy Logic: Introduction and new directions", Prentice Hall, USA, 2000.
13. K. Ogata, "Ingeniería de Control Moderna", 3ª Edici6n, Prentice Hall Hispanoamericana, Spain, 1998.
14. H. Paul and C. Yang, "Sistemas de control en ingeniería", Prentice Hall Iberia, Spain, 1999.
15. M. Sugeno, "On Stability of Fuzzy Systems Expressed by Fuzzy Rules with Singleton Consequents", IEEE Trans. Fuzzy Systems, vol. 7, no. 2, 1999.
16. L.X. Wang, " A Course in Fuzzy Systems and Control", Prentice Hall, 1997.
17. J. Yen and R. Langari, "Fuzzy Logic: Intelligence, Control, and Information", Prentice Hall, New Jersey, USA, 1998.
18. H.Ying, "Practical Design of Nonlinear Fuzzy Controllers with Stability Analysis for Regulating Processes with Unknown Mathematical Models", Automatica, vol. 30, no. 7, pp. 1185-1195, 1994.
19. L.A. Zadeh, "The concept of a linguistic variable and its application to approximate reasoning", J. Information Sciences, vol. 8, pp. 43-80,1975.
20. L.A. Zadeh, "Fuzzy Logic = Computing With Words", IEEE Trans. Fuzzy Systems, vol. 4, no. 2, pp. 103-111, 1996.

A Comparative Study of Controllers Using Type-2 and Type-1 Fuzzy Logic

Roberto Sepulveda and Patricia Melin

[1]CITEDI-IPN and [2]Tijuana Institute of Technology, Tijuana, Mexico

Abstract. Uncertainty is an inherent part in controllers used for real-world applications. The use of new methods for handling incomplete information is of fundamental importance in engineering applications. This paper deals with the design of controllers using type-2 fuzzy logic for minimizing the effects of uncertainty produced by the instrumentation elements. We simulated type-1 and type-2 fuzzy logic controllers to perform a comparative analysis of the systems' response, in the presence of uncertainty.

1 Introduction

Uncertainty affects decision-making and appears in a number of different forms. The concept of information is fully connected with the concept of uncertainty. The most fundamental aspect of this connection is that the uncertainty involved in any problem-solving situation is a result of some information deficiency, which may be incomplete, imprecise, fragmentary, not fully reliable, vague, contradictory, or deficient in some other way [1]. The general framework of fuzzy reasoning allows handling much of this uncertainty, fuzzy systems employ type-1 fuzzy sets, which represents uncertainty by numbers in the range [0, 1]. However, when something is uncertain, like a measurement, it is difficult to determine its exact value, and of course type-1 fuzzy sets makes more sense than using crisp sets. However, it is not reasonable to use an accurate membership function for something uncertain, so in this case what we need is another type of fuzzy sets, those, which are able to handle these uncertainties, the so called type-2 fuzzy sets [2]. So, the amount of uncertainty in a system can be reduced by using type-2 fuzzy logic because it offers better capabilities to handle linguistic uncertainties by modeling vagueness and unreliability of information.

R. Sepulveda and P. Melin: *A Comparative Study of Controllers Using Type-2 and Type-1 Fuzzy Logic*, StudFuzz **208**, 151–178 (2007)
www.springerlink.com

Recently, we have seen the use of type-2 fuzzy sets in fuzzy logic systems to deal with uncertain information. So we can find some papers emphasizing on the implementation of a type-2 Fuzzy Logic System (FLS) [3]; in others, it is explained how type-2 fuzzy sets let us model and minimize the effects of uncertainties in rule-base FLSs [4]. Some research works are devoted to solve real world applications in different areas, for example, in signal processing type-2 fuzzy logic is applied in prediction in Mackey-Glass chaotic time-series with uniform noise presence [5]. In medicine, an expert system was developed for solving the problem of Umbilical Acid-Base (UAB) assessment [6]. In industry, type-2 fuzzy logic and neural networks was used in the control of non-linear dynamic plants [7,8].

This work deals with the advantages of using type-2 fuzzy sets in the implementation of a Fuzzy Logic Controller (FLC), for a real system. It is a fact, that in the control of real systems, the instrumentation elements (instrumentation amplifier, sensors, digital to analog, analog to digital converters, etc.) introduce some sort of unpredictable values in the information that has been collected. So, the controllers designed under idealized conditions tend to behave in an inappropriate manner. Since, uncertainty is inherent in the design of controllers for real world applications, we are presenting how to deal with it using type-2 FLC to diminish the effects of imprecise information. We are supporting this statement with experimental results, qualitative observations, and quantitative measures of errors. For quantifying the errors, we utilized three widely used performance criteria, these are: Integral of Square Error (ISE), Integral of the Absolute value of the Error (IAE), and Integral of the Time multiplied by the Absolute value of the Error (ITAE) [9].

This paper is organized as follows: section 2 presents an introductory explanation of type-1 and type-2 FLCs and the performance criteria for evaluating the transient and steady state closed-loop response in a computer control system. In section 3, we are showing details of the implementation of the feedback control system used in this work, we are presenting some experimental results and a performance comparison between type-1 and type-2 fuzzy logic controllers. Finally, we have the conclusions.

2 Fuzzy Controllers

In the 40's and 50's, many researchers proved that many dynamic systems can be mathematically modeled using differential equations. These previous works represent the foundations of the Control theory, which, in addition with the Transform theory, provided an extremely powerful means of analyzing and designing control systems [10]. These theories were being developed until the 70's, when the area was called System theory to indicate its definitiveness [11]. Its principles have been used to control a very big amount of systems taking mathematics as the main tool to do it during many years. Unfortunately, in too many cases this approach could not be sustained because many systems have unknown parameters or highly complex and nonlinear characteristics that make them not to be amenable to the full force of mathematical analysis as dictated by the Control theory.

Soft computing techniques have become a research topic, which is applied in the design of controllers [12]. These techniques have tried to avoid the above-mentioned drawbacks, and they allow us to obtain efficient controllers, which utilize the human experience in a more related form than the conventional mathematical approach. In the cases in which a mathematical representation of the controlled systems cannot be obtained, the process operator should be able to express the relationships existing in them, that is, the process behavior.

A FLS, described completely in terms of type-1 fuzzy sets is called a type-1 fuzzy logic system (type-1 FLS). It is composed by a knowledge base that comprises the information given by the process operator in form of linguistic control rules, a fuzzification interface, who has the effect of transforming crisp data into fuzzy sets, an inference system, that uses them in conjunction with the knowledge base to make inference by means of a reasoning method, and a defuzzification interface, which translate the fuzzy control action so obtained to a real control action using a defuzzification method [10].

In our paper, the implementation of the fuzzy controller in terms of type-1 fuzzy sets, has two input variables such as the error $e(t)$, the difference between the reference signal and the output of the process, as well as the error variation $\Delta e(t)$,

$$e(t) = r(t) - y(t) \tag{1}$$

$$\Delta e(t) = e(t) - e(t-1) \tag{2}$$

so the control law can be represented as in Fig. 1.

A FLS described using at least one type-2 fuzzy set is called a type-2 FLS. Type-1 FLSs are unable to directly handle rule uncertainties, because they use type-1 fuzzy sets that are certain. On the other hand, type-2 FLSs, are very useful in circumstances where it is difficult to determine an exact certainties, and measurement uncertainties [2].

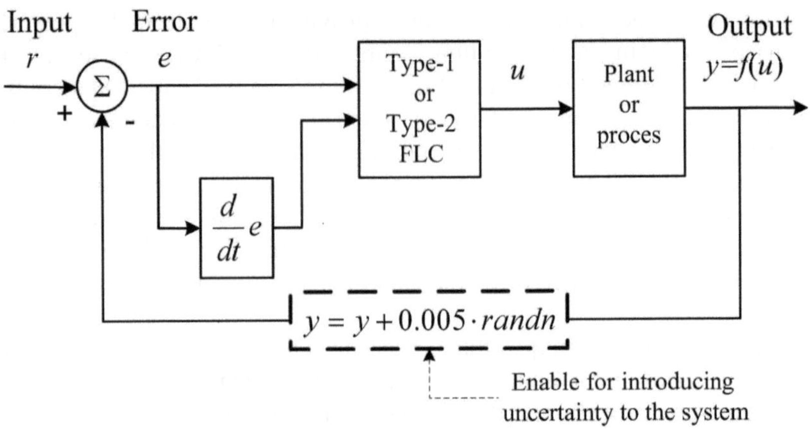

Fig. 1. System used for obtaining the experimental results.

It is known that type-2 fuzzy sets let us to model and to minimize the effects of uncertainties in rule-based FLS. Unfortunately, type-2 fuzzy sets are more difficult to use and understand than type-1 fuzzy sets; hence, their use is not widespread yet. In [4] were mentioned at least four sources of uncertainties in type-1 FLSs:

1. The meanings of the words that are used in the antecedents and consequents of rules can be uncertain (words mean different things to different people).
2. Consequents may have histogram of values associated with them, especially when knowledge is extracted from a group of experts who do not all agree.
3. Measurements that activate a type-1 FLS may be noisy and therefore uncertain.
4. The data used to tune the parameters of a type-1 FLS may also be noisy.

All of these uncertainties translate into uncertainties about fuzzy set membership functions. Type-1 fuzzy sets are not able to directly model such uncertainties because their membership functions are totally crisp. On the other hand, type-2 fuzzy sets are able to model such uncertainties because their membership functions are themselves fuzzy. A type-2 membership grade can be any subset in [0,1], the primary membership, and corresponding to each primary membership, there is a secondary membership (which can also be in [0,1]) that defines the possibilities for the primary membership. A type-1 fuzzy set is a special case of a type-2 fuzzy set; its secondary membership function is a subset with only one element, unity.

Similar to a type-1 FLS, a type-2 FLS includes fuzzifier, rule base, fuzzy inference engine, and output processor. The output processor includes type-reducer and defuzzifier; it generates a type-1 fuzzy set output (from the type-reducer) or a crisp number (from the defuzzifier). A type-2 FLS is again characterized by IF-THEN rules, but its antecedent or consequent sets are now type-2. Type-2 FLSs, can be used when the circumstances are too uncertain to determine exact membership grades such as when training data is corrupted by noise. In our case, we are simulating that the instrumentation elements (instrumentation amplifier, sensors, digital to analog, analog to digital converters, etc.) are introducing some sort of unpredictable values in the collected information.

In the case of the implementation of the type-2 FLC, we have the same characteristics as in type-1 FLC, but we used type-2 fuzzy sets as membership functions for the inputs and for the output.

For evaluating the transient closed-loop response of a computer control system we can use the same criteria that normally are used for adjusting constants in PID (Proportional Integral Derivative) controllers. These are [9]:

1. Integral of Square Error (ISE).

$$\mathrm{ISE} = \int_0^\infty [e]^2\, dt \tag{3}$$

2. Integral of the Absolute value of the Error (IAE).

$$\mathrm{IAE} = \int_0^\infty |\, e\,|\, dt \tag{4}$$

3. Integral of the Time multiplied by the Absolute value of the Error (ITAE).

$$\mathrm{ITAE} = \int_0^\infty t\,|\, e\,|\, dt \tag{5}$$

The selection of the criteria depends on the type of response desired, the errors will contribute different for each criterion, so we have that large errors will increase the value of ISE more heavily than to IAE. ISE will favor responses with smaller overshoot for load changes, but ISE will give longer settling time. In ITAE, time appears as a factor, and therefore, ITAE will penalize heavily errors that occurs late in time, but virtually ignores errors that occurs early in time. Designing using ITAE will give us the shortest settling time, but it will produce the largest overshoot among the three criteria considered. Designing considering IAE will give us an intermediate results, in this case, the settling time will not be so large than using ISE nor so small than using ITAE, and the same applies for the overshoot response. The selection of a particular criterion is depending on the type of desired response.

3 Simulation Results

We are showing in Fig. 1, the feedback control system that was used for achieving the results of this paper. It was implemented in Matlab where the controller was designed to follow the input as closely as possible. The plant was modeled using equation (6)

$$y(i)=0.2\cdot y(i-3)\cdot 0.7y(i-2)+0.9\cdot y(i-1)+0.005u(i-1)+0.5\cdot u(i-2) \tag{6}$$

The controller's output was applied directly to the plant's input. Since we are interested in comparing the performance between type 1 and type 2 FLC system, we tested the controller in two ways:

1. One is considering the system as ideal, that is, we did not introduce in the modules of the control system any source of uncertainty. See experiments 1, and 2.
2. The other one is simulating the effects of uncertain modules (subsystems) response introducing some uncertainty. See experiments 3, and 4.

For both cases, as is shown in Fig. 1, the system's output is directly connected to the summing junction, but in the second case, the uncertainty was simulated introducing random noise with normal distribution (the dashed square in Fig. 1). We added noise to the system's output $y(i)$ using equation (7), which in turn was introduced to the summing junction of the controller system.

$$y(i)=y(i)+0.05\cdot randn \tag{7}$$

We tested the system using as input, a unit step sequence free of noise, $r(i)$. For evaluating the system's response and compare between type 1 and type 2 fuzzy controllers, we used the performance criteria ISE, IAE, and ITAE. In table I, we summarized the values obtained for each criterion considering 400 units of time. For calculating ITAE we considered a sampling time $T_s = 0.1\,\text{sec}$.

For Experiments 1, 2, 3, and 4 the reference input r is stable and noisy free. In experiments 3 and 4, although the reference appears clean, the feedback at the summing junction is noisy since we introduced deliberately noise for simulating the overall existing uncertainty in the system, in consequence, the controller's inputs e (error), and $\frac{\Delta}{\Delta t}e$ contains uncertainty data.

For each input of the type-1 FLC, we defined three type-1 fuzzy Gaussian membership functions: negative, zero, positive. The universe of discourse for these membership functions is in the range [-10 10]; their mean is -10, 0 and 10 respectively, and their standard deviation are 9, 2 and 9 respectively.

For the output, we have five type-1 fuzzy Gaussian membership functions: NG, N, Z, P and PG. They are on the interval [-10 10], their means are -10, -4.5, 0, 4, and 10 respectively; and their standard deviations are 4.5, 4, 4.5, 4 and 4.5 respectively.

In the type-2 FLC, for each input we defined three type-2 fuzzy Gaussian membership functions: negative, zero, positive. In this case the fuzzy membership functions have uncertain mean and fixed standard deviation on the interval [-10 10]. For the upper membership functions we have -10.5, -1, and 9.5 uncertain means; for the lower membership functions we have -9.5, 1, and 10.5 uncertain means respectively; for the fixed standard deviations 9, 2 and 9 respectively.

For computing the output we have five type-2 fuzzy Gaussian membership functions with uncertain mean and fixed standard deviations: NG, N, Z, P and PG, on the interval [-10 10]. For the upper membership functions we have -10.25, -4.75, -0.25, 3.75 and 9.75 uncertain means; for the lower membership functions we have -9.75, -4.25, 0.25, 4.25 and 10.25 uncertain means respectively. The fixed standard deviations: 4.5, 4, 4.5, 4 and 4.5 respectively.

Experiment 1. Ideal system using a type-1 FLC.

In this experiment, we did not add uncertainty data to the system, the system response is illustrated in Fig. 2. Note that the settling time is in about 140 units of time; i.e., the system trends to stabilize with time and the output will follow accurately the input. In Table I, we listed the obtained values of ISE, IAE, and ITAE for this experiment. We are showing in Fig. 3, 4 and 5 the ISE, IAE, and ITAE behavior of this experiment.

Experiment 2. Ideal system using a type-2 FLC.

Here, we used the same test conditions of Experiment 1, but in this case, we implemented the controller's algorithm with type-2 fuzzy logic, its output sequence is illustrated in Fig. 2, and the corresponding performance criteria are listed in Table I. By visual inspection, we can observe that the output system response of Experiment 1, and this one, are very similar, they are almost overlapped.

Using the performance criteria we can get a quantitative comparison, where we can observe small differences favoring Experiment 1, i.e., the results obtained using a type-1 FLC. We can observe in Fig. 3, 4, and 5 that using a type-1 FLC we got the lower errors.

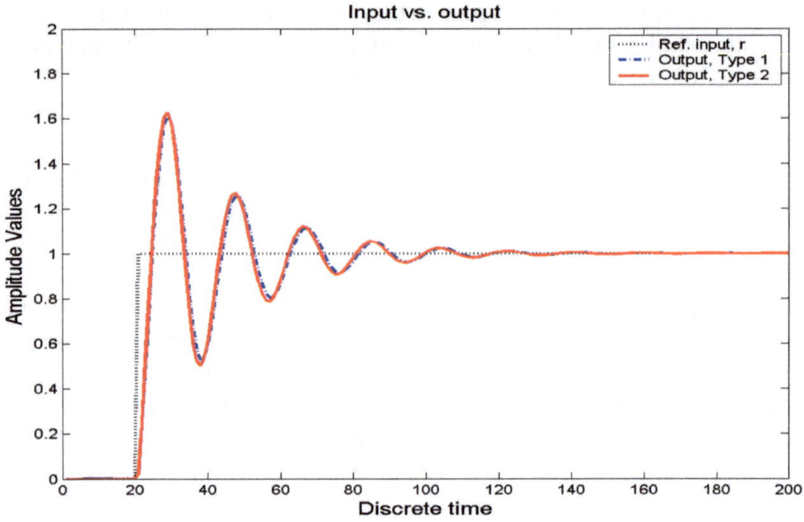

Fig. 2. This graphic shows the system's response to a unit step sequence.

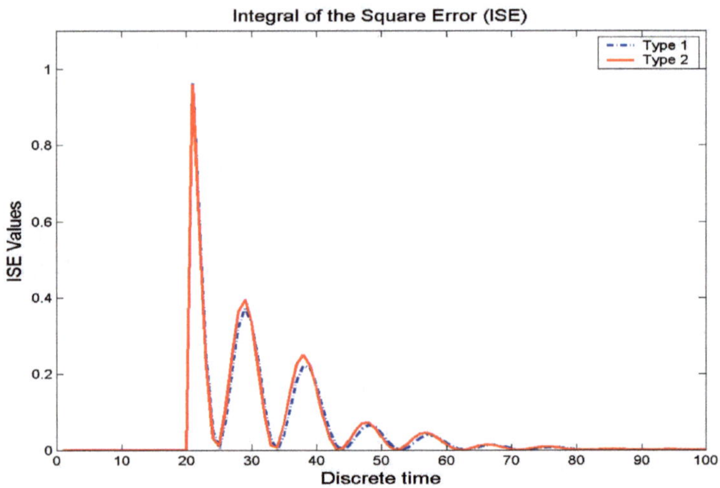

Fig. 3. In uncertainty absence, the ISE values are very similar for type-1 and type-2 FLCs.

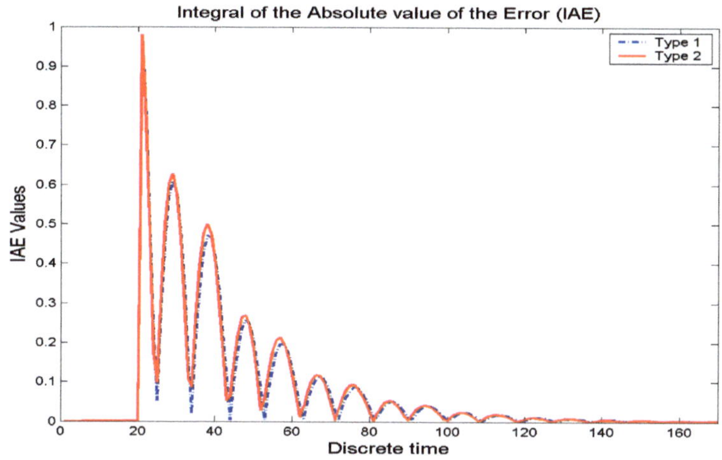

Fig. 4. In uncertainty absence, the IAE values obtained at the plant's output are very similar for type-1 and type-2 FLCs.

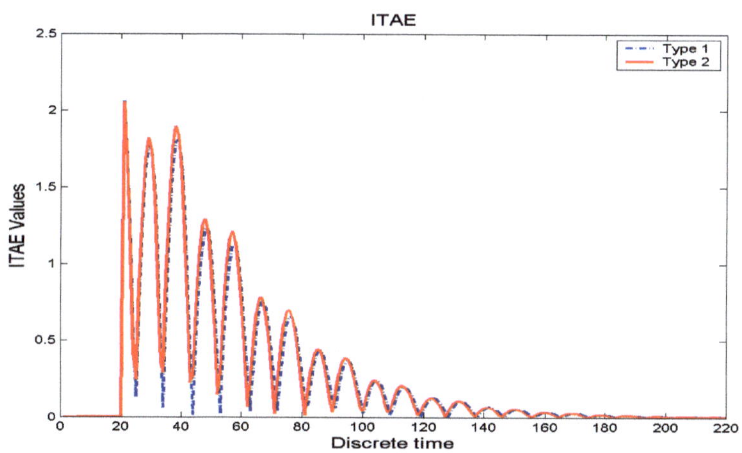

Fig. 5. In uncertainty absence, the ITAE values obtained at the plant's output are very similar for type-1 and type-2 FLCs, in accordance with Figure 13, it is evident a type-1 FLC works a little better.

Experiment 3. System with uncertainty using a type-1 FLC.

In this case, we simulated using equation (7), the effects of uncertainty introduced to the system by transducers, amplifiers, and any other element that in real world applications affects expected values. We are showing in Fig. 6, the system's response output. In Fig. 7 and 8 are plotted the performance criteria ISE, IAE, ITAE.

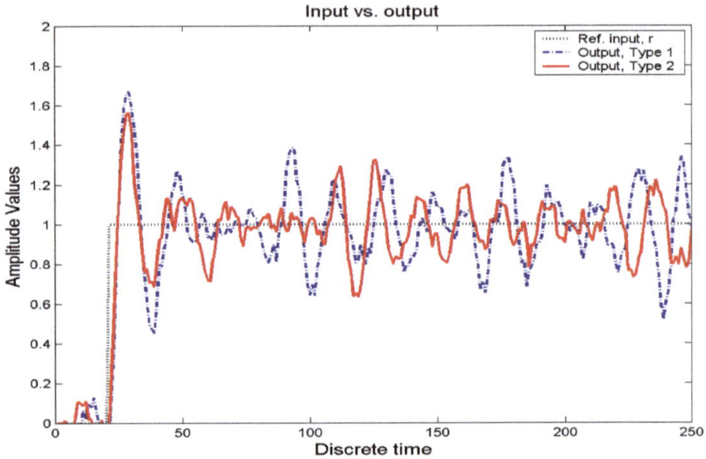

Fig. 6. This graphic was obtained with uncertainty presence; compare the system's outputs produced by type-1 and type-2 FLCs.

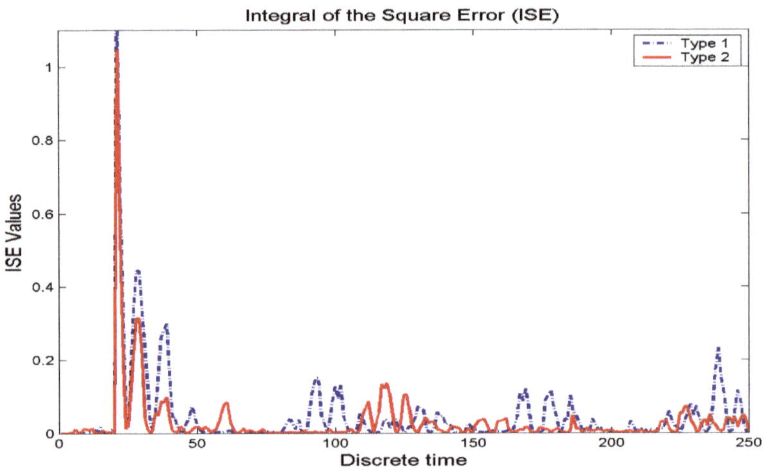

Fig. 7. Here we can see that a type-2 FLC produces lower overshoot errors, quantitatively the ISE overall error of using type-2 is 9.5516 against 15.1143 of the overall error produced by the type-1 FLC.

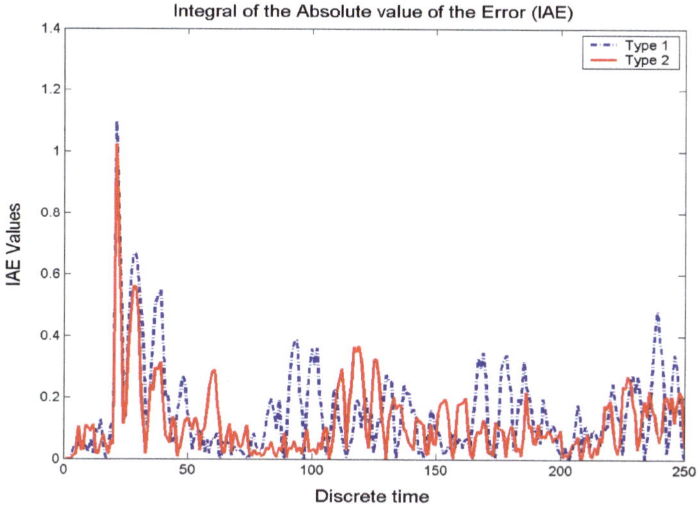

Fig. 8. In accordance with Fig. 6, IAE confirms that we obtained the best system response using a type-2 FLC with uncertainty presence.

Experiment 4. System with uncertainty using a type-2 FLC. In this experiment, we introduced uncertainty in the system, in the same way as in Experiment 3. In this case, we used a type-2 FLC and we improved those results obtained with a type-1 FLC (Experiment 3).

4 Conclusions

We observed and quantified using performance criteria such as ISE, IAE, and ITAE that in systems without uncertainties (ideal systems) is a better choice to select a type-1 FLC since it works a little better than a type-2 FLC, and it is easier to implement it. It is known that type-1 FLC can handle nonlinearities, and uncertainties up to some extent.

References

1. George J. Klir, Bo Yuan, *Fuzzy Sets and Fuzzy Logic: Theory and Applications*, Ed. Prentice Hall USA, 1995.
2. Jerry M. Mendel, Uncertain Rule-Based Fuzzy Logic Systems: Introduction and new directions, Ed. Prentice Hall, USA, 2000.

3. Nilesh N. Karnik, Jerry M. Mendel, Qilian Liang, Type-2 Fuzzy Logic Systems, IEEE Transactions on Fuzzy Systems, Vol. 7, No. 6, December 1999.
4. Jerry M. Mendel and Robert I. Bob John, Type-2 Fuzzy Sets Made Simple, IEEE Transactions on Fuzzy Systems, Vol. 10, No. 2, April 2002.
5. Jerry M. Mendel , Uncertainty, fuzzy logic, and signal processing, Ed. Elsevier Science B. V. 2000.
6. Turhan Ozen and Jonathan Mark Garibaldi, *Investigating Adaptation in Type-2 Fuzzy Logic Systems Applied to Umbilical Acid-Base Assessment*, European Symposium on Intelligent Technologies, Hybrid Systems and their implementation on Smart Adaptive Systems (EUNITE 2003), Oulu, Finland, 10-12 July 2003.
7. P.Melin and O. Castillo, *Intelligent control of non-linear dynamic plants using type-2 fuzzy logic and neural networks,* Proceedings of the Annual Meeting of the North American Fuzzy Information Processing Society, 2002.
8. P.Melin and O. Castillo, A new method for adaptive model-based control of non-linear plants using type-2 fuzzy logic and neural networks telligent control of non-linear dynamic plants using type-2 fuzzy logic and neural networks, Proceedings of the 12th IEEE International conference on Fuzzy Systems, 2003.
9. Pradee B. Deshpande, Raymond H. Ash, *Computer Process Control with Advanced Control Applications*, Ed. Instrument Society of America, USA, 1988.
10. O. Cordón, F. Herrera, E. Herrera-Viedma, M. Lozano, *Genetic Algorithms and Fuzzy Logic in Control Processes*, Technical Report #DECSAI-95109, Universidad de Granada, Spain, 1995.
11. Mamdani, E. H., Twenty years of Fuzzy Control: Experiences Gained and Lessons Learn. Fuzzy Logic Technology and Applications, R. J. Marks II (Eds.), IEEE Press, 1993.
12. J.-S. R. Jang, C.-T. Sun, E. Mizutani. Neuro-Fuzzy and Soft Computing, A Computational Approach to Learning and Machine Intelligence, Matlab Curriculum Series. Prentice Hall.

Evolutionary Computing for Topology Optimization of Type-2 Fuzzy Controllers

Oscar Castillo, Gabriel Huesca, and Fevrier Valdez

Department of Computer Science, Tijuana Institute of Technology, Tijuana, Mexico

Abstract. We describe in this paper the use of hierarchical genetic algorithms for fuzzy system optimization in intelligent control. In particular, we consider the problem of optimizing the number of rules and membership functions using an evolutionary approach. The hierarchical genetic algorithm enables the optimization of the fuzzy system design for a particular application. We illustrate the approach with the case of intelligent control in a medical application. Simulation results for this application show that we are able to find an optimal set of rules and membership functions for the fuzzy system.

1 Introduction

We describe in this paper the application of a Hierarchical Genetic Algorithm (HGA) for fuzzy system optimization (Man et al. 1999). In particular, we consider the problem of finding the optimal set of rules and membership functions for a specific application (Yen and Langari 1999). The HGA is used to search for this optimal set of rules and membership functions, according to the data about the problem. We consider, as an illustration, the case of a fuzzy system for intelligent control.

Fuzzy systems are capable of handling complex, non-linear and sometimes mathematically intangible dynamic systems using simple solutions (Jang et al. 1997). Very often, fuzzy systems may provide a better performance than conventional non-fuzzy approaches with less development cost (Procyk and Mamdani 1979). However, to obtain an optimal set of fuzzy membership functions and rules is not an easy task. It requires time, experience and skills of the designer for the tedious fuzzy tuning exercise. In principle, there is no general rule or method for the fuzzy logic set-up, although a heuristic and iterative procedure for modifying the membership

O. Castillo et al.: *Evolutionary Computing for Topology Optimization of Type-2 Fuzzy Controllers*, StudFuzz **208**, 163–178 (2006)
www.springerlink.com

functions to improve performance has been proposed. Recently, many researchers have considered a number of intelligent schemes for the task of tuning the fuzzy system. The noticeable Neural Network (NN) approach (Jang and Sun 1995) and the Genetic Algorithm (GA) approach (Homaifar and McCormick 1995) to optimize either the membership functions or rules, have become a trend for fuzzy logic system development.

The HGA approach differs from the other techniques in that it has the ability to reach an optimal set of membership functions and rules without a known fuzzy system topology (Tang et al. 1998). During the optimization phase, the membership functions need not be fixed. Throughout the genetic operations (Holland 1975), a reduced fuzzy system including the number of membership functions and fuzzy rules will be generated (Yoshikawa et al. 1996). The HGA approach has a number of advantages:

1. An optimal and the least number of membership functions and rules are obtained
2. No pre-fixed fuzzy structure is necessary, and
3. Simpler implementing procedures and less cost are involved.

We consider in this paper the case of automatic anesthesia control in human patients for testing the optimized fuzzy controller. We did have, as a reference, the best fuzzy controller that was developed for the automatic anesthesia control (Karr and Gentry 1993, Lozano 2003), and we consider the optimization of this controller using the HGA approach. After applying the genetic algorithm the number of fuzzy rules was reduced from 12 to 9 with a similar performance of the fuzzy controller. Of course, the parameters of the membership functions were also tuned by the genetic algorithm. We did compare the simulation results of the optimized fuzzy controllers obtained with the HGA against the best fuzzy controller that was obtained previously with expert knowledge, and control is achieved in a similar fashion. Since simulation results are similar, and the number of fuzzy rules was reduced, we can conclude that the HGA approach is a good alternative for designing fuzzy systems. We have to mention that Type-2 fuzzy systems are considered in this research work, which are more difficult to design and optimize.

2 Genetic Algorithms for Optimization

In this paper, we used a floating-point genetic algorithm (Castillo and Melin 2001) to adjust the parameter vector θ, specifically we used the Breeder Genetic Algorithm (BGA). The genetic algorithm is used to optimize the fuzzy system for control that will be described later (Castillo and Melin 2003). A BGA can be described by the following equation:

$$BGA=(P_g^0, N, T, \Gamma, \Delta, HC, F, term) \quad (1)$$

where: P_g^0=initial population, N=the size of the population, T=the truncation threshold, Γ=the recombination operator, Δ=the mutation operator, HC=the hill climbing method, F=the fitness function, *term*=the termination criterion.

The BGA uses a selection scheme called truncation selection. The %T best individuals are selected and mated randomly until the number of offspring is equal the size of the population. The offspring generation is equal to the size of the population. The offspring generation replaces the parent population. The best individual found so far will remain in the population. Self-mating is prohibited (Melin and Castillo 2002). As a recombination operator we used "extended intermediate recombination", defined as: If $x=(x_i,...x_n)$ *and* y $y=(y_1,...,y_n)$ are the parents, then the successor $z=(z_1,...,z_n)$ is calculated by:

$$z_i=x_i+\alpha_i(y_i-x_i) \qquad i=1,\dots n \quad (2)$$

The mutation operator is defined as follows: A variable x_i is selected with probability p_m for mutation. The BGA normally uses $p_m = 1/n$. At least one variable will be mutated. A value out of the interval $[-range_i, range_i]$ is added to the variable. $range_i$ defines the mutation range. It is normally set to $(0.1 \times searchinterval_i)$. $searchinterval_i$ is the domain of definition for variable x_i. The new value z_i is computed according to

$$z_i=x_i\pm range_i\cdot\delta \quad (3)$$

The + or – sign is chosen with probability 0.5. δ is computed from a distribution which prefers small values. This is realized as follows

$$\delta=\sum_{i=0}^{15}\alpha_i 2^i \qquad \alpha_i \in 0,1 \quad (4)$$

Before mutation we set α_i=0. Then each α_i is mutated to 1 with probability p_δ=1/16. Only α_i=1 contributes to the sum. On the average there will be just one α_i with value 1, say α_j. Then δ is given by

$$\delta=2^{-j} \quad (5)$$

The standard BGA mutation operator is able to generate any point in the hypercube with center x defined by $x_i \pm range_i$. But it generates values much more often in the neighborhood of x. In the above standard setting, the mutation operator is able to locate the optimal x_i up to a precision of $ramge_i \cdot 2^{-150}$.

To monitor the convergence rate of the LMS algorithm, we computed a short term average of the squared error $e^2(n)$ using

$$ASE(m) = \frac{1}{K} \sum_{k=n+1}^{n+K} e^2(k) \tag{6}$$

where $m=n/K=1,2,....$ The averaging interval K may be selected to be (approximately) K=10N. The effect of the choice of the step size parameter Δ on the convergence rate of LMS algorithm may be observed by monitoring the ASE(m).

2.1 Genetic Algorithm for Optimization

The proposed genetic algorithm is as follows:

1. We use real numbers as a genetic representation of the problem.
2. We initialize variable i with zero (i=0).
3. We create an initial random population P_i, in this case (P_0). Each individual of the population has n dimensions and, each coefficient of the fuzzy system corresponds to one dimension.
4. We calculate the normalized fitness of each individual of the population using linear scaling with displacement (Melin and Castillo 2002), in the following form:
$$f_i^{'} = f_i + \frac{1}{N} \sum |f_i| + \left| \min_i (f_i) \right| \qquad \forall i$$
5. We normalize the fitness of each individual using:
$$F_i = \frac{f_i^{'}}{\sum_{i=1}^{N} f_i^{'}} \qquad \forall i$$
6. We sort the individuals from greater to lower fitness.
7. We use the truncated selection method, selecting the %T best individuals, for example if there are 500 individuals and, then we select 0.30*500=150 individuals.

8. We apply random crossover, to the individuals in the population (the 150 best ones) with the goal of creating a new population (of 500 individuals). Crossover with it self is not allowed, and all the individuals have to participate. To perform this operation we apply the genetic operator of extended intermediate recombination as follows:

   If $x=(x_1,...,x_n)$ and $y=(y_1,...,y_n)$ are the parents, then the successors $z=(z_1,...,z_n)$ are calculated by, $z_i=x_i+\alpha_i(y_i-x_i)$ *for* $i=1,...,n$ where α is a scaling factor selected randomly in the interval $[-d,1+d]$. In intermediate recombination $d=0$, and for extended $d>0$, a good choice is $d=0.25$, which is the one that we used.
9. We apply the mutation genetic operator of BGA. In this case, we select an individual with probability $p_m=1/n$ (where n represents the working dimension, in this case n=25, which is the number of coefficients in the membership functions). The mutation operator calculates the new individuals z_i of the population in the following form: $z_i=x_i\pm range_i\,\delta$ we can note from this equation that we are actually adding to the original individual a value in the interval: $[-range_i, range_i]$ the range is defined as the search interval, which in this case is the domain of variable x_i, the sign ± is selected randomly with probability of 0.5, and is calculated using the following formula,

$$\delta = \sum_{i=0}^{m-1} \alpha_i 2^{-i} \qquad \alpha_i \in 0,1$$

   Common used values in this equation are $m=16$ y $m=20$. Before mutation we initiate with $_i=0$, then for each $_i$ we mutate to 1 with probability $p=1/m$.
10. Let i=i+1, and continue with step 4.

3 Evolution of Fuzzy Systems

Ever since the very first introduction of the fundamental concept of fuzzy logic by Zadeh in 1973, its use in engineering disciplines has been widely studied. Its main attraction undoubtedly lies in the unique characteristics that fuzzy logic systems possess. They are capable of handling complex, non-linear dynamic systems using simple solutions. Very often, fuzzy systems provide a better performance than conventional non-fuzzy approaches with less development cost.

However, to obtain an optimal set of fuzzy membership functions and rules is not an easy task. It requires time, experience, and skills of the operator for the tedious fuzzy tuning exercise. In principle, there is no general rule or method for the fuzzy logic set-up. Recently, many researchers have considered a number of intelligent techniques for the task of tuning the fuzzy set. Here, another innovative scheme is described (Tang et al. 1998). This approach has the ability to reach an optimal set of membership functions and rules without a known overall fuzzy set topology. The conceptual idea of this approach is to have an automatic and intelligent scheme to tune the membership functions and rules, in which the conventional closed loop fuzzy control strategy remains unchanged, as indicated in Figure 1.

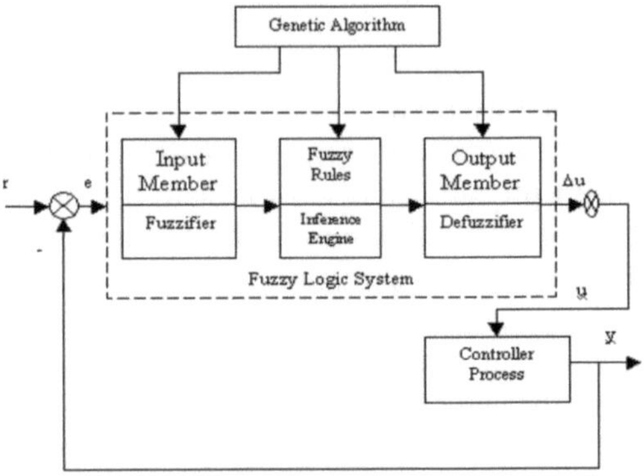

Fig. 1 Genetic algorithm for a fuzzy control system.

In this case, the chromosome of a particular system is shown in Figure 2. The chromosome consists of two types of genes, the control genes and parameter genes. The control genes, in the form of bits, determine the membership function activation, whereas the parameter genes are in the form of real numbers to represent the membership functions.

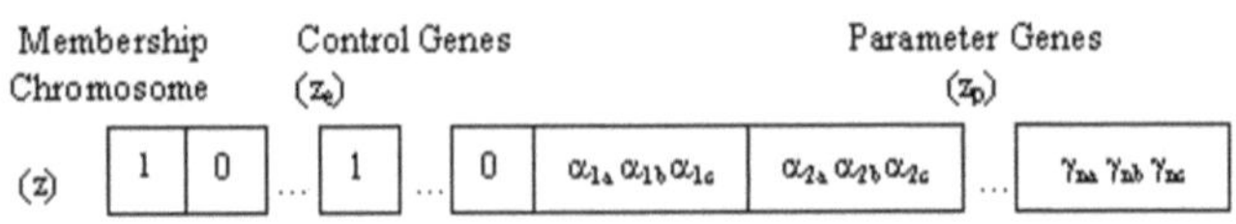

Fig. 2 Chromosome structure for the fuzzy system.

To obtain a complete design for the fuzzy control system, an appropriate set of fuzzy rules is required to ensure system performance. At this point it should be stressed that the introduction of the control genes is done to govern the number of fuzzy subsets in the system. Once the formulation of the chromosome has been set for the fuzzy membership functions and rules, the genetic operation cycle can be performed. This cycle of operation for the fuzzy control system optimization using a genetic algorithm is illustrated in Figure 3. There are two population pools, one for storing the membership chromosomes and the other for storing the fuzzy rule chromosomes. We can see this in Figure 3 as the membership population and fuzzy rule population, respectively. Considering that there are various types of gene structure, a number of different genetic operations can be used. For the crossover operation, a one-point crossover is applied separately for both the control and parameter genes of the membership chromosomes within certain operation rates. There is no crossover operation for fuzzy rule chromosomes since only one suitable rule set can be assisted.

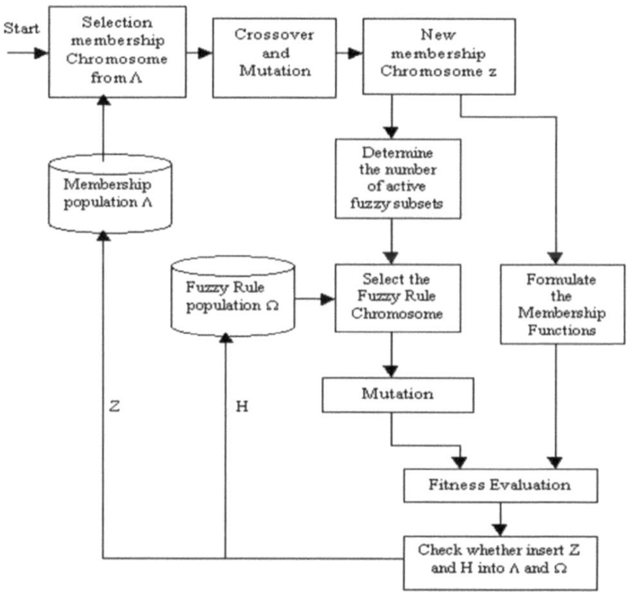

Fig. 3. Genetic cycle for fuzzy system optimization.

Bit mutation is applied for the control genes of the membership chromosome. Each bit of the control gene is flipped if a probability test is satisfied (a randomly generated number is smaller than a predefined rate). As for the parameter genes, which are real number represented, random mutation is applied.

The fitness function can be defined in this case as follows:

$$f_i = \Sigma / y(k) - r(k) / \qquad (7)$$

where Σ indicates the sum for all the data points in the training set, and y(k) represents the real output of the fuzzy system and r(k) is the reference output. This fitness value measures how well the fuzzy system is approximating the real data of the problem.

4 Type-2 Fuzzy Logic

The concept of a type-2 fuzzy set, was introduced by Zadeh (Melin and Castillo 2002) as an extension of the concept of an ordinary fuzzy set (henceforth called a "type-1 fuzzy set"). A type-2 fuzzy set is characterized by a fuzzy membership function, i.e., the membership grade for each element of this set is a fuzzy set in [0,1], unlike a type-1 set (Castillo and Melin 2001, Melin and Castillo 2002) where the membership grade is a crisp number in [0,1]. Such sets can be used in situations where there is uncertainty about the membership grades themselves, e.g., an uncertainty in the shape of the membership function or in some of its parameters. Consider the transition from ordinary sets to fuzzy sets (Castillo and Melin 2001). When we cannot determine the membership of an element in a set as 0 or 1, we use fuzzy sets of type-1. Similarly, when the situation is so fuzzy that we have trouble determining the membership grade even as a crisp number in [0,1], we use fuzzy sets of type-2.

Example: Consider the case of a fuzzy set characterized by a Gaussian membership function with mean m and a standard deviation that can take values in $[\sigma_1,\sigma_2]$, i.e.,

$$\mu(x)=\exp\{-½[(x-m)/\sigma]^2\}; \quad \sigma \in [\sigma_1,\sigma_2] \tag{8}$$

Corresponding to each value of σ, we will get a different membership curve (Figure 4). So, the membership grade of any particular x (except x=m) can take any of a number of possible values depending upon the value of σ, i.e., the membership grade is not a crisp number, it is a fuzzy set. Figure 4 shows the domain of the fuzzy set associated with x=0.7.

The basics of fuzzy logic do not change from type-1 to type-2 fuzzy sets, and in general, will not change for any type-n (Castillo and Melin 2003). A higher-type number just indicates a higher "degree of fuzziness". Since a higher type changes the nature of the membership functions, the operations that depend on the membership functions change; however, the basic principles of fuzzy logic are independent of the nature of membership functions and hence, do not change. In Figure 5 we show the general structure of a type-2 fuzzy system. We assume that both antecedent and consequent sets are type-2; however, this need not necessarily be the case in practice.

The structure of the type-2 fuzzy rules is the same as for the type-1 case because the distinction between type-2 and type-1 is associated with the nature of the membership functions. Hence, the only difference is that now some or all the sets involved in the rules are of type-2. In a type-1 fuzzy system, where the output sets are type-1 fuzzy sets, we perform defuzzification in order to get a number, which is in some sense a crisp (type-0) representative of the combined output sets. In the type-2 case, the output sets are type-2; so we have to use extended versions of type-1 defuzzification methods. Since type-1 defuzzification gives a crisp number at the output of the fuzzy system, the extended defuzzification operation in the type-2 case gives a type-1 fuzzy set at the output. Since this operation takes us from the type-2 output sets of the fuzzy system to a type-1 set, we can call this operation "type reduction" and call the type-1 fuzzy set so obtained a "type-reduced set". The type-reduced fuzzy set may then be defuzzified to obtain a single crisp number; however, in many applications, the type-reduced set may be more important than a single crisp number. Type-2 sets can be used to convey the uncertainties in membership functions of type-1 fuzzy sets, due to the dependence of the membership functions on available linguistic and numerical information.

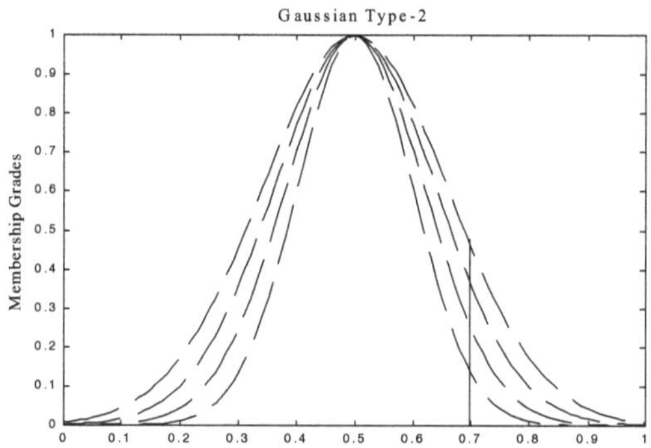

Fig. 4. A type-2 fuzzy set representing a type-1 set with uncertain deviation.

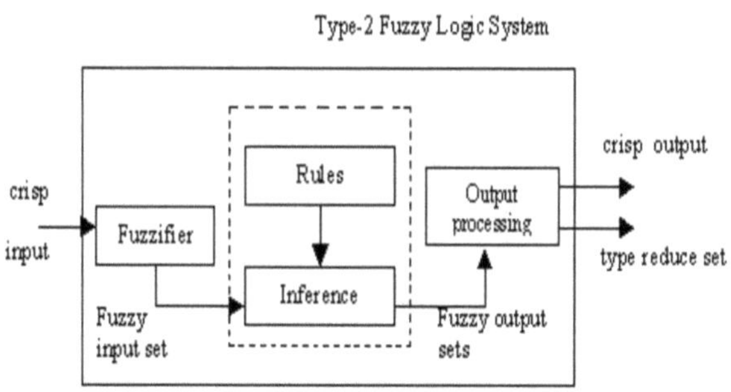

Fig. 5. Structure of a type-2 fuzzy system.

5 Application to Intelligent Control

We consider the case of controlling the anesthesia given to a patient as the problem for finding the optimal fuzzy system for control (Lozano 2003). The complete implementation was done in the MATLAB programming language. The fuzzy systems were build automatically by using the Fuzzy Logic Toolbox, and genetic algorithm was coded directly in the MATLAB language. The fuzzy systems for control are the individuals used in the genetic algorithm, and these are evaluated by comparing them to the ideal control given by the experts. In other words, we compare the performance of the fuzzy systems that are generated by the genetic algorithm, against the ideal control system given by the experts in this application.

5.1 Anesthesia Control Using Fuzzy Logic

The main task of the anesthesist, during and operation, is to control anesthesia concentration. In any case, anesthesia concentration can't be measured directly. For this reason, the anesthesist uses indirect information, like the heartbeat, pressure, and motor activity. The anesthesia concentration is controlled using a medicine, which can be given by a shot or by a mix of gases. We consider here the use of isoflurance, which is usually given in a concentration of 0 to 2% with oxygen. In Figure 6 we show a block diagram of the controller.

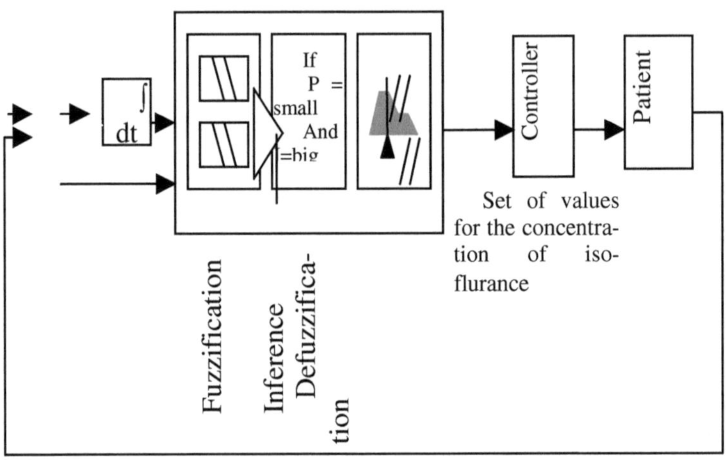

Fig. 6. Architecture of the fuzzy control system.

The air that is exhaled by the patient contains a specific concentration of isoflurance, and it is re-circulated to the patient. As consequence, we can measure isoflurance concentration on the inhaled and exhaled air by the patient, to estimate isoflurance concentration on the patient's blood. From the control engineering point of view, the task by the anesthesist is to maintain anesthesia concentration between the high level W (threshold to wake up) and the low level E (threshold to success). These levels are difficult to be determine in a changing environment and also are dependent on the patient's condition. For this reason, it is important to automate this anesthesia control, to perform this task more efficiently and accurately, and also to free the anesthesist from this time consuming job. The anesthesist can then concentrate in doing other task during operation of a patient.

The first automated system for anesthesia control was developed using a PID controller in the 60's. However, this system was not very succesful due to the non-linear nature of the problem of anesthesia control. After this first attempt, adaptive control was proposed to automate anesthesia control, but robustness was the problem in this case. For these reasons, fuzzy logic was proposed for solving this problem.

5.2 Characteristics of the Fuzzy Controller

In this section we describe the main characteristics of the fuzzy controller for anesthesia control. We will define input and output variable of the fuzzy system. Also, the fuzzy rules of fuzzy controller previously designed will be described.

The fuzzy system is defined as follows:

1. Input variables: Blood pressure and Error
2. Output variable: Isoflurance concentration
3. Nine fuzzy if-then rules of the optimized system, which is the base for comparison
4. 12 fuzzy if-then rules of an initial system to begin the optimization cycle of the genetic algorithm.

The linguistic values used in the fuzzy rules are the following:

PB = Positive Big
PS = Positive Small
ZERO = zero
NB =Negative Big
NS = Negative Small

We show below a sample set of fuzzy rules that are used in the fuzzy inference system that is represented in the genetic algorithm for optimization.

if Blood pressure is NB and error is NB then conc_isoflurance is PS
if Blood pressures is PS then conc_isoflurance is NS
if Blood pressure is NB then conc_isoflurance is PB
if Blood pressure is PB then conc_isoflurance is NB
if Blood pressure is ZERO and error is ZERO then conc_isoflurance is ZERO
if Blood pressure is ZERO and error is PS then conc_isoflurance is NS
if Blood pressure is ZERO and error is NS then conc_isoflurance is PS
if error is NB then conc_isoflurance is PB
if error is PB then conc_isoflurance is NB
if error is PS then conc_isoflurance is NS
if Blood pressure is NS and error is ZERO then conc_isoflurance is NB
if Blood pressure is PS and error is ZERO then conc_isoflurance is PS.

5.3 Genetic Algorithm Specification

The general characteristics of the genetic algorithm that was used are the following:
NIND = 40; % Number of individuals in each subpopulation.
MAXGEN = 300; % Maximum number of generations allowed.
GGAP = .6; %"Generational gap", which is the percentage from the complete population of new individuals generated in each generation.
PRECI = 120; % Precision of binary representations.
SelCh = select('rws', Chrom, FitnV, GGAP); % Roulette wheel method for selecting the indivuals participating in the genetic operations.
SelCh = recombin('xovmp',SelCh,0.7); % Multi-point crossover as recombination method for the selected individuals.
ObjV = FuncionObjDifuso120_555(Chrom, sdifuso); Objective function is given by the error between the performance of the ideal control system given by the experts and the fuzzy control system given by the genetic algorithm.

5.4 Representation of the Chromosome

In Table 1 we show the chromosome representation, which has 120 binary positions. These positions are divided in two parts, the first one indicates the number of rules of the fuzzy inference system, and the second one is divided again into fuzzy rules to indicate which membership functions are active or inactive for the corresponding rule.

Table 1. Binary Chromosome Representation.

Bit assigned	Representation
1 a 12	Which rule is active or inactive
13 a 21	Membership functions active or inactive of rule 1
22 a 30	Membership functions active or inactive of rule 2
...	Membership functions active or inactive of rule...
112 a 120	Membership functions active or inactive of rule 12

6 Simulation Results

We describe in this section the simulation results that were achieved using the hierarchical genetic algorithm for the optimization of the fuzzy control system, for the case of anesthesia control. The genetic algorithm is able to evolve the topology of the fuzzy system for the particular application. We used 300 generations of 40 individuals each to achieve the minimum error. We show in Figure 7 the final results of the genetic algorithm, where the error has been minimized. This is the case in which only nine fuzzy rules are needed for the fuzzy controller. The value of the minimum error achieved with this particular fuzzy logic controller was of 0.0064064, which is considered a small number in this application.

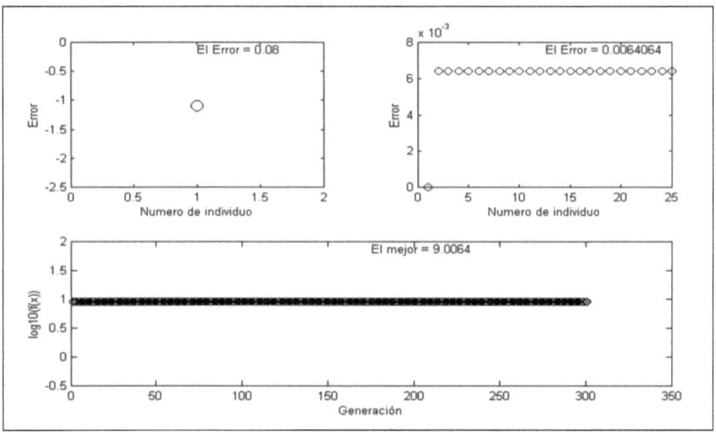

Fig. 7. Plot of the error after 300 generations of the HGA.

In Figure 8 we show the simulation results of the fuzzy logic controller produced by the genetic algorithm after evolution. We used a sinusoidal input signal with unit amplitude and a frequency of 2 radians/second, with a transfer function of [1/(0.5s +1)]. In this figure we can appreciate the comparison of the outputs of both the ideal controller (1) and the fuzzy controller optimized by the genetic algorithm (2). From this figure it is clear that both controllers are very similar and as a consequence we can conclude that the genetic algorithm was able to optimize the performance of the fuzzy logic controller. We can also appreciate this fact more clearly in Figure 9, where we have amplified the simulation results from Figure 8 for a better view.

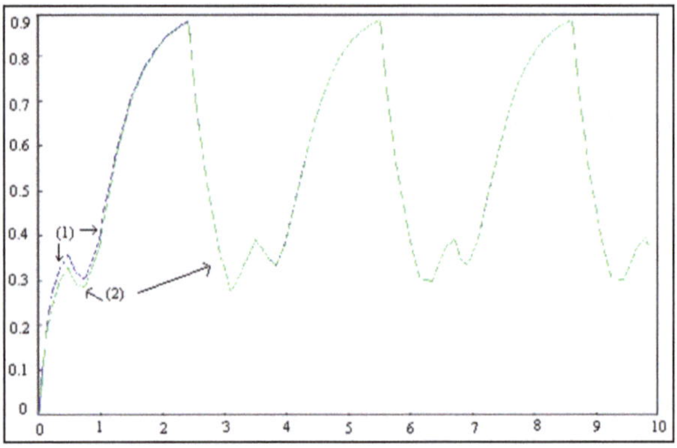

Fig. 8. Comparison between outputs of the ideal controller (1) and the fuzzy controller produced with the HGA (2).

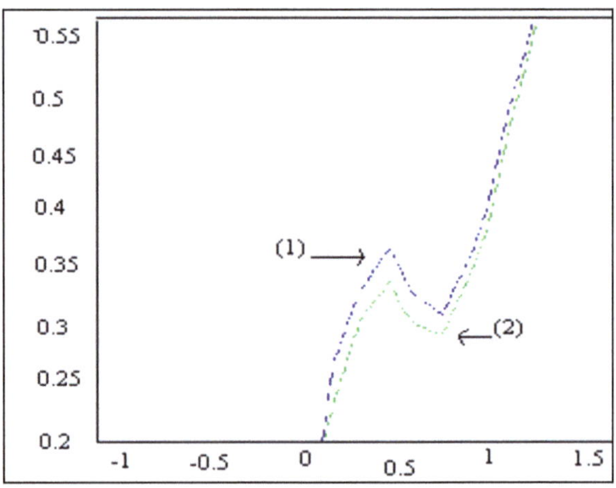

Fig. 9. Zoom in of figure 8 to view in more detail the difference between the controllers.

Finally, we show in Figure 10 the block diagram of the implementation of both controllers in Simulink of MATLAB. With this implementation we are able to simulate both controllers and compare their performances.

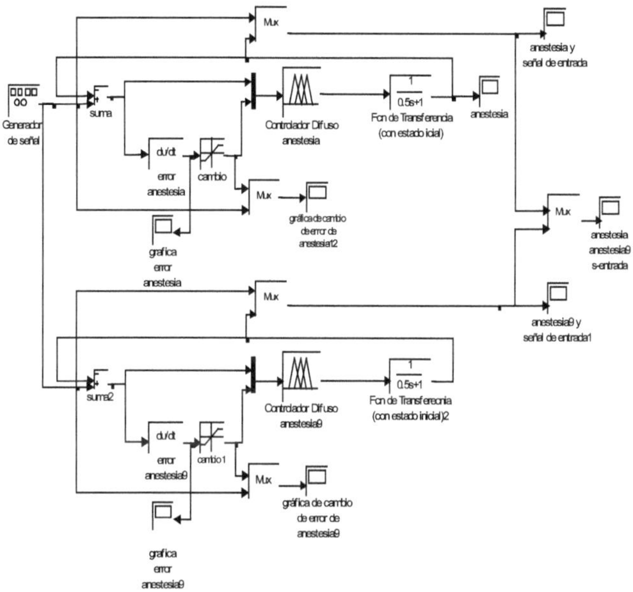

Fig. 10. Implementation in Simulink of MATLAB of both controllers for comparison of their performance.

7 Conclusions

We consider in this paper the case of automatic anesthesia control in human patients for testing the optimized fuzzy controller. We did have, as a reference, the best fuzzy controller that was developed for the automatic anesthesia control (Karr and Gentry 1993, Lozano 2003), and we consider the optimization of this controller using the HGA approach. After applying the genetic algorithm the number of fuzzy rules was reduced from 12 to 9 with a similar performance of the fuzzy controller. Of course, the parameters of the membership functions were also tuned by the genetic algorithm. We did compare the simulation results of the optimized fuzzy controllers obtained with the HGA against the best fuzzy controller that was obtained previously with expert knowledge, and control is achieved in a similar fashion.

Acknowledgments

We would like to thank the Research Grant Committee of COSNET for the financial support given to this project (under grant 424.03-P). We would also like to thank CONACYT for the scholarships given to the students that work in this research project (Gabriel Huesca and Fevrier Valdez).

References

O. Castillo and P. Melin (2001), "Soft Computing for Control of Non-Linear Dynamical Systems", Springer-Verlag, Heidelberg, Germany.

O. Castillo and P. Melin (2003), "Soft Computing and Fractal Theory for Intelligent Manufacturing", Springer-Verlag, Heidelberg, Germany.

J. Holland, (1975), "Adaptation in natural and artificial systems" (University of Michigan Press).

A. Homaifar and E. McCormick (1995), "Simultaneous design of membership functions and rule sets for fuzzy controllers using genetic algorithms", *IEEE Trans. Fuzzy Systems*, vol. 3, pp. 129-139.

J.-S. R. Jang and C.-T. Sun (1995) "Neurofuzzy fuzzy modeling and control", *Proc. IEEE*, vol. 83, pp. 378-406.

J.-S. R. Jang, C.-T. Sun, and E. Mizutani (1997), "Neuro-fuzzy and Soft Computing, A computational approach to learning and machine intelligence", , Prentice Hall, Upper Saddle River, NJ.

C.L. Karr and E.J. Gentry (1993), "Fuzzy control of pH using genetic algorithms", *IEEE Trans. Fuzzy Systems*, vol. 1, pp. 46-53.

A. Lozano (2003), "Optimización de un Sistema de Control Difuso por medio de algoritmos genéticos jerarquicos", Thesis, Dept. of Computer Science, Tijuana Institute of Technology, Mexico.

K.F. Man, K.S. Tang, and S. Kwong (1999), "Genetic Algorithms: Concepts and Designs", Springer Verlag.

P. Melin and O. Castillo (2002), "Modelling, Simulation and Control of Non-Linear Dynamical Systems", Taylor and Francis, London, Great Britain.

T.J. Procyk and E.M. Mamdani (1979), "A linguistic self-organizing process controller" *Automatica*, vol. 15, no. 1, pp 15-30.

K.-S. Tang, K.-F. Man, Z.-F. Liu and S. Kwong (1998), "Minimal fuzzy memberships and rules using hierarchical genetic algorithms", *IEEE Trans. on Industrial Electronics*, vol. 45, no. 1.

J. Yen, and R. Langari (1999), "Fuzzy Logic: intelligence, control and information", Prentice Hall, Inc.

T. Yoshikawa, T. Furuhashi and Y. Uchikawa (1996), "Emergence of effective fuzzy rules for controlling mobile robots using NA coding method", *Proc. ICEC'96*, Nagoya, Japan, pp. 581

Part III Robotic Applications

Decision Trees and CBR for the Navigation System of a CNN-based Autonomous Robot

Giovanni Egidio Pazienza[1], Elisabet Golobardes-Ribé[1], Xavier Vilasís-Cardona[1], and Marco Balsi[2]

[1]Enginyeria i Arquitectura La Salle, Universitat "Ramon Llull", Pg. Bonanova 8, 08022 Barcelona, Spain, gpazienza@salleurl.edu,
[2]Dipartimento di Ingegneria Elettronica, Università "La Sapienza", Via Eudossiana 18, 00184 Roma, balsi@uniroma1.it

Abstract. In this paper we present a navigation system based on decision trees and CBR (Case-Based reasoning) to guide an autonomous robot. The robot has only real-time visual feedback, and the image processing is performed by CNNs to take advantage of the parallel computation. We successfully tested the system on a SW simulator.

1 Introduction

Among the challenges in autonomous robotics, we find navigation in unstructured environments with vision based algorithms. The choice for visual feedback is likely to be based more on human perception analysis, for which sight is the primary navigation input, rather than on technological criteria. Still, dealing with images allows for a highly valuable verbalisation of the processing steps guiding to the heuristic development of procedures and algorithms. Yet, we find in recent references such as [1, 2] promising results, at the price, however, of a large amount of computing effort. A compromise is then to be found between real autonomy, robot resources and real-time operation. We have been working for some time to study the capacity of Cellular Neural Networks (CNNs) [4, 5] to break the robot vision compromise [9, 10, 3]. CNNs represent a convenient tool for real-time image processing for three main reasons: they are, by construction, a massively parallel system with the capability of universal Turing machines [6] which can be implemented directly on analog or digital VLSI devices [7, 8]. Although CNN hardware exists, Cellular Neural Networks are often emulated using general purpose processors or programmable hardware. In

G.E. Pazienza et al.: *Decision Trees and CBR for the Navigation System of a CNN-Based Autonomous Robot*, StudFuzz **208**, 181–201 (2007)
www.springerlink.com

this manner, the limitations in availability and image sizes of CNN chips can be overcome. This has been our strategy to develop a CNN-based vision system to control an autonomous robot with only real-time visual feedback in an unknown environment. We have built a robot platform on purpose for this problem, where all processing tasks, such as CNN emulation, are carried out by a DSP, while its unique sensor is a B/W camera. Robot is fully autonomous, while its dimensions are kept small, 25x16x10cm, in order to achieve the goal of real-time operation with limited resources. Our robot guidance solution splits into two processes: visual processing and guidance operation. The first is based exclusively on CNN calculations to extract relevant parameters from the image, whereas the second is in charge of finding the trajectory to be followed by the robot. The fact that both CNN emulation and the navigation system run on the DSP, places an upper bound to the complexity of the algorithm, because it has to be executed in real-time. Actually, our efforts have been devoted more to the design of suitable CNN algorithms than to the refinement of the guidance procedures, for which we resorted to adapting a well known solution of the truck backer-upper based on a fuzzy controller [14]. In this scheme, we align the robot to a desired straight trajectory by determining the steering angle from the distance and the angle of the axis of the cart with respect to the targeted line. These two last parameters are extracted from the CNN processing of the image. This simple strategy has allowed us to solve problems such as line following, obstacle avoidance and tracking. Still, the guidance system has often required more attention and tuning than initially expected. Actually, the first step of our programme was proving that the image processing for driving a robot in a maze made of black lines could be performed using CNNs [9]. During the test phase, we noticed that the CNN visual system was effective and quite fast, but the development of an adequate guidance system turned out to be a not trivial task. In fact, as the camera of the robot is inclined to enlarge the visual field, the scene may be misrepresented due to the perspective distortion; therefore, it was not possible to use a guidance algorithm based exclusively on geometric considerations. Using a fuzzy system can be effective, but the design procedure involves tedious heuristic tuning of membership functions. Moreover, as in the robot used in the experiment the DSP is in charge of both vision algorithm and guidance operations, it is better to get the second as simple as possible to dedicate resources to the image elaboration. In order to obtain a more efficient solution, in this paper we present an alternative guiding algorithm for the line following problem based on decision trees, integrated by a Case-Based Learning scheme. It will be shown that this system improves performance, meeting the requirement of real-time operation. The chapter is structured as follows: first, we shall review the essentials of Cellular Neural Networks and then sketch our CNN vision algorithm; then, we detail the new navigation system based on decision trees inserted in a Case-Based learning structure and compare it with the old fuzzy-reasoning-based guidance system; finally, we comment on results and give conclusions.

2 Cellular Neural Networks for Image Processing

Cellular Neural Networks (CNNs) [4, 5] are arrays of dynamical arti• cial neurons with local connections only. This essential point has made hardware implementation of large networks possible on a single VLSI chip [7, 8]. Our solution refers to the Discrete-Time CNN model (DTCNN hereafter) [11] because it performs better under emulation without loss of generality. The DTCNN core operation is described by the following system of iterative equations:

$$x_{ij}(n+1) = \mathrm{Sign}\left(\sum_{kl \in N(ij)} A_{k-i,l-j} \; x_{ij}(n) + \sum_{kl \in N(ij)} B_{k-i,l-j} u_{ij}(n) + I \right) \quad (1)$$

where:

- n is (discrete) time;
- xij is the state of the cell (neuron) in position ij, that corresponds to the image pixel in the same position;
- uij is the input to the same cell, representing the luminosity of the corresponding image pixel, suitably normalised;
- A is a matrix representing the interaction between cells, which is local (as speci• ed by the fact that summations are taken over the set N of indexes of neighbour cells) and space-invariant (as implied by the fact that weights depend on the dierence between cell indexes, rather than on their absolute values);
- xij is the state of the cell (neuron) in position ij, that corresponds to the image pixel in the same position;
- uij is the input to the same cell, representing the luminosity of the corresponding image pixel, suitably normalised;
- B is a matrix representing forward connections issuing from a eighbourhood of inputs. I is a bias. $N(i, j)$ is the set of indexes corresponding to cell ij itself and a small neighbourhood e.g. cell ij and its 8 nearest neighbours). Due to the locality of the computation, implied by the summation over this neighbourhood, and spaceinvariance, implied by

the differences of indexes in matrices *A* and *B* in equation (1), it is sufficient to define *A* and *B* for a few instances of the indexes, so that they can be represented by small matrices.

The operation performed by the network is fully defined by the so-called cloning template {A, B, I} (see examples in Table 1). Under suitable conditions and with time-invariant input u, a steady state is reached. In the DTCNN model, states are binary to allow for faster convergence of the function but this does not affect the structure of the cloning templates with respect to the continuous-time model. Images to be processed are fed to the network as initial state and/or input, and the result is the steady state value, which realistically comes after some time steps (ranging normally from 1 to several tens according to the task).

3 Generalities on the Problem

3.1 Brief Description of the Robot Pasqual8

Although in this paper we present only a SW simulation of our system, in a near future we plan to implement it on a small autonomous robot called Pasqual8 we have already been employing to test other CNN algorithms [3]. Therefore, in this chapter we wish to give a rough idea of the hardware of the robot, above all to consider its limitations that must be taken into account during the development of the navigation algorithm.

The main part of Pasqual8 is the processing board, hosting some RAM, a TMS320C32 DSP from Texas Instruments, which emulates the CNNs, and a FLEX10K20 FPGA from Altera whose task is to interface a digital B/W camera that is the only sensor of the robot. The control algorithm, which is in charge of calculating the correct steering angle using data coming from the image processing, is implemented on the same DSP where the CNN emulator runs. Therefore, the less complex the control algorithm is, the faster the CNN emulation results because the two processes cannot be done in parallel. With a 40 MHz clock, the board is able to process about 4 64x64 pixel images per second. To sum up, Pasqual8 is not an elaborate robot but our aim is to show how such a simple hardware with no sensors, apart from a B/W camera, can execute complex tasks. Nevertheless, this simplicity constitutes a boundary for the complexity of both the visual processing and the guidance algorithm.

3.2 A Fuzzy-Based Navigation System for Pasqual8

For the sake of simplicity, at the beginning we chose to implement a quite standard fuzzy controller inspired by the solution to the truck backer-upper problem [14]. The controller is fed with the the distance X_c from the centre of the front wheels to the line to be followed, and with the angle A_{th} made by the axis of the robot and the line (see figure 1). In the most-simple version, this fuzzy system has 15 rules that allow to compute the steering angle A_{st}. The consequent sets are simplified to singletons for simpler on-line implementation, and the values for the singletons are adjusted according to the robot platform steering capabilities. There is no systematic way to determine such values, and usually they must be tuned experimentally. For example, we observed that a good solution is to choose 7 consequents for A_{st} that are 0 for Zero and ± 15, 30, 60 for Small, Medium, Large Left/Right respectively. When a line is in sight, the robot calculates the values assumed by of Xc and Ath and applies the fuzzy algorithm in order to align itself on the line as fast as possible.

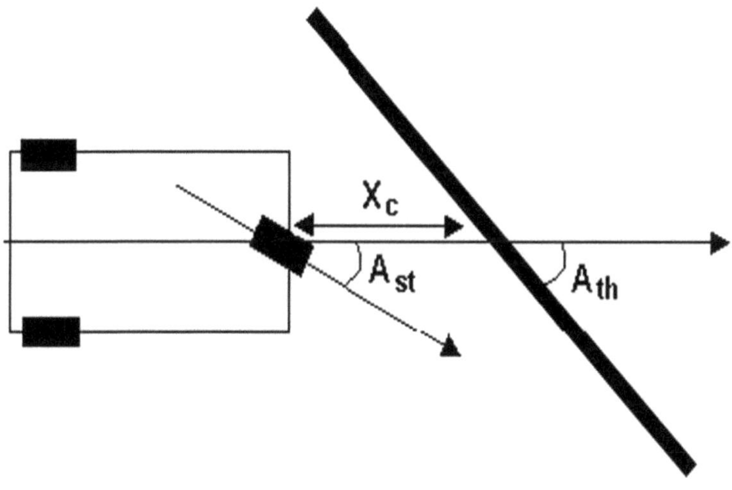

Fig. 1. Representation of the alignment positioning parameters.

As mentioned in the introduction, some aspects of the fuzzy system are improvable. In the first place, the image processing needed to get X_c and A_{th} is not trivial, and several templates must be used to obtain the correct parameters; consequently, the final CNN algorithm results rather slow and it may be the bottleneck of the whole process. Then, establishing fuzzy sets is a heuristic process and it may be long and not accurate. In addition, this mechanism is not efficient because the robot calculates the steering angle every time the camera takes a new image although the previous information could be stored and used again. Last but not least, the parameter fuzzyfication and the application of fuzzy algorithm implies an excessive use of the DSP that can affect the image processing. As a result, this system results to be unsatisfactory and it must be substituted by a faster method, with a simpler CNN processing, so simple as to be implementable on a DSP, and, if possible, based on experience.

4 A New Approach to the Problem

4.1 CNN Visual Processing

It follows from section 2 that the image processing is based on the action of a sequence of templates over the image in the manner of a 'CNN Universal Machine' algorithm: all templates used in the algorithm come from the standard library [12]. As we mentioned in the introduction, our purpose is to drive an autonomous robot in a maze made of black lines using only the information coming from a B/W camera. Basically, the aim of the visual algorithm is to get useful features from the image of a line in order to estimate its slope. For this purpose we devise the necessary sequence of actions to be taken on the image to obtain the required result, while checking they can be performed by CNN templates. We said previously that one of the drawbacks of the fuzzy navigation system is the complexity of the CNN visual algorithm needed to get the parameters required. One of the aims of the new navigation system is to simplify the CNN algorithm, and in this section we show the new visual processing whose usefulness will be clearer after the reading of section 4.3.

First of all, we start by preprocessing the image taken by the camera with the so-called Small Object Remover template to make a preliminary cleaning and binarisation (Figure 2 (a) and (b)).

Next step consists in obtaining the line position and orientation in order to establish the right angle the robot must turn to keep staying on the line. If the image contains at most one line, its direction can be extracted by projecting the line onto the horizontal and vertical axis by means of the so-called Hole Detection cloning template. Operations of such templates, with reference to horizontal holes (vertical direction), at an intermediate and

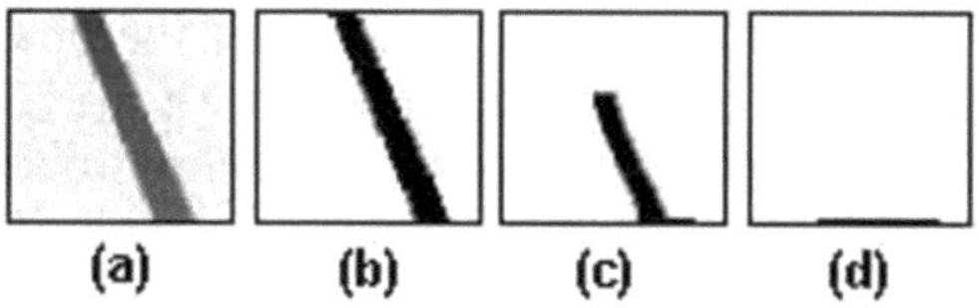

Fig. 2. Image taken from the actual camera with size 64x64 pixels (a), cleaning and binarisation (b), hole detection intermediate result (c), final result (d)

final stage of processing are depicted in Figure 2(c,d). It is apparent that besides getting the desired projections, we can gain from intermediate results also the information about which extreme is closer to the lower border of the image. As shown in the next section, these data are sufficient to compute direction and position of the line with the maximum precision allowed by image definition.

Table 1. Templates used in the processing stage

	A	B	I
Small object remover	$\begin{pmatrix}1&1&1\\1&2&1\\1&1&1\end{pmatrix}$	0	0
Vertical hole detector	$\begin{pmatrix}0&0&0\\1&2&-1\\0&0&0\end{pmatrix}$	0	0
Horizontal hole detector	$\begin{pmatrix}0&1&0\\0&2&0\\0&-1&0\end{pmatrix}$	0	0
Gradient	$\begin{pmatrix}0&0&0\\0&1&0\\0&0&0\end{pmatrix}$	$\begin{pmatrix}b&b&b\\b&0&b\\b&b&b\end{pmatrix}$	-0.1
ApproxVertical Line Detector	$\begin{pmatrix}0&0&0\\0&2&0\\0&0&0\end{pmatrix}$	$\begin{pmatrix}-1&-1&-1&-1&-1\\0.5&1&-1&1&0.5\\1&1&5&1&1\\0.5&1&-1&1&0.5\\-1&-1&-1&-1&-1\end{pmatrix}$	-13

4.2 Navigation as a Classification Problem

As shown in section 4.1, after the CNN processing two images are available, each containing a projection of the line onto an axis. Moreover, it it is known which extreme of the line is closer to the lower border and this information can be used to establish if the angle the line forms with the horizontal axis is positive or negative. Although we worked in the following examples with 15x15 pixel images, the algorithm we are going to illustrate is applicable to images of any size. There is no loss of generality in taking into consideration small images, but by working this way, the system can be tested in a more accurate way. Moreover, we take into account 3 pixel-width lines to simplify the test phase, but also this hypothesis can be removed without problems. The main idea of this method is to classify all possible examples according to the slope of the line in the image. It means that we must extract useful information from all the images in the database, trying to codify it in some way, and finally associate a label (that is an angle) to that image. In this way we no more try to measure the slope of a line geometrically, but we deal with the topic exactly as a classification problem. Generally speaking, the intrinsic simplicity of this method allows faster processing with respect to the fuzzy system. Moreover, in this case the perspective distortion can be discarded without consequences because the training samples will be affected by the same amount of distortion of the images actually seen by the robot. Finally, as we will see in section 5, the system is able to improve with experience.

4.3 Use of the Decision Trees

A way to implement this new approach to the navigation problem, is employing the decision trees. The decision tree algorithm we use is ID3 [15], which is a method to perform supervised batch inductive learning and classification tasks. It is simple yet powerful, and these characteristics make it suitable for our system. The first step to design the classification system consists in creating the training set. There are about 600 possible ways in which a 3 pixelwidth line can appear in a 15x15 pixel image: we stored them in a database and processed the images with CNNs in order to get the projections onto the axes. Next step is choosing an efficient representation of features (examples) for the classification system, for instance deciding the number of attributes for each feature and how to relate them with the image. The first idea was associating each pixel of the projections with an attribute, creating a feature of 30 (15+15) attributes. As images are binary, the resulting tree is binary as well. Unfortunately, this way of operating has some drawbacks. Firstly, the size of features grows quickly as the image gets bigger (for example a 64x64 pixel image would require 128 attributes) and this might cause problems with data storage; secondly, searching in binary trees is not efficient because of their depth; thirdly, a mechanism to relate each pixel with its neighbors lacks, and so the information about the width of the line is lost. To sum up, we need to code the information in a more efficient way. We can overcome the problem by clustering the pixels, adding their values and considering the sum as one attribute. An example of this method is illustrated in Figure 3. Thanks to the clustering, a feature has less attributes than before, so the same information can be stored and recalled quicker. Moreover, clustering allows to relate different pixels and this may result useful in many situations. For instance, pixels belonging to a projection usually have at least one neighbour, whereas pixels with no neighbours come probably from a spot in the image. We can discern this behaviour by codifying the information as illustrated before. A further advantage of clustering is that trees are no more binary, therefore searching in the tree is faster than before. Note that all the attributes of a feature range in the same interval (from 0 to the number of features clustered) so a normalization of values is not necessary. By clustering 3 pixels in each projection, every image can be represented by a feature of 11 attributes: 5 for the horizontal projection, 5 for the vertical projection and 1 for the inclination of the line.

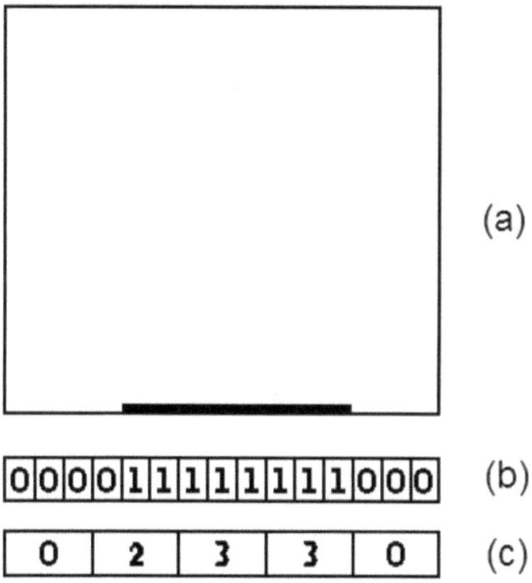

Fig. 3. The horizontal projection of a line (a), its coding in a one-pixel one attribute feature (b), clustering of three pixels (c)

The last attribute is necessary because two different images can have the same projections. For example, Figure 4 shows three images with the correspondent features below. Line (a) and line (c) have the same projections but different slope. The only way to discern between them is looking at attribute 11, which is 1 if the angle of the line with respect to the horizontal is below 90° and 2 otherwise. Once defined how to build the features, we must focus our attention on the relation between decision trees and robot guidance. Every image in the training set is associated with a class that, in our system, corresponds to an angle. Like in the fuzzy system, we considered 7 possible angles the robot can steer associating them with labels Zero and Small, Medium, Large Left or Right respectively. The values of the classes (that is, the angles) depend on the position of the camera and the perspective distortion. When a line is in sight, the CNN-based vision system processes the image and passes the results to the controller, that looks in the tree for the class the line belongs to and transmits the corresponding angle to the actuator.

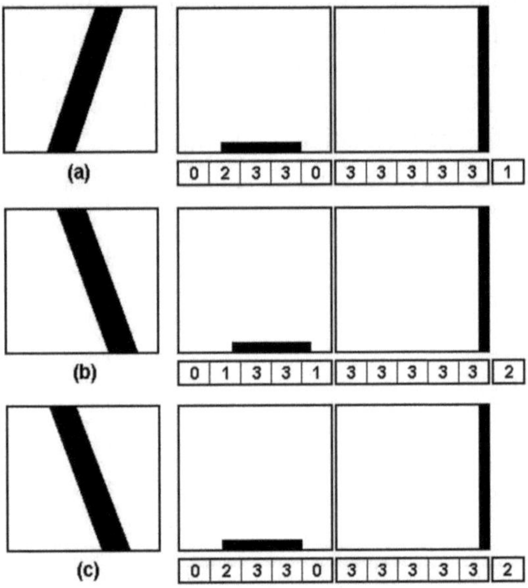

Fig. 4. Each row contains a different line, its projections on the axes and the correspondent feature

5 Incremental Learning

5.1 Case-Based Reasoning

In the previous section we assumed that all possible images were known and stored in a database, but this hypothesis is not always verified. On the contrary, usually the robot does not know what it is going to see. Moreover, broadly speaking, a robot navigation system needs to be able to learn and operate in real-time in order to be capable of continual adaptation to changes in the environment. In conclusion, we wish our system to be able to learn incrementally by acquiring new knowledge. A powerful method to implement this behaviour is using Case-Based Reasoning (CBR hereafter), which integrates in one system two different characteristics: machine learning and problem solving capabilities. CBR uses a human-inspired philosophy: it tries to solve new cases by using previously solved ones. The process of solving new cases also updates the system providing new information and new knowledge to the system. This new knowledge can be used for solving other future cases. The basic method can be easily described in terms of its four phases (see Figure 5). The first phase retrieves old solved cases similar to the new one. Then, in the second phase, the system tries to reuse the solutions of the previously retrieved cases for solving the new case. Next, the third phase revises the proposed solution. Finally, the fourth phase retains the useful information obtained while solving the new case. Basically, a new problem is matched against cases in the database, and one or more similar cases are retrieved. A solution suggested by the matching cases is then reused and tested for success. Unless the retrieved case is a close match, the solution may have to be revised producing a new case that can be retained.

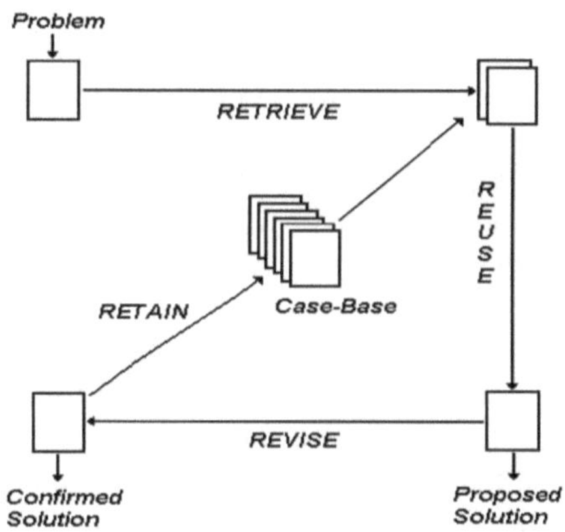

Fig. 5. Cycle of CBR

One key point in the whole algorithms is the concept of most similar case used in the retrieval phase of CBR. The definition of this concept determines which stored cases are retrieved as the bases of the classification process of the new cases being solved. In this work, the notion of similarity between two cases is computed using functions based on distance metrics. The most used similarity function is the Nearest Neighbour Algorithm

(NNA). The NNA computes the similarity between two cases using a global similarity metric. In this work, the practical implementation of the NNA is the Minkowski metric (see Table 5.1). This metric admits several values for r parameter. We use r=1 (Hamming distance), r=2 (Euclidean distance), and r=3 (Cubic distance). Case x and Case y are two cases (whose similarity is computed), F is the number of features that describes the case, and x_i, y_i are the value of the ith feature of cases Case x and Case y respectively. Detailed descriptions of CBR can be found in [16–18].

Table 2. Similarity functions based on distance metric

Metric	Function
Minkowski	$Similarity(Case\_x, Case\_y) = \sqrt[r]{\sum_{i=1}^{F} \lvert x_i - y_i \rvert^r}$

5.2 Application of CBR to Our Navigation System

In order to apply the CBR algorithm, we split the database into two sets: a training set and a test set. In this way, we can simulate an incomplete knowledge of the possible images the robot can see. In this phase we can appreciate the benefit of knowing all the possible ways the line can be seen by the robot; thanks to this fact, we can test our system reliably. Now the database is constituted by the training set, whereas the test set is used to measure the reliability of CBR algorithm. As far as the retrieve phase is concerned, we used the Minkowski metric of first order (Hamming distance) to determine the similarity between two attributes. In spite of its simplicity, it is the most suitable method for this task because all the attributes, except for the one indicating the slope of the line, have the same importance and range in the same interval. We must compare a new case only with the samples in the training set that have the same value of attribute 11, otherwise we may make a mistake. For example, considering the Hamming distance, the feature corresponding to image (c) in Figure 4 is as similar to the first as to the second (in both the case it has just one pixel of difference), but the image itself is far more similar to image (b) than to image (a). In our system, the reuse phase consists in finding in the database the set of cases similar to the new case, and assigning to it the class most present in the set. For the nature of the problem, our system does not need a revision of the solution proposed. Finally, the new feature (that is, the new case and the class deduced) is added to the database and a new tree is created.

6 Extensions of the Navigation System to Other Tasks

6.1 Obstacle Avoidance

The navigation system presented in this paper can be also used to avoid obstacles and move the robot in a corridor without colliding with the walls. In order to do this new task, it is necessary to change the way in which the sequence of monocular images is processed, allowing to localize the object (or the wall) through the identification of significant parameters. For example, if there is a great contrast between the ground and the object, the gradient of the brightness assumes high values on the border of the object. This hypothesis is often true, therefore we can detect the obstacle by just calculating the gradient of brightness and looking for its maximum. We outline the new CNN image processing in Figure 6.1. Panel (a) is shows a wall seen from the robot point of view. It is apparent how noisy the image is, so the first template to be applied is the Small Object Remover that cleans and binarises the image. Subsequently, in order to detect borders, we use the Gradient template, whose result is depicted in Figure 6.1 (c).

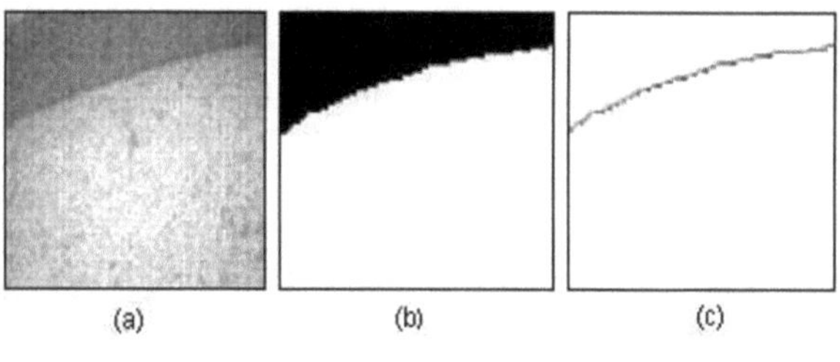

Fig. 6. Example of image processing: junction between a wall and the floor (a), binarisation (b), image obtained by application of Gradient Template (c)

Thanks to the addition of just one template in the algorithm, the robot is able to recognize the border of the wall that is nothing but a line. Consequently, we can use the navigation system showed before to establish the right steering angle, but now there is a difference with respect to the previous case: we do not want to follow a line, but to avoid it. It means that is necessary to change the values of the classes, and increment slightly the turning angles to avoid hitting the wall. The process is exactly the same when an object is involved, but in this case the border line will not take up the whole visual field.

6.2 Crossing Detection

A further extension of this navigation system is the case of crossing detection. Also in this example we must change the visual processing for another one, and the proposed solution is illustrated in Figure 6.2. If necessary the image can be preprocessed as usual employing the Small Object Remover template to get a quasi-noiseless crossing. Then, in order to discern the different lines that constitute the crossing according to their slope, we employ a set of directional filters. Its way of operating can be understood looking at the Figure 6.2, where the action of the ApproxHorizontal Line Detector template is represented, whose purpose is detecting approximately horizontal lines. By modifying the positions of the elements of the B template, namely rotating B, the template can be sensitized to other directions as well. As the bandwidth of a single filter is about 90º, the sequence of 4 templates covers the whole visual field. In Figure 6.2 (b) and 6.2 (c) are depicted the images resulting after the application of the ApproxHorizontal filter and an ApproxDiagonal filter.

It is apparent that the lines extracted are incomplete that is, they are affected by noise, so their projections may not match exactly with those used to train the system. Anyway, the mechanism of clustering of several pixels, explained in 4.3, makes our system quite robust, so that this small amount of noise does not represent a problem. At the end of the processing all the lines belonging to the crossing are detected, and the slope of each one can be determined as explained in the previous sections.

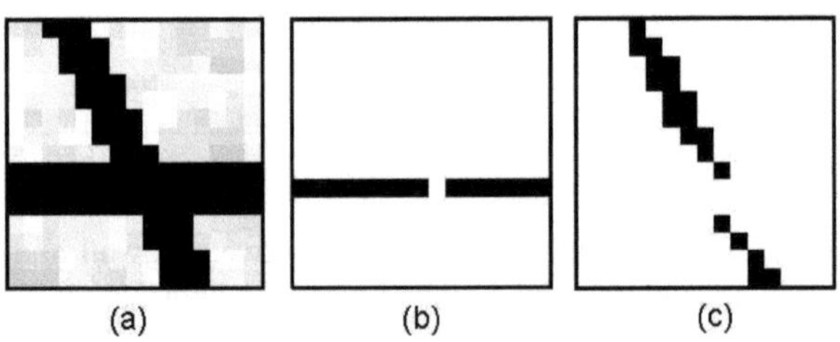

Fig. 7. Crossing detection: original image (a), after the application of the Approx-Horizontal Line Detector (b), ApproxDiagSWNE Line Detector (c)

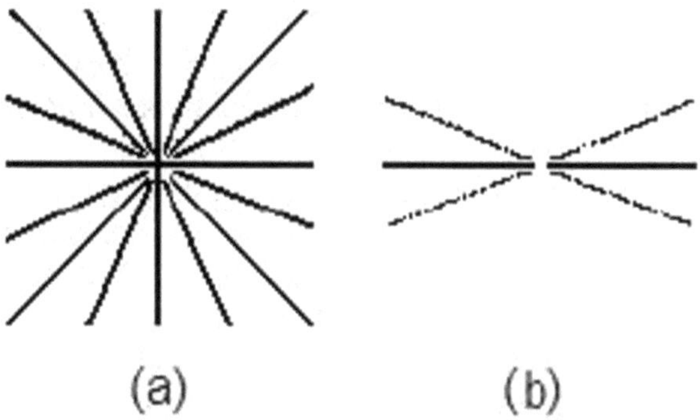

Fig. 8. ApproxHorizontal Line Detector: example of the effects of the template on lines with various orientations. Original image (a), result (b)

7 Results

7.1 Vision Algorithm

The vision algorithm resulted robust enough to the effects of a small change of the brightness, and little spots on the ground are usually eliminated by the Small Object Remover template. We can foretell that when the algorithm will be implemented on Pasqual8, the CNN processing is still the bottleneck of the whole algorithm in spite of the fact that it implies only 3 templates. This is a consequence of the CNNs emulation on a DSP, but we are working to move the CNN emulation from the DSP to the FPGA so as to speed up the image processing. For example, in [19] is shown that the FPGA can emulate CNNs up to 9 times faster than this DSP can do (Pasqual8 can process about 4 images per second, as mentioned in section 3.1). This means that moving the emulation to the FPGA, we would be able to enlarge the images or increment the resolution going on processing the same number of images per second, or decide to use the same class of images but processing up to 36 images per second.

7.2 Navigation System

An exhaustive study of performances using a fuzzy-based guide system can be found in [10], so we will analyse in detail only the guidance system based on decision trees. As we said in 4.3, we took into consideration 15x15 pixel images in which a 3 pixel-width line is drawn. We found heuristically that there are about 600 possible ways to focus such a kind of line, and all the possibilities, properly classified according to the slope of the line, were stored in a database. Then, we split the database into a training set and a test set, building a decision tree with the samples of the train set. Finally, we tested the system varying the dimension of the training set, observing how many cases belonging to the test set were classified correctly. Due to the small number of instances, we implemented a k-fold cross-validation test, with k ranging from 5 to 50. When cases in the test set were more than examples of the training set, we used an holdout test, varying the size of the test set from 55% to 95% of total instances. The test was repeated removing the CBR mechanism (that is, classifying the examples using only the tree) so to observe how much it influences the final result. Results are depicted in the Figure 9. This result can be improved in a very simple way. Sometimes a leaf has the value ”unknown” when there are no examples in the training set corresponding to that path in the tree. In this case, we could assign to that example the value of a neighbour leaf. This method is not effective as the CBR, but can increase the amount of matchings. Obviously, the number of well-classified cases becomes larger when the percentage of training set samples on the total increases. In particular, when the training set is composed of 95% of the total examples, more than 85% of the test set samples are classified correctly.

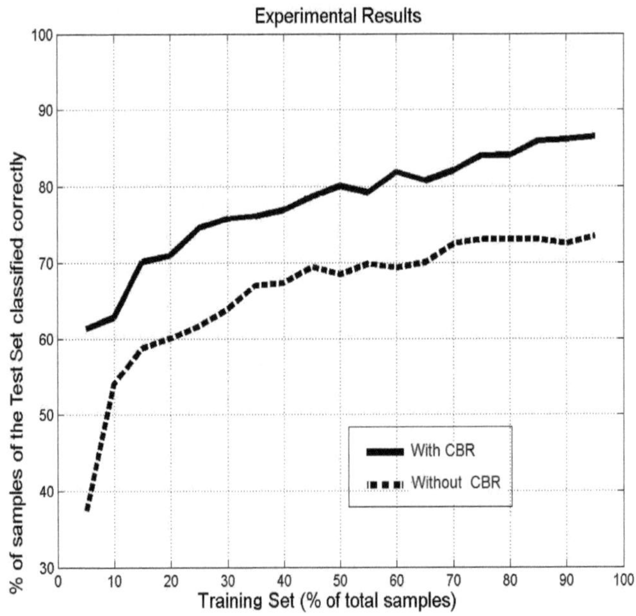

Fig. 9. Experimental results. On the x-axis there is the percentage of the total examples belonging to the training set; on the y-axis there is the percentage of the samples in the test set classified correctly

There are two reasons because not all the test set is well-classified: first, when the number of test samples is low, few errors can affect considerably the final result; second, due to the small quantity of total examples, the system can suffer from overtraining. This problem can be overcome just adding more samples (for example noisy images well-classified) to the original database so to characterize better the training set. When the examples of the training set constitutes 66% of the total, 4 out of 5 cases in the test set are classified correctly. Furthermore, when the training set is far smaller than the test set, the percentage of well-classified instances is about 70%. It means that when the system is fed with just 120 samples chosen randomly, it is capable to classify correctly about 350 of the remaining examples. The dashed line in the figure represents the same system without the CBR mechanism. As only the decision tree is used, performances are worse than the previous system, and the difference is larger as the training set gets smaller. Looking at the figure, we can realize that also when the training set is very large, the CBR improves the performances of the system significantly. The CBR is absolutely necessary when the training set is a small part of the whole set of examples, because without it only 35% of the instances are classified correctly. We never obtained trees deeper than 7 levels, which it means that searching is efficient and fast. The last significative thing is that when the system makes a mistake during the classification, it assigns to the example a class neighbour of the correct one. For instance, if the right class for an example is "Large Right", the system could classify erroneously the example as "Medium Right", but it will to that example a class like "Medium Left". This aspect is very important during the guide of the robot: it means the turning angle will never be very different from the correct one.

7.3 Robot Simulator

Finally, we tested the behaviour of the robot by means of a simulator that takes into consideration mechanical characteristics of the robot like velocity, height, inclination and visual field of the camera to make the simulation as reliable as possible. The simulator is fed with a map containing the path that the robot has to follow. It able to see the environment from the robot point of view, and processes the images as the robot would do. As we said in 4.3, the values associated to the classes depend on the position of the camera. For example, in this case the camera was inclined by 15° with respect to the flor and an adequate set of angles, found experimentally, was (0, ±15, ±30, ±50). Figure 10 shows a test path drawn in black and the trajectory of the robot in blue. It is possible to appreciate that the robot, starting from a random point, is able to find the line and walk on it. Due to its mechanic characteristics, the robot cannot steer more than 50º, so, when it finds an angle of 90º, it needs to turn more than one time to be aligned with the line after the bend. It can be possible to appreciate that the robot is almost always on the line, apart from a couple of points in which it turns 15º instead of going straight ahead. The robot cannot steer more than 50º so, when it finds a right angle, it needs turn just before the bend and not in correspondence with it.

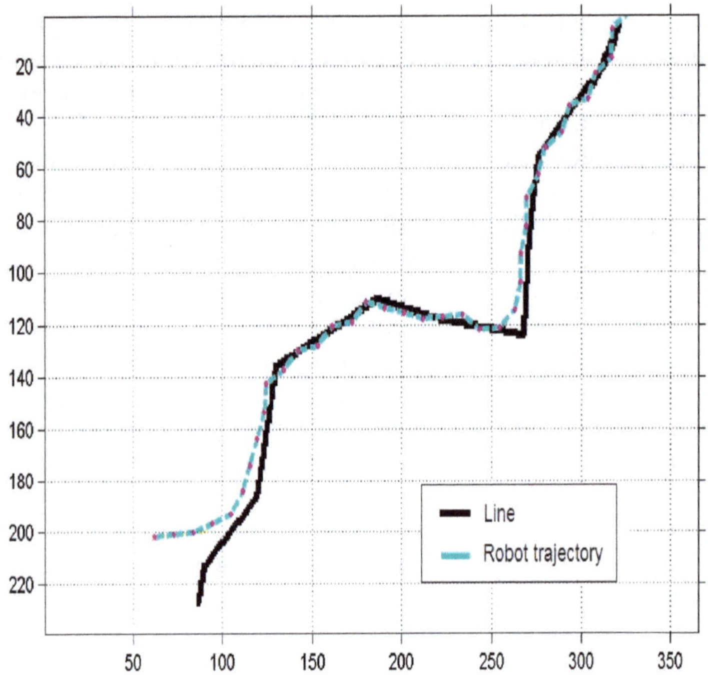

Fig. 10. Simulation of the line following.

Finally, in the Figure 11 is shown another simulation in which all the parameters are set exactly as before, but the values of the angles (that is, the classes) are: 0, ±5, ±10, ±30. It is evident that these values are not adequate for this problem, although the robot is able to reach the line and begin to follow it. The problem comes when the angle gets bigger, because the robot cannot turn more than 60º degrees.

To test the algorithm with obstacles and crossings we can use the same simulator, just changing the CNN processing as showed before.

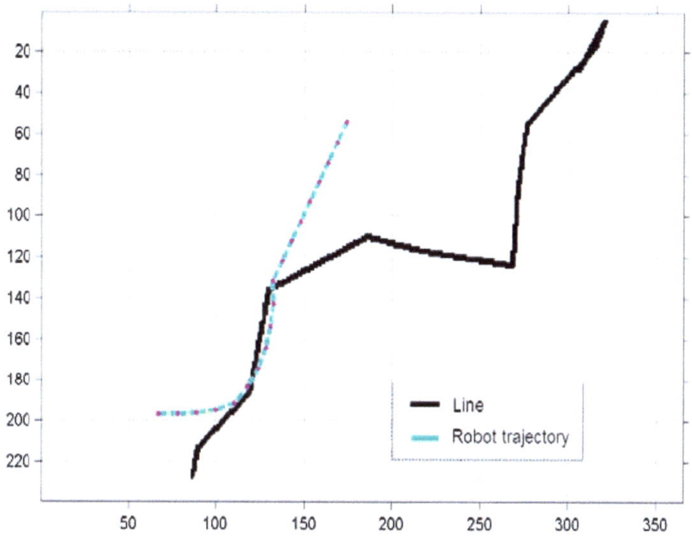

Fig. 11. Line following with wrong values for the classes

8 Conclusions

We have shown how a combination of soft computing techniques provides a powerful solution to develop a purely visual robot guidance system. On one hand Cellular Neural Networks prove to be sufficient to extract the relevant parameters from images in an efficient and simple manner. Though at present our CNN implementation lies on a DSP processor, we are currently involved in a project concerning the emulation of CNNs on an FPGA that should allow a faster processing of the image. However CNNs only show their real capabilities in hardware form.

As far as the navigation system is concerned, we can say that decision trees are a good tool to deal with the problem. Results are encouraging and can improved just reclassifying the images in the database. Building a decision tree is quite fast in comparison with other methods, and the CBR mechanism does not take much time. While in the fuzzy system perspective distortion causes some imperfections of the guide algorithm, using decision trees this do not happens, because both images in the train set and images in test set are affected by the same amount of distortion. Finally, the tree can be stored in the limited memory of an embedded system without any problem.

In conclusion, the architecture proved effective for the task considered. In the near future we will implement the same algorithm on Pasqual8, and our aim is to apply our studies to real life problems.

Acknowledgments

This work is partially supported by Funitec under projects PGR-PR2003-03, PGR-PR2004-03.

References

1. De Souza G.N., Avinash C. Kak A.C., Vision for Mobile Robot Navigation: A Survey, IEEE Trans. on Pattern Analysis and Machine Intelligence, vol. 24, no. 2, (2002) 237–267
2. Weng J., Chen S., Visual Learning with Navigation as an Example, IEEE Intelligent Systems, (Sep/Oct 2000) 63–71
Bertozzi M., Broggi A.,GOLD: A Parallel Real-Time Stereo Vision System for Generic Obstacle and Lane Detection. IEEE Trans. on Image Proc., vol. 7, no. 1, (1998), 62–81
4. Pomerleau D.: RALPH: Rapidly Adapting Lateral Position Handler. IEEE Symposium on Intelligent Vehicles, vol. 7, no. 1, (1995) 506–511
5. Chua L.O., L. Yang, Cellular Neural Networks: Theory. IEEE Trans. Circuits and Systems, vol. 35, no. 10, (1988) 1257–1272
6. Chua L.O., Roska T., The CNN Paradigm. IEEE Trans. Circuits and Systems—I, Fundamental Theory and Appl., vol. 40, no. 3, (1993) 147–156
7. Vilasís-Cardona X., Luengo S., Solsona J., Apicella G., Maraschini A., Balsi M., Guiding a Mobile Robot with Cellular Neural Networks. Int. J. of Circ. Th. Appl., vol. 30, (2002) 611–624
8. Balsi M., Vilasís-Cardona X., Robot Vision Using Cellular Neural Networks. Autonomous Robotic Systems, Physica-Verlag (2002) 431–450
9. Pazienza G.E., Giangrossi P., Tortella S., Balsi M., Vilas´ıs-Cardona X., Tracking for a CNN Guided Robot. ECCTD '05 Proc., Cork, Ireland (2005)
10. Roska T., Chua L.O., The CNN Universal Machine: an Analogic Array Computer. Machine Learning, vol. 1, (1986) 81–106
11. Liñan L., Espejo S., Domínguez-Castro R., Rodríguez-Vázquez A., ACE4K: An Analog I/O 64x64 Visual MicroProcessor Chip with 7-bit Analog Accuracy. Int. J. of Circ. Th. Appl., vol. 30, no. 2-3, (2002) 89–116
12. Kananen A., Paasio A, Laiho M, Halonen K, CNN Applications from the Hardware Point of View: Video Sequence Segmentation. Int. J. of Circ. Th. Appl., vol. 2, (2002) 117–137
13. Harrer H., Nossek J.A., Discrete-time Cellular Neural Networks., Int. J. of Circ. Th. Appl., vol. 20, (1992) 453–467

14. Roska T., Kék L., Nemes L., Zarándy À., Szolgay P., CSL-CNN Software Library, Version 7.3. Hungarian Academy of Sciences, Budapest, Hungary, (1999)
15. Kosko B., Neural Fuzzy Systems, Prentice Hall Englewood Cliffs, NJ, (1992)
16. Quinlan J.R., Induction of decision trees, IEEE Trans. Circuits and Systems—II, Analog and Digital Signal Processing, vol. 40, no. 3, (1993) 163–173
17. Aamodt A., Plaza E., Case-Based Reasoning Foundational Issues, Methodological Variations, and System Approaches, AI Communications, vol. 7(1), (1994) 39–59
18. Riesbeck C.K., Schank R.C., Inside Case-Based Reasoning, Lawrence Erlbaum Associates, Hillsdale, NJ, US, (1989)
19. Kolodner J., Case-Based Reasoning, Morgan Kaufmann Publishers, Inc., (1993)
20. López García J.C.,Moreno-Armendáriz M.A., Riera-Baburés Jordi , Vilasís-Cardona X., Real-Time Vision by FPGA Implemented CNNs, ECCTD '05 Proc., Cork, Ireland, (2005)

Intelligent Agents in Distributed Fault Tolerant Systems

Arnulfo Alanis Garza, Juan José Serrano, Rafael Ors Carot, José Mario García Valdez

[1]Dpto. de Sistemas Computacionales, Instituto Tecnológico de Tijuana (México) Calzada Tecnológico S/N, Unidad Tomas Aquino, {alanis,ocastillo,mario}@tectijuana.mx

[2]D. Inf. de Sistemas y Computadoras, Camí de Vera, s/n, 46022 VALÈNCIA, ESPAÑA, 00+34 96387,Universidad Politécnica de Valencia (España) {jserrano,rors}@disca.upv.es

Abstract. Intelligent Agents have originated a lot of discussion about what they are, and how they are different from general programs. We describe in this paper a new paradigm for intelligent agents. This paradigm helped us deal with failures in an independent and efficient way. We proposed three types of agents to treat the system in a hierarchic way. A new way to visualize fault tolerant systems (FTS) is proposed, in this paper with the incorporation of intelligent agents, which as they grow and specialized create the Multi-Agent System (MAS). The MAS contains a diversified range of agents, which depending on the perspective will be specialized or evolutionary (from our initially proposal) they will be specialized for the detection and possible solution of errors that appear in an FTS). The initial structure of the agent is proposed in [1] and it is called a reflected agent with an internal state and in the Method MeCSMA [2].

1 Introduction

At the moment, the approach using agents for real applications, has worked with movable agents, which work at the level of the client-server architecture. However, in systems where the requirements are higher, as in the field of the architecture of embedded industrial systems, the idea is to innovate in this area by working with the paradigm of intelligent agents. Also, it is a good idea in embedded fault tolerant systems, where it is a new and good strategy for the detection and resolution of errors.

A.A. Garza et al.: *Intelligent Agents in Distributed Fault Tolerant Systems*, StudFuzz **208**, 203–213 (2007)
www.springerlink.com

The main goals of the present research work were the following:

- To create a new visualization tool of the application of intelligent agents, in the fault tolerant systems for embedded systems.
- To create a model, that will help the programmers to create profiles in embedded circuits, according to utility, by means of, Intelligent Agents

The reflected agent with an internal state sets out the general structure of the recovery Intelligent Agent for Fault tolerant Systems in Distributed Systems, whit three types of intention agents.

1.1 Where do Agents Come From?

Agents have their origins in four different research areas: robotics, artificial intelligence, distributed systems, and computer graphics.

Agents working in robotics and artificial intelligence were originally strongly interrelated. Robots such as SHAKEY were programmed to exhibit autonomous behavior in well-defined environments, and laid the groundwork for AI planning systems to this day. The first software agent was probably ELIZA [12], a program which could engage in a conversation with a user. Another influential program, SHRDLU [13], allowed a person to have a conversation with a simulated robot.

The notion of multi-agent systems was brought to the fore-front by Marvin Minsky in his work on the "Society of Mind" [14]. His vision was that a complex system such as the human mind should be understood as a collection of relatively simple agents, each of which was a specialist in a certain narrow domain. Through structures called K-lines, agents would activate each other whenever their context became relevant.

The work of Minsky showed remarkable vision, but was ahead of its time since software complexity had not yet reached the level where the advantages of such structures would have a practical impact.

However, the idea of decomposing a complex system into simple agents found willing takers in robotics. Frustrated with the complexity of robots built around general and thus large homogeneous software systems, Rodney Brooks [18] proposed a radically different design. In his view, intelligent and complex behavior would be emergent in the interplay of many simple behaviors. Each behavior is a simple agent whose activation is decided by a control architecture. Complex general vision systems were replaced by simple detectors specialized in particular situations, and actions were taken based on very simple rules. Brooks showed that using this approach, one could very easily build robust autonomous robots, which had not been possible otherwise [9] [10] [11].

1.2 Agents

Let's first deal with the notion of intelligent agents. These are generally defined as "software entities", which assist their users and act on their behalf. Agents make your life easier, save you time, and simplify the growing complexity of the world, acting like a personal secretary, assistant, or personal advisor, who learns what you like and can anticipate what you want or need. The principle of such intelligence is practically the same of human intelligence. Through a relation of collaboration-interaction with its user, the agent is able to learn from himself, from the external world and even from other agents, and consequently act autonomously from the user, adapt itself to the multiplicity of experiences and change its behavior according to them. The possibilities offered for humans, in a world whose complexity is growing exponentially, are enormous [1][4][5][6].

2 Distributed Artificial Intelligence

Distributed Artificial Intelligence (DAI) systems can be defined as cooperative systems where a set of agents act together to solve a given problem. These agents are often heterogeneous (e.g., in Decision Support System, the interaction takes place between a human and an artificial problem solver).

Its metaphor of intelligence is based upon social behavior (as opposed to the metaphor of individual human behavior in classical AI) and its emphasis is on actions and interactions, complementing knowledge representation and inference methods in classical AI.

This approach is well suited to face and solve large and complex problems, characterized by physically distributed reasoning, knowledge and data managing. In DAI, there is no universal definition of agent, but Ferber's definition is quite appropriate for drawing a clear image of an agent: "An agent is a real or virtual entity, which is emerged in an environment where it can take some actions, which is able to perceive and represent partially this environment, which is able to communicate with the other agents and which possesses an autonomous behaviour that is a consequence of its observations, its knowledge and its interactions with the other agents".

DAI systems are based on different technologies like, e.g., distributed expert systems, planning systems or blackboard systems. What is now new in the DAI community is the need for methodology for helping in the development and the maintenance of DAI systems. Part of the solution relies on the use of more abstract formalisms for representing essential DAI properties (in fact, in the software engineering community, the same problem led to the definition of specification languages) [7][8].

3 FIPA (The Foundation of Intelligence Physical Agents)

FIPA specifications represent a collection of standards, which are intended to promote the interoperation of heterogeneous agents and the services that they can represent.

The life cycle [9] of specifications details what stages a specification can attain while it is part of the FIPA standards process. Each specification is assigned a specification identifier [10] as it enters the FIPA specification life cycle. The specifications themselves can be found in the Repository [11].

The Foundation of Intelligent Physical Agents (FIPA) is now an official IEEE Standards Committee.

4 FIPA ACL Message

A FIPA ACL message contains a set of one or more message elements. Precisely which elements are needed for effective agent communication will vary according to the situation; the only element that is mandatory in all ACL messages is the performative, although it is expected that most ACL messages will also contain sender, receiver and content elements.

If an agent does not recognize or is unable to process one or more of the elements or element values, it can reply with the appropriate not-understood message.

Specific implementations are free to include user-defined message elements other than the FIPA ACL message elements specified in Table 1. The semantics of these user-defined elements is not defined by FIPA, and FIPA compliance does not require any particular interpretation of these elements.

Some elements of the message might be omitted when their value can be deduced by the context of the conversation. However, FIPA does not specify any mechanism to handle such conditions, therefore those implementations that omit some message elements are not guaranteed to interoperate with each other

The full set of FIPA ACL message elements is shown in Table 1 without regard to their specific encodings in an implementation. FIPA-approved encodings and element orderings for ACL messages are given in other specifications. Each ACL message representation specification contains precise syntax descriptions for ACL message encodings based on XML, text strings and several other schemes.

A FIPA ACL message corresponds to the abstract element message payload identified in the [15]

Table 1. FIPA ACL Message Elements

Element	Category of Elements
`performative`	Type of communicative acts
`sender`	Participant in communication
`receiver`	Participant in communication
`reply-to`	Participant in communication
`content`	Content of message
`language`	Description of Content
`encoding`	Description of Content
`ontology`	Description of Content
`protocol`	Control of conversation
`conversation-id`	Control of conversation
`reply-with`	Control of conversation
`in-reply-to`	Control of conversation
`reply-by`	Control of conversation

He following terms are used to define the ontology and the abstract syntax of the FIPA ACL message structure:
Frame. This is the mandatory name of this entity, that must be used to represent each instance of this class.
Ontology. This is the name of the ontology, whose domain of discourse includes their elements described in the table.
Element. This identifies each component within the frame. The type of the element is defined relative to a particular encoding. Encoding specifications for ACL messages are given in their respective specifications.
Description. This is a natural language description of the semantics of each element. Notes are included to clarify typical usage.
Reserved Values. This is a list of FIPA-defined constants associated with each element. This list is typically defined in the specification referenced.

All of the FIPA message elements share the frame and ontology shown in Table 2.

Table 2. FIPA ACL Message Frame and Ontology

Frame	FIPA-ACL-Message
Ontology	FIPA-ACL

5 Proposed Method

Let DS denote a distributed system made up of a set of Nodes N = { Ni }, where each Ni can be formed by several Devices (De) [Di, z]. On the other hand, a DS also contains a set of Tasks to execute, T = { Tj }.

Definition 1: N = {Ni}, where i is the number of nodes of the distributed system.

Definition 2: T = {Tj}, where j is the number of tasks that are executed in the system.

Definition 3: De = [Di, z], where z is the number of devices that will be monitored by Ni from these definitions, it can be made the following one:

Definition 4: Let a distributed system DS be pair <N, T>

This is where we equiped this DS with certain characteristics of failure tolerance.

This is where the use of the DAI paradigm, applied to the Fault Tolerant System (FTS) as a DS can represent a new approach with the implementation of Intelligent Agents.

IAFT = {ANi,AT j,AS} will now define the Fault tolerant Agents, that work a DS.

The Node Agent (ANi) € Ni, whose mission is related to the tolerance to failures at node level (What works and what not within the node).

The Task Agent (ATj) € ATj, whose mission is related to the tolerance to failures at task level (like recovering the tasks of the possible errors that can suffer)

System Agent (AS) € DS, whose mission is the related to the tolerance to failures at the system level (what tasks must be executed in the system and on what nodes)

With it a fault tolerant DS is defined as:

Definition 5: A Distributed Fault Tolerant System DFTS is the pair <DS, IAFT>, DSTF is defined as {DS, IAFT}

6 Control of Conversation

In this section we describe the control of conversation between agents. In table 3 we show the protocol. In this table 4 we sow the conversation identifier of the node agent. In table 5 we show the reply of an agent.

Table 3. Protocol

Element	Description	Reserved Values
Protocol TCP/IP	Denotes the interaction protocol that the sending agent is employing with this ACL message	See [16]

Table 4. Conversation Identifier of Node Agent (ANi)

Element	Description	Reserved Values
• (AN*i*).Phase.Detection y (AN*i*).{Input-Error (*i*,*j*).Error} • (AN*i*).Phase.Location y (AN*i*).Input-Error(*i*,*j*).Error • (AN*i*).Phase.Isolation y (AN*i*).Device[D*i*,*m*].Incorrect • (AN*i*).Phase.Recunfiguration • (AN*i*).Phase.Recunfiguration y **AN*i*T*j*. Recovered**	Introduces an expression (a conversation identifier) which is used to identify the ongoing sequence of communicative acts that together form a conversation.	

Table 5. Reply With

Element	Description	Reserved Values
• (AN*i*).State.Suspect • (AN*i*).{Test[D*i* *k*]} • (AN*i*).{Device[D*i*,*m*]. Incorrect} • (AN*i*).{Test [D*i*,*l*]} **(AN*i*S). low** y (AN*i*).State.low • (AN*i*).Actions-Isolation-Device(*m*) • **AN*i*T*j*.A-to Recover** y (AN*i*).Phase. recovery • (AN*i*).Phase.Detection y (AN*i*).State.Correcto.	Introduces an expression that will be used by the responding agent to identify this message.	

7 Considerations

The agent counts on a AID, which is "intelligent Agents as a new paradigm of Distributed Fault tolerant Systems for industrial control" to as Architecture of Reference fipa/Data minimum of an agent is specified in the norms of Fipa (, says: Aid- the agent must have a unique name globally).

The agent contains descriptions of transport in the development of his documentation, which fulfills the specifications of fipa (Architecture of Reference fipa/Data minimum of an agent, says: Localizer one or but descriptions of the transport that as well, contains the type of transport by ej. Protocol), but does not specify the protocol that uses like type of transport, this in phase of analysis.

It concerns the communication and cooperation between agents, the document "intelligent Agents as New Paradigm of Distributed Fault tolerant Systems for Industrial Control" says to us that the communication between the agents occurs of ascending or descendent form depending on the type of agent. A little superficial explanation occurs, without specifying for example that type of language of communication between agents uses, or KQML or the Fipa-acl.

8 Conclusions

We described in this paper our approach for building multi-agents system for achieving fault tolerant control system in industry. The use of the paradigm of intelligent agents has enabled the profile generation of each of the possible failures in an embedded industrial system. In our approach, each of the intelligent agents is able to deal with a failure and stabilize the system in an independent way, and that the system has a behavior that is transparent for the use application as well as for the user.

Reference

1. Stuart Russell and Peter Norvig, Artificial Intelligence to Modern Aproach, Pretence artificial Hall series in intelligence, Chapter Intelligent Agent, pages. 31-52.
2. A.Alanis, Of Architectures for Systems Multi-Agentes, (Master Degree thesis in computer sciences), Tijuana Institute of Technology, November, 1996.
3. Michael J. woodridge, Nicholas R. Jennings. (Eds.), Intelligence Agents, Artificial Lecture Notes in 890 Subseries of Lectures Notes in Computer Science, Amsterdam, Ecai-94 Workshop on Agent Theories, Architectures, and languages, The Netherland, Agust 1994 Proceedings, ed. Springer-Verlag, págs. 2-21.
4. P.R. Cohen ET al.?An Open Agent Architecture, working Notes of the AAAI Spring symp.: Software Agent, AAAI Press, Cambridge, Mass., 1994 págs. 1-8.
5. Bratko I. Prolog for Programming Artificial Intelligence, Reding, Ma. Addison-Wesley, 1986.

6. Or Etzioni, N. Lesh, and R. Segal?Bulding for Softbots UNIX? (preliminary report). Tech. Report 93-09-01. Univ. of Washington, Seattle, 1993.
7. Elaine Rich, Kevin Knight, Artificial intelligence, SecondEdition, Ed. Mc Graw-Hill, págs. 476-478.
8. N. Jennings, M. Wooldridge: Intelligent agents: Theory and practice. The Knowledge Engineering Review 10, 2 (1995), 115– [10] Durfee et al. 89.
9. E. H. Durfee, V. R. Lesser, D. D. Corkill: Trends in cooperative distributed problem solving. IEEE Transactions on Knowledge and Data Engineering KDE-1, 1(March 1989), 63–83.
10. http://www.fipa.org/specifications/lifecycle.html
11. http://www.fipa.org/specifications/identifiers.html
12. http://www.fipa.org/specifications/index.html
13. M. Yokoo, T. Ishida, K. Kuwabara: Distributed constraint satis-faction for DAI problems. In Proceedings of the 1990 Distributed AI Workshop (Bandara, TX, Oct. 1990).
14. J. Weizenbaum: ELIZA – a computer program for the study of natural language communication between man and machine. Communications of the Association for Computing Machinery 9, 1(Jan. 1965), 36–45.
15. T. Winograd: A procedural model of language understanding. In Computer Models of Thought and Language, R.Schank and K. Colby, Eds. W.H.Freeman, New York, 1973, pp. 152–186.
16. FIPA Abstract Architecture Specification. Foundation for Intelligent Physical Agents, 2000. http://www.fipa.org/specs/fipa00001/
17. FIPA Interaction Protocol Library Specification. Foundation for Intelligent Physical Agents, 2000. http://www.fipa.org/specs/fipa00025/

Genetic Path Planning with Fuzzy Logic Adaptation for Rovers Traversing Rough Terrain

Mahmoud Tarokh

Department of Computer Science, San Diego State University, San Diego, CA 92182-7720, U.S.A., tarokh@sdsu.edu

Abstract. The paper develops a genetic algorithm approach to path planning for a mobile robot operating in rough environments. Path planning consists of a description of the environment using a fuzzy logic framework, and a two-stage planner. A global planner determines the path that optimizes a combination of terrain roughness and path curvature. A local planner uses sensory information, and in case of detection of previously unknown and unaccounted for obstacles, performs an on-line planning to get around the newly discovered obstacle. The adaptation of the genetic operators is achieved by adjusting the probabilities of the genetic operators based on a diversity measure of the population and traversability measure of the path. Path planning for an articulate rover in a rugged Mars terrain is presented to demonstrate the effectiveness of the proposed path planner.

1 Introduction

Path planning for a mobile robot is defined as determining a route from the start to the goal for successfully navigating the robot around obstacles in some optimal manner [1]. The optimality is usually taken as finding a short path, but research work has also considered the shortest motion time [2]. It is well known that path panning in its general form is an NP-complete problem, and thus the problem is usually solved using heuristic approaches; most notably genetic and evolutionary algorithms.

There are two groups of path planners - local and global. The local planners (e.g. [3]-[5]) consider a subset of the environment in the vicinity of the robot to plan the path. Therefore they are often fast but lack the necessary broad perspective and can often get stuck in traps. Global planners (e.g. [6]-[8]) conceive their paths after a complete survey of the whole environment and are therefore much slower, but are generally much more

M. Tarokh: *Genetic Path Planning with Fuzzy Logic Adaptation for Rovers Traversing Rough Terrain*, StudFuzz **208**, 215–228 (2007)
www.springerlink.com

trap resistant than the local planners. However, detailed information about the entire environment is often not available, and this information become known only when the robot moves and surveys its surroundings.

Most approaches to mobile path planning concentrate on a binary representation of the terrain where a given region of the terrain is considered either free or occupied by obstacles. An evolutionary algorithm has been shown to be effective for this environment [6]. However, the binary setting not appropriate for path planning of rovers that can climb over some obstacles or rocks if such a traversal results in a more optimal path.

In previous papers [9]-[10], we developed the concept of path impedance that will take into account such parameters as the height and size of obstacles, terrain slope, and concentration of obstacles. Most of this information can be obtained using imaging techniques. However, extracting this information from the images of scene is associated with considerable uncertainty and vagueness. In addition, relating path impedance to the terrain characteristics, such as height and size of obstacles, is best done by a rule base approach rather than analytical expressions. These considerations naturally lead to the fuzzy logic framework, which is explored in this paper.

An important requirement for a path planner is to cope with partially known environments, where the rover may encounter previously unknown and unaccounted for obstacles. This requires the rover to deviate from the planned path and perform collision avoidance as the obstacles are detected. One of the contributions of this paper is to enhance a genetic path planner with such a capability. A further contribution of the paper is to devise a scheme for adaptation of the genetic operators as the environment is learned. The adaptation consists of adjusting the probabilities of the genetic operators based on diversity and traversability of paths population in the genetic process.

2 Terrain and Path Representation

The terrain is represented by a regular grid consisting of square cells. The size of a cell depends on the dimensions of the rover and the desired resolution of the terrain description. The terrain roughness within a cell depends on a number of parameters such as the height of the tallest obstacle in the cell, and the size or surface area of the cell occupied by obstacles. Stereo vision and region growing techniques can be used to determine obstacle height and surface area using the image. Despite the availability of vision processing software, exact determination of the heights and sizes of rocks affecting roughness is not possible. These parameters can be found approximately due to errors, misinterpretations and ambiguity involved in extracting information from images. We employed fuzzy logic and approximate reasoning framework, which involves defining five fuzzy sets for the height, i.e. VL (very low), LO (low), MD (medium), HI (high) and VH (very high). Similarly, five fuzzy sets are defined for the obstacle size, i.e. VL (very large), LG (large), MD (medium), SM (small) and TI (tiny). Height and size of obstacles in a cell are fuzzified to obtain fuzzy variables $\widetilde{h}_i$ and $\widetilde{s}_i$, respectively, where $i = 1,2,...,n$ and n is the number of cells. A rule matrix is designed to relate these quantities to fuzzy roughness $\widetilde{\rho}_i$ of the cell using rules

$$\begin{aligned} &\textit{if } \widetilde{h}_i \textit{ Height is HeightSet and } \widetilde{s}_i \textit{ Size is SizeSet} \\ &\quad \textit{then } \widetilde{\rho}_i \textit{ is RoughnessSet} \end{aligned} \tag{1}$$

where *HeightSet* and *SizeSet* can take on one of the defined fuzzy sets defined above, and *RoughnessSet* is associated with one of the five fuzzy sets, VH, HI, MD, LO, VL. There will be $5 \times 5 = 25$ rules of the form (1). The fuzzy variable $\widetilde{\rho}_i$ can be defuzzified to obtain the crisp roughness ρ_i for each cell, which are used in the genetic planner [10]. In this paper, however, we do not defuzzify the roughness, but use its fuzzy value in conjunction with the path curvature, to be defined shortly, in order to obtain a fitness function for the path.

A path is represented by a sequence of way-points connecting the start to the goal. In a computer, the path is stored in a linked list data structure consisting of a number of nodes each of which stores the information about a way-point, and is linked to the next node (way-point) in the list. The way-points $W_k, k = 1,2,...,m$, where m is the number of waypoints, are specified by their (x_k, y_k) coordinates on the terrain. The generation and evolution of a path refers to the creation and modification of the way-points. These way-points in turn specify the terrain cells that the path traverses over. A cell that is located on a path, will be referred to as a *path cell*, and has two main attributes as follows:

(a) The roughness of the cell, which provides information on the heights, sizes and concentration of obstacles on a cell as described above, and

(b) The curvature or jaggedness of a path cell that is obtained using the information about the way-points. Specifically, the curvature ζ_k of the way-point W_k is defined

$$\zeta_k = \frac{d_k}{D_k}, \quad k = 1,2,\ldots,m \tag{2}$$

where d_k is the perpendicular distance from W_k to the line segment joining the previous way-point W_{k-1} to the next way-point W_{k+1} and D_k is the distance between W_{k-1} and W_{k+1}. Note that ζ_k is a dimensionless quantity, and that (2) also gives the curvature of the path cell that contains a way-point. The curvature is fuzzified using three fuzzy sets, HI (high), MD (medium) and LO (low) to obtain the fuzzy curvature $\tilde{\zeta}_k$.

The cell impedance η_i combines the roughness and curvature and is a measure of the difficulty of the traversal. The fuzzy cell impedance $\tilde{\eta}_i$ is obtained from

$$\begin{aligned} &\textit{if } \tilde{\rho}_i \textit{ is RoughnessSet and } \tilde{\zeta}_k \textit{ is CurvatureSet} \\ &\qquad \textit{then } \tilde{\eta}_i \textit{ is ImpedanceSet} \end{aligned} \tag{3}$$

where *RoughnessSet* and *CurvatureSet* can take on one of the fuzzy sets defined above and *ImpedanceSet* can assume VH, HI, MD, LO or VL. It is noted that other path attributes such as slope can easily be included in the above formulation of the path impedance. There are $5 \times 3 = 15$ fuzzy rules of the form (3). A cell whose impedance belongs to VL (very high) becomes intraversable. The scope (base width) of VL is determined based on the mobility characteristics of the particular rover being used. We identify a path as being traversable if every cell on the path is traversable, otherwise the whole path becomes intraversable. In the genetic evolutionary process, these two types of paths are treated separately. Although, traversable paths have priority over intraversable paths, the latter are not automatically discarded since they may prove to produce good off-springs later on during the evolutionary process.

The *path impedance* is obtained by the union of the clipped fuzzy sets resulting from the rules (3). The defuzzified value of the union set gives the crisp value of the path impedance, which will be used as the fitness value of the path in the evolutionary path planning.

3 Genetic Path Planning

Path planning consists of two phases - global and local. Prior to the motion, the rover takes images of the environment using cameras (e.g. see [11]). The images provide a panoramic view of the environment based on which the terrain representation discussed above is obtained. The acquired terrain will be used for global planning. As the rover moves, previously un-represented obstacles may be detected by the rover's proximity sensors, e.g. sonar. Using this information, the rover must perform a local planning to avoid the newly detected obstacles.

Global Planning

The path planner starts by creating several random paths between start and goal locations on the terrain. Each path consists of a random number of way-points between start and goal. These initial paths in general go though rough or impassable regions on the terrain, and must be improved. This improvement is achieved by applying certain genetic operators to a randomly selected path from the population. Each genetic operator has a particular role in bringing about a change in the path. For example, the *replace* operator replaces an undesirable way-point (a way-point on a rough region) with a random and potentially better way-point. The selection of particular operator is based on the probability assigned to it.

After a genetic operation is performed and new paths are generated, a fitness proportion selection is employed to select the path for the next generation. The population goes through generations and is thus evolved. After each generation, the quality of paths is either improved or in the worst case remains unchanged. The evolution is continued until an acceptable path is found, or until a preset number of generations is performed.

In order to evolve paths from one generation to the next, several operators have been devised. Two of these operators, namely crossover and mutation, are commonly used in genetic algorithms. Others are specifically designed for the path planner. Operators are applied to way-points, and as a result of changes in way-points, the path cells are also changed.

The *cross-over* operator randomly selects two paths from the population, say P_1 and P_2, and divides each path into two path segments about a randomly selected way-point. Denoting these paths by $P_1 = (P_{11}, P_{12})$ and $P_2 = (P_{21}, P_{22})$ where P_{ij} is the j-th segment of the path i, then two new paths are formed as $P_1 = (P_{11}, P_{22})$ and $P_2 = (P_{21}, P_{12})$. The *mutate* operator randomly selects a path and a way-point in this path. It then changes the (x, y) coordinates of the selected way-point with random values within the terrain. Mutate operator can produce a significant change in the path.

The *replace* operator is applied to an intraversable path. It replaces an intraversable way-point with one or more way-points whose location and number are random. If there are more than one intraversable way-points, one of them is selected randomly for replacement.

The *swap* operator interchanges the locations of two randomly selected way-points on a randomly selected path. The swap operator can be applied to both traversable and intraversable paths. It has the possibility of either removing or introducing a "zig-zag" which could avoid an obstacle.

The role of *smooth* operator is to reduce sharp turns. The way-point with the highest curvature, say W_k, is selected and two new way-points are inserted, one on a randomly selected cell between the previous way-points W_{k-1} and W_k and the other on a cell between W_k and the next way-point W_{k+1}. After this insertion, the way-point W_k is removed. The effect of this operation is the smoothing of a sharp turn. This operator is applied to traversable paths only.

The *pull-out* operator is intended to pull out a path segment from inside an intraversable region to its surrounding traversable region. Pull-out is more elaborate than the other operators, and details of its implementation is omitted here for the sake of brevity.

3.2 Local On-line Planning

As the rover moves along the path found by the global planner, its proximity sensors may detect obstacles that were not previously accounted for in the global planning. In such a case, three heuristics strategies of maneuvering around, ancestral knowledge and partial replanning are attempted in that order. These are described in the following paragraphs.

The *maneuvering around* strategy aims at repairing the path segment that crosses an obstacle by randomly inserting new way-points in a region around the obstacle in an attempt to circumvent the detected obstacle. This region is constructed using one of the two methods shown in Fig 1, where the original path segment is ABCD. In the first method (top diagram in Fig. 1), the region is the equilateral triangle BEF whose side BE is perpendicular to the obstacle intersecting segment BC, and whose side length is equal to BC, i.e. BE=EF=FE=BC. An identical region, not shown in the figure, is constructed on the other side of the path ABCD. In each region a number of random waypoints (three in our experiments) are inserted (one waypoint is shown in Fig. 1 as G). For each of the inserted waypoints, the corresponding modified path segment (ABGCD in Fig. 1) is checked for collision, and if any is collision free the new waypoint is selected.

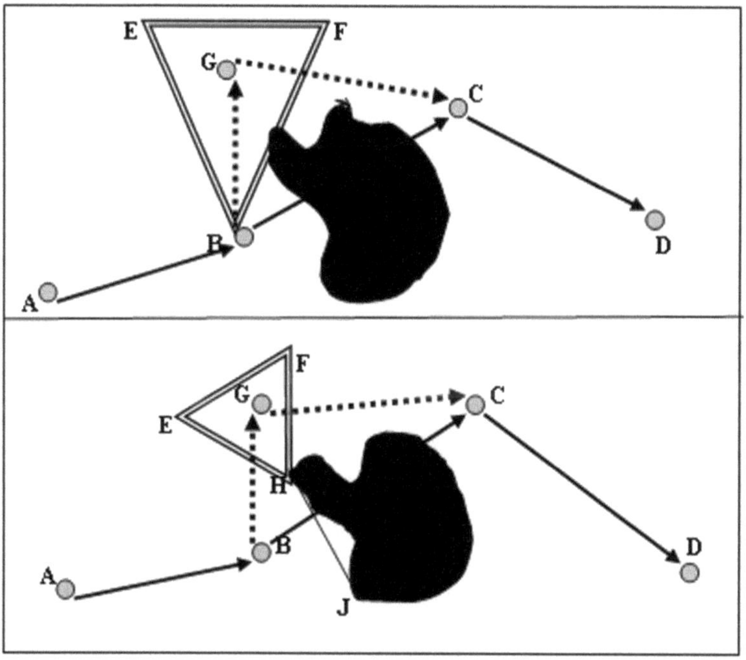

Fig. 1. Strategies of maneuvering around a previously undetected obstacle.

In case none of the inserted waypoints produces a collision free modified path segment, the second method, shown in bottom of Fig.1, is attempted. In this method the line JH is drawn between two extreme points of the obstacle as seen by the rover. An equilateral triangular region is now formed with the vertex at the extreme point H such that JH bisects the angle EHF. The side length of the triangle is chosen to be equal to JH. Similarly another triangular region is formed with its vertex placed at J. New waypoints are inserted into randomly in these regions, and for each the modified path segment (ABGCD) is tested to find a collision free path segment around the obstacle.

In any of the above two maneuvering method, once the obstacle is circumvented, the remaining path to the goal as determined by the global planner is followed. In difficult environments, the above maneuvering around methods may not produce collision free path segments. In such cases we apply a strategy that we will call *ancestral knowledge*. During the genetic evolution of the path in the global planning stage, a link is retained to the previous state of a path, i.e. its ancestral chain. Upon the discovery of a previously unaccounted obstacle, the ancestral chain is traversed until a path that is both traversable and does not intersect the new obstacle is found. This path is then followed from the nearest waypoint to the current rover position and finally to the goal.

In extremely complex environment with substantial previously undetected obstacles, the above strategies of maneuvering around and ancestral knowledge may not yield a collision free path. In such a case a *replanning* from the current rover position to the goal is performed using the global genetic planning by considering the newly detected obstacle information.

3.3 Adaptation of Genetic Operators

A population of paths has two fundamental characteristics, namely diversity and traversability, which can be used to gage the effectiveness of a particular genetic operator in evolving better (fitter) paths. Diversity is the degree of variability of the path and is defined as the variance of the fitness of the path in the population. Suppose that there are p paths in the population, and denote their fitness by $F_j, j = 1,2,...,p$. The diversity of the population is then defined as

$$\delta = \frac{1}{n}\sqrt{\sum_{j=1}^{n}(F_j^2 - F_{ave}^2)} \tag{4}$$

where F_{ave} is the average fitness value of the population. The traversability is defined as

$$\tau = \frac{N_T}{N_i + 1} \tag{5}$$

where N_T and N_I are the number of traversable and intraversable paths in the population. These two population characteristics, i.e. diversity and traversability, are used to adjust the genetic operator's probability. For example, a population with low diversity and low traversability (high impedance) indicates that the mutation operator must be given higher probability, and cross over to be assigned a lower probability. This is due to the fact that in this situation, the crossover of intraversable paths produces other intraversable paths and a substantial change is needed which is achieved by mutation. On the other hand, a population with both high traversability and diversity has ideal characteristics and paths need only to be smoothed using the smooth operator.

Based on the above argument, we propose the following rule matrix for the boosting (increasing) an operators probability, where diversity and traversability are fuzzified each with three fuzzy sets LO (low), MD (medium) and HI (high). Note that when an operator probability increased, other operators' probabilities are reduced proportionally to keep the total probability equal to one.

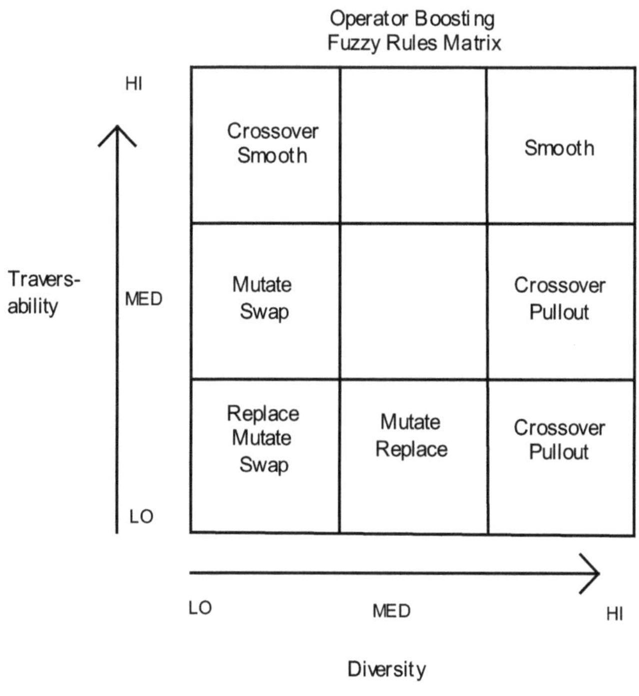

Fig. 2. The rule matrix for adjusting operators' probabilities

The upper left box in the rule matrix, for example, describes the following fuzzy rule

if diversity is LO and traversability is HI
then boost Replace, Mutate, Swap

The magnitude of the boost in the operator's probability is proportion to the truth values of the rule antecedent (i,e. "*if diversity is LO and traversability is HI*", in the above example). The truth value is obtained using standard fuzzy logic computation.

3 Results

In this section, we present the results of applying the method discussed above to an image of the Mars terrain to plan two paths. The software developed for image processing and obtaining the height and sizes of the rocks is described in [10]. Figure 3(a) shows the image of a 10 by 10 meters terrain obtained from the JPL Mars yard, and its contour obtained by applying certain image processing techniques [10]. The darker areas in the contour map show higher elevations. The 512 by 512 pixel contour image was divided into 32 by 32 cells, and the roughness of each cell was determined by the method described in Section 2. The number of cells can be increased for a higher resolution, if required.

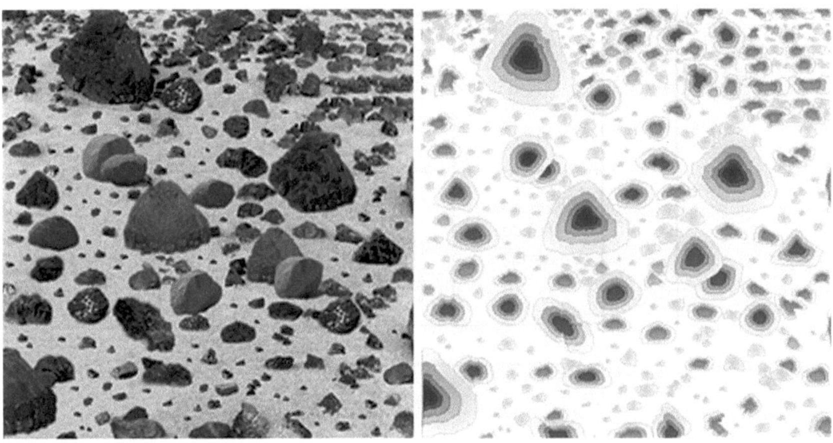

Fig. 3. Image of Mars terrain and its contour map

The genetic algorithm described in Section 3 was then applied. A population size of five paths was chosen, and these paths went through the genetic evolution by applying one of the genetic operators. The fitness of each path was evaluated using the procedure described in Section 3. Finally the genetic operators were adapted during the evolution as discussed in Section 4. The initial intraversable paths were quickly evolved into traversable paths, and as the evolution continued these paths in turn changed into shorter ones passing through less rock concentrated areas and avoiding larger rocks. Near optimal paths were usually found after 200 to 400 iterations (generations), thus good paths were found very quickly. Figure 4 shows the results of changes in the diversity and traversability as the evolution progresses. It is seen that the population initially has a high diversity and very low traversability due to random paths that were generated. However, as the evolution progresses, both these quantities converge to satisfactory values.

Figure 5 shows typical adaptations of the genetic operator probabilities as functions of generations. It is seen that depending on the traversability and diversity, the operator probabilities change with generations according to the table in Figure 2. Generally, mutation has a higher probability at the beginning but its probability is reduced towards the end of evolution where traversable paths are found and smoothing gets a higher chance of being applied.

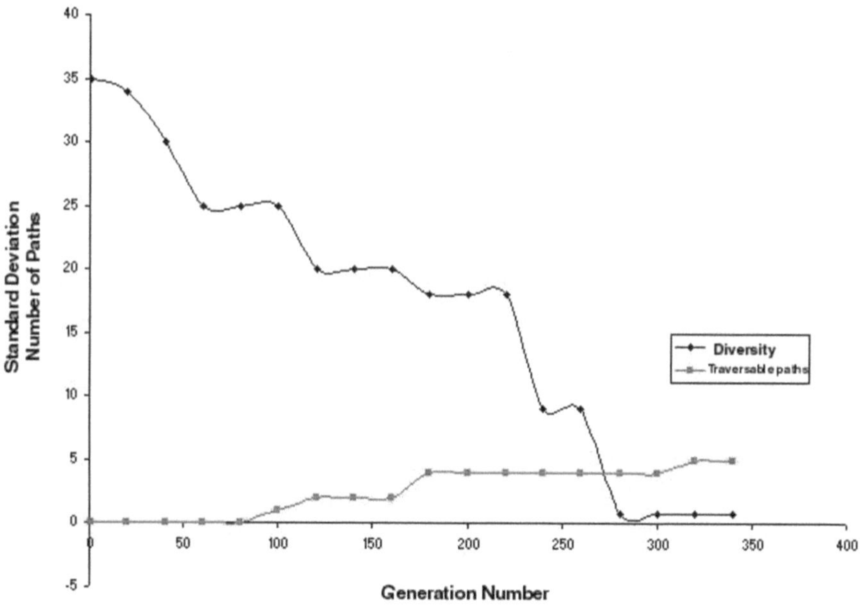

Fig. 4. Changes in diversity and traversability with generations.

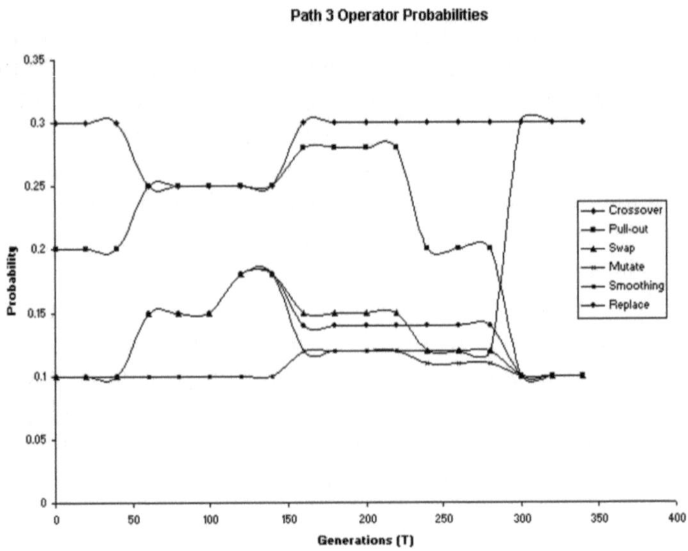

Fig. 5. Adaptation of the genetic operators probabilities versus generation number.

Figure 6 shows two typical path planned by the genetic algorithm. In the first case, the start and goal points are at left and right of the terrain, respectively. In the second case, the start and goal points are located in the upper left corner and lower center of the terrain. Note that the genetic planner produces the waypoints, and in Fig. 6 these waypoints are connected by straight line segments. To obtain smoother paths, these waypoints can be connected by cubic polynomials or another suitable interpolation method. It is noted from Fig. 6 that the path sometimes traverses over small rocks to achieve shorter path lengths. A closer examination shows that all paths are in fact traversable by the rover (in this case NASA's Rocky 7 rover [11]).

Finally, experiments consisting of 100 path planning trials were conducted in which obstacles were placed randomly on the rover path to simulate previously undetected obstacles in the global planning stage. The local planner then used strategies discussed in Section 3.2 to maneuver around the newly detected obstacle. In more than 70% of cases, the first maneuvering around strategy (Fig. 1) was able to suitably modify the path. In the remaining cases, the two other maneuvering around strategies, ancestral knowledge method and re-planning were used and successfully found traversable paths.

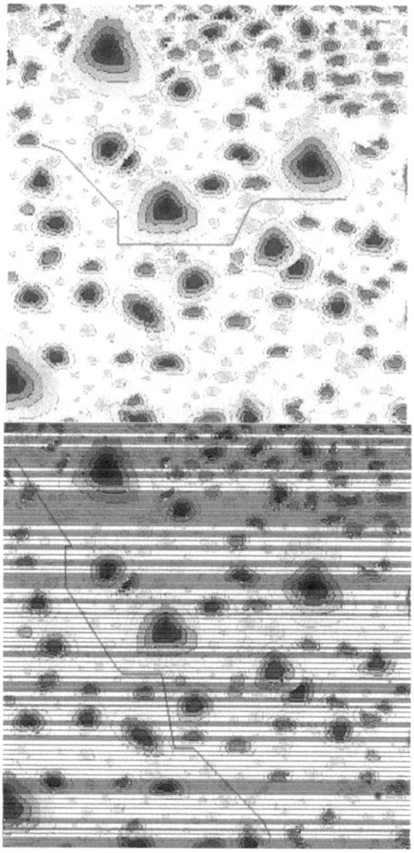

Fig. 6 Two typical planned paths.

5 Conclusions

The main contributions of this paper are the development of a genetic path planner, a two-phase planning method, and an adaptation procedure for adjusting the probabilities of the genetic operators. The planner uses a fuzzy logic description of the terrain topology to come up with a roughness measure for the terrain. This description captures and copes with the uncertainties in sensory terrain data acquisition. The global path planner finds an optimal path that avoids rough areas, and quickly converges to a solution due to the adaptation of the genetic operators. The local planner resolves the situations when on-board sensors discover new obstacles, and performs on-line strategies to circumvent the newly detected obstacles.

Several strategies are described for on-line collision avoidance to deal with a variety of difficult situations. Extensive simulation tests have been carried out for path planning in rocky Mars environments, and the results demonstrate the effectiveness of the proposed method.

References

1. Latomb, J.-C. : Robot Motion Planning. Kluwer Academic Publishers (1991).
2. Shiller, Z., Gwo, Y-R.: Dynamic motion planning of autonomous vehicles. IEEE Trans. Automation and Robotics, Vol. 2, (1991) 241-249.
3. Vadakkepat, P., Chen, T.K. : Evolutionary artificial potential fields and their application in real time robot path planning, proc. Congress on Evolutionary Computation, San Diego, CA. (2000), pp. 256-264.
4. Gallardo, D., Colomnia, O. : A genetic algorithm for robust motion planning, 11th Int. Conf. on Industrial and Engineering Applications of Artificial Intelligence and Expert Systems, Castellon, Spain, (1998) 1150121.
5. Cazangi, R. R., Figuieredo, M. : Simultaneous emergence of conflicting basic behaviors and their coordination in an evolutionary autonomous navigation systems. Proc. IEEE Conf. on Evolutionary Computation, (2002).
6. Xiao, J., Michalewicz, Z., Zhang, L., Trojanowski, K. : Adaptive evolutionary planner/navigator for mobile robots. IEEE Trans. Evolutionary Computation, Vol. 1, No. 1 (1997) 18-28.
7. Hocaoglu, C. Sanderson, A.C. : Planning multiple paths with evolutionary speciation. IEEE Trans. Evolutionary Computation, vol. 5, No. 3, (2001) 169-191.
8. Sugihara, K. Smith, J.: Genetic algorithms for adaotive motion planning of an autonomous mobile robot. Proc. IEEE Int. Conf. on Computational Intelligence in Robotics and Automation, Monterey, CA, (1997), 138-146.
9. Tarokh, M., Shiller, Z., Hayati, S. : A comparison of two traversability based path planners for planetary rovers. Proc. 5th Int. Symposium on Artificial Intelligence, Robotics and Automation in Space (1999) 151-165.
10. Tarokh, M., Chan, R.W., Song, C. : Path planning of rovers using fuzzy logic and genetic algorithm. Proc. World Automation Conf., ISORA-026, Hawaii (2000) 1-7.
11. Hayati, S., et al. : The Rocky 7 rover: A Mars science craft prototype. Proc. IEEE Int. Conf. Robotics and Automation (1997) 2458-2464.

Chattering Attenuation Using Linear-in-the-Parameter Neural Nets in Variable Structure Control of Robot Manipulators with Friction

Ricardo Guerra, Luis T. Aguilar, and Leonardo Acho

Instituto Politécnico Nacional, Centro de Investigación y Desarrollo de Tecnología Digital, 2498 Roll Dr. #757 Otay Mesa, San Diego CA 92154 USA, rguerra{laguilar, leonardo}@citedi.mx

Abstract. Variable structure control is a recognized method to stabilize mechanical systems with friction. Friction produces non linear phenomena, such as tracking errors, limit cycles, and undesired stick-slip motion, degrading the performance of the closed-loop system. The main drawback of variable structure control is the presence of chattering, which is not suitable in mechanical systems. In this paper, we design a variable structure controller complemented with Linear-in-the-Parameter neural nets to attenuate chattering. Experimental validation applied to a three degree of freedom robot mechanical manipulator is shown to support the results.

1 Introduction

Friction is the resistance to motion, during sliding or rolling, that is experienced whenever one solid body moves tangentially over another with which it is in contact. Friction is undesirable in mechanical systems because can lead to tracking errors, limit cycles, and undesired stick-slip motion (cf. [1]). Control strategies for friction compensation have been proposed in [1]-[7], among others. In these papers the authors propose friction model based controllers to mitigate the friction effects. It is well-known that the phenomenon of friction is not yet completely understood and it is hard to model [8], therefore stabilization of mechanical systems through a feedback law with an imprecise friction compensation term may result in a considerable degree of uncertainty, thus not producing the expected motion. If the uncertainties are bounded, discontinuous robust control methods ([8]) provide simple and straightforward solutions to the friction compensation design, however, the system exhibits an infinitely fast switching of the input control called chattering ([9]) inducing fatigue in mechanical

R. Guerra et al.: *Chattering Attenuation Using Linear-in-the-Parameter Neural Nets in Variable Structure Control of Robot Manipulators with Friction* , StudFuzz **208**, 229–241 (2007)
www.springerlink.com

parts and the system could be damaged in a short time. For instance, in [8] and [10], friction compensator design involves chattering behavior where the chattering controller deals with the friction model uncertainties, which is a desired property in friction compensation.

This paper is intended to provide a solution to mechanical problems due to chattering without losing robustness properties given by variable structure controllers [11]. Neural nets have been used extensively in feedback control (see, for instance, [12]-[15]). Also, adaptive control theory has evolved as a powerful methodology for designing nonlinear feedback controllers for systems with uncertainties [16]. Using the advantage of chattering control to deal with uncertainty in the friction model ([8] and [10]), and utilizing a linear-in-the-parameter (LIP) neural net, a chattering friction compensation design is proposed, where a dynamic adaptation law for the parameters of the LIP neural nets is designed to attenuate the amplitude of the chattering once the control objective is achieved. In this way, chattering appears only when it is needed.

To the best knowledge of the authors, the chattering attenuation problem for the class of Variable Structure Control (VSC) introduced in this paper has not been reported. On the other hand, few results have appeared in research papers dealing with the chattering problem for sliding mode control: Parra-Vega *et al.* [17], for example, showed that adaptive and non-adaptive cases of variable structure robot control undergo chattering attenuation. Bartolini *et al.* [18] demonstrated that it is possible to eliminate chattering by generating a second-order sliding mode control using the first derivative of the control law as a control input instead of the actual control law. Another alternative used in control applications is to replace the signum function with a smooth approximation (*e.g.* tanh, sigmoid function, among others).

This paper is organized as follows: Section 2 presents the problem statement along with the dynamic model of mechanical manipulators and the previous result on chattering control developed by Orlov *et al.* [8]; Section 3 presents the neural nets chattering controller applied to a n-degrees-of-freedom robot manipulator where it is assumed that joint positions are the only information available for feedback, along with its stability analysis; Section 4 provides experimental results made for a three degrees of freedom mechanical manipulator using the neural nets chattering controller described in Section 3; and Section 5 presents some conclusions.

The following notations will be adopted throughout this paper. $\lambda_{\min}(A)$ and $\lambda_{\max}(A)$ denote the minimum and maximum eigenvalues of a symmetric positive definite matrix $A \in R^{n\times n}$, respectively, and $\|x\| = \sqrt{x^T x}$ represents the Euclidean norm of vector $x \in R^n$.

2 Problem Statement

In the present paper we study controlled n-link mechanical manipulators described by interconnected second-order differential equations of the form [8]:

$$M(q)\ddot{q} + C(q,\dot{q})\dot{q} + G(q) + F(\dot{q}) = \tau \tag{1}$$

where $q \in R^n$ is the position vector, $\tau \in R^n$ is the control input, $M(q)$, $C(q,\dot{q})$, $G(q)$ are smooth functions of appropriate dimensions, $M(q) = M^T(q) > 0$,

$$F(\dot{q}) = K_b\dot{q} + K_f \operatorname{sgn}(\dot{q}) \tag{2}$$

$$\operatorname{sgn}(\dot{q}) = [\operatorname{sgn}(\dot{q}_1), \operatorname{sgn}(\dot{q}_2), \ldots, \operatorname{sgn}(\dot{q}_n)]^T \tag{3}$$

$$\operatorname{sgn}(z) = \begin{cases} 1 & if \quad z > 0 \\ [-1,1] & if \quad z = 0 \quad \forall z \in R \\ -1 & if \quad z < 0 \end{cases} \tag{4}$$

and $K_b = diag\{k_{b_i}\}$ and $K_f = diag\{k_{f_i}\}$ are positive definite and diagonal matrices. Throughout, the precise meaning of solutions of the system (1) with discontinuous functions $F(\dot{q})$ and $\tau(q,\dot{q})$ are defined in Filippov's sense [8].

From the physical point of view, the position q represents the generalized coordinates, the control input τ is the vector of external torques, $M(q)$ is the inertia matrix, $C(q,\dot{q})\dot{q}$ is the vector of Coriolis and centripetal torques, $G(q)$ is the vector of gravitational forces, $F(\dot{q})$ represents the friction torques, where k_{b_i} and k_{f_i}, $i = 1,2,\ldots,n$ are the constant coefficients of viscous and Coulomb frictions, respectively. Because frictions are uncoupled among joints, we have assumed that the matrices K_b and K_f are diagonal.

Consider the following control law:

$$\tau = G(q) - K_d\dot{x} - K_p e - K_\alpha \operatorname{sgn}(e) \tag{5}$$

$$\dot{x} = -Lx + K_d e \tag{6}$$

where $L \in R^{n \times n}$ is a symmetric positive definite matrix, $K_d \in R^{n \times n}$ is a symmetric positive semi-definite matrix, $K_p = diag\{k_{p_i}\} \in R^{n \times n}$ is a diagonal positive definite matrix, $K_\alpha = diag\{k_{\alpha_i}\} \in R^{n \times n}$ is a diagonal matrix such that $K_\alpha > K_f$, and $e = q - q_d$ represents the position error with respect to the constant desired position q_d. Equation (6) is a first-order linear compensator used to replace the velocity feedback (cf. [19]).

The control law (5)-(6), that belongs to the *variable structure controllers* family, is called a *chattering controller* because it generates no sliding mode, except at the origin, while exhibiting an infinite number of switches in a finite time interval ([8]).

Theorem 1 ([8]). *Let the friction manipulator (1)-(4) be driven by the switched position feedback controller (5)-(6) with the assumptions given above. Then, the closed loop system (1)-(6) is globally asymptotically stable at the equilibrium point* $(\dot{q}, e, x) = 0$.

The switched term in (5), represented by $K_\alpha \operatorname{sgn}(e)$, can be interpreted as LIP neural nets with dendrite weights equal to one, and with firing thresholds (the so called 'bias' terms) equal to zero. The cell inputs are the components of the vector e. The outputs are the components of the switched term. Because the dendrite weights are positive the neural nets correspond to *excitatory* synapses. Here, the activation functions are the so called *symmetric hard limit.* Representing the activation function by $\sigma(\cdot)$ we have:

$$y_i = k_{\alpha_i} \operatorname{sgn}(e_i) = k_{\alpha_i} \sigma(e_i); \quad i = 1,2,\ldots,n \tag{7}$$

where y_i are the outputs of the LIP neural nets.

The problem to tackle is to find a training rule for each k_{α_i} such that the closed-loop system be globally asymptotically stable at the equilibrium point $(\dot{q}, e, x) = 0$ with the property that each k_{α_i} converges to zero as the system approaches the equilibrium point.

3 Neural Nets Chattering Controller

Considering the following control law:

$$\tau = G(q) - K_d \dot{x} - K_p e - \delta(t) K_\alpha sign(e) \tag{8}$$

$$\dot{x} = -Lx + K_d e \tag{9}$$

$$\dot{\delta} = -\alpha \log(1+\delta) + k_r \frac{(1+\delta)}{\log(1+\delta)+1} \|e\|^2 \tag{10}$$

where $\alpha, k_r \in R^+$ and $\delta(t) \in R$ is the adaptive term that will regulate the amplitude of the chattering term. In fact, the above controller is a LIP neural net with *dynamic* training implemented with a point of view similar to [14] and [15].

Again, $L \in R^{n\times n}$ is a symmetric positive definite matrix, $K_d \in R^{n\times n}$ is a symmetric positive semi-definite matrix, $K_p = diag\{k_{p_i}\} \in R^{n\times n}$ is a diagonal positive definite matrix, $K_\alpha = diag\{k_{\alpha_i}\} \in R^{n\times n}$ is a diagonal matrix such that $K_\alpha > K_f$, and $e = q - q_d$ represents the position error with respect to the constant desired position q_d.

Lemma 1 [20]: *Suppose the ordinary differential equation in (10) has initial condition* $\delta(t_0) \geq 0$, *then* $\delta(t) \geq 0$ *for all* $t \geq t_0$.

Our main result follows.

Theorem 2. *Let the friction manipulator (1)-(4) be driven by the switched position feedback controller (8)-(10) with the assumptions given above. Suppose that* $K_f = \beta K_\alpha$ *with* $\beta < 1$ *and* $0 \leq \delta(t) \leq \beta + 1$ *for all* $t \geq t_0$. *Then, the closed-loop system (1)-(4) and (8)-(10) is globally asymptotically stable at the equilibrium point* $(\dot{q}, e, x, \delta) = (\dot{e}, e, x, \delta) = 0$ *if* $\delta(t) \geq 0$ *and*

$$0 < k_r < \frac{1}{\beta+1} \lambda_{\min}\left(\begin{bmatrix} K_d L K_d & -\frac{1}{2} K_d LL \\ -\frac{1}{2}(K_d LL)^T & LLL \end{bmatrix} \right). \tag{11}$$

Remark 1: *Because $\delta = 0$ is an equilibrium point of the closed-loop system (1)-(4) and (8)-(10), the chattering amplitude vanishes as $t \to \infty$.*

Proof. To this end, we follow the same line of reasoning given in [8]. Let us introduce the Lyapunov candidate function

$$\begin{aligned} V(\dot{q},e,x,\delta) = & \frac{1}{2}\dot{q}^T M(q)\dot{q} + \frac{1}{2}e^T K_p e + \frac{1}{2}(K_d e - Lx)^T (K_d e - Lx) \\ & + k_{\alpha_1}|e_1| + \ldots + k_{\alpha_n}|e_n| + (1+\delta)\log(1+\delta). \end{aligned} \tag{12}$$

This Lyapunov function is similar to the one proposed in [8] but the last term involves the dynamic adaptation law. This last term was also utilized in [20]. The time derivative of (12), along the trajectories of the closed loop system (1)-(4) and (8)-(10) yields:

$$\begin{aligned} \dot{V}(\dot{q},e,x,\delta) = & \dot{q}^T M(q)\ddot{q} + \frac{1}{2}\dot{q}^T \dot{M}(q)\dot{q} + e^T K_p \dot{e} + \dot{q}^T K_\alpha \operatorname{sgn}(e) \\ & + (K_d e - Lx)^T (K_d \dot{e} - L\dot{x}) + \dot{\delta}(\log(1+\delta)+1). \end{aligned} \tag{13}$$

Employing the well-known property $\dot{q}^T[\frac{1}{2}\dot{M}(q) - C(q,\dot{q})]\dot{q} = 0$, for all $q \in R^n$, and substituting the control law (8) into (13) we have

$$\begin{aligned} \dot{V}(\dot{q},e,x,\delta) = & -\dot{q}^T K_b \dot{q} - \dot{q}^T K_f \operatorname{sgn}(\dot{q}) - e^T K_d \dot{x} - \dot{q}^T K_\alpha (\delta - 1)\operatorname{sgn}(e) \\ & + (K_d e - Lx)^T (K_d \dot{e} - L\dot{x}) + \dot{\delta}(\log(1+\delta)+1). \end{aligned} \tag{14}$$

From (9), the above equation is simplified to

$$\begin{aligned} \dot{V}(\dot{q},e,x,\delta) = & -\dot{q}^T K_b \dot{q} - \dot{q}^T K_f \operatorname{sgn}(\dot{q}) - \dot{q}^T K_\alpha (\delta - 1)\operatorname{sgn}(e) \\ & - (K_d e - Lx)^T L (K_d e - Lx) + \dot{\delta}(\log(1+\delta)+1). \end{aligned} \tag{15}$$

Invoking (10), the above equation is reduced to

$$\begin{aligned}\dot{V}(\dot{q},e,x,\delta) &= -\dot{q}^T K_b \dot{q} - \dot{q}^T K_f \operatorname{sgn}(\dot{q}) - \dot{q}^T K_\alpha (\delta - 1)\operatorname{sgn}(e) \\ &\quad - (K_d e - Lx)^T L (K_d e - Lx) + k_r (1+\delta)\|e\|^2 \\ &\quad - \alpha \log(1+\delta)(\log(1+\delta)+1) \\ &= -\dot{q}^T K_b \dot{q} - \dot{q}^T K_f \operatorname{sgn}(\dot{q}) - \dot{q}^T K_\alpha (\delta - 1)\operatorname{sgn}(e) \\ &\quad - \begin{bmatrix} e \\ x \end{bmatrix}^T \begin{bmatrix} K_d L K_d & -\frac{1}{2} K_d L L \\ -\frac{1}{2}(K_d L L)^T & LLL \end{bmatrix} \begin{bmatrix} e \\ x \end{bmatrix} \\ &\quad + k_r (1+\delta)\|e\| - \alpha \log(1+\delta)(\log(1+\delta)+1) \\ &\le -\dot{q}^T K_b \dot{q} - \dot{q}^T K_f \operatorname{sgn}(\dot{q}) - \dot{q}^T K_\alpha (\delta - 1)\operatorname{sgn}(e) \\ &\quad - (\lambda_{\min}(Q) - k_r(1+\delta))\|e\|^2 - \lambda_{\min}(Q)\|x\|^2 \\ &\quad - \alpha \log(1+\delta)(\log(1+\delta)+1)\end{aligned}$$

where

$$Q = \begin{bmatrix} K_d L K_d & -\frac{1}{2} K_d L L \\ -\frac{1}{2}(K_d L L)^T & LLL \end{bmatrix} \tag{16}$$

From (11) we forward to

$$\begin{aligned}\dot{V}(\dot{q},e,x,\delta) &\le -\dot{q}^T K_b \dot{q} - \dot{q}^T K_f \operatorname{sgn}(\dot{q}) - \dot{q}^T K_\alpha (\delta - 1)\operatorname{sgn}(e) \\ &\quad - \lambda_{\min}(Q)\|x\|^2 - \alpha \log(1+\delta)(\log(1+\delta)+1) \\ \dot{V}(\dot{q},e,x,\delta) &\le -\dot{e}^T K_b \dot{e} - \dot{e}^T K_f \operatorname{sgn}(\dot{e}) - \dot{e}^T K_\alpha (\delta - 1)\operatorname{sgn}(e) \\ &\quad - \lambda_{\min}(Q)\|x\|^2 - \alpha \log(1+\delta)(\log(1+\delta)+1).\end{aligned}$$

By virtue of

$$\begin{aligned}& -\dot{e} K_f \operatorname{sgn}(\dot{e}) - \dot{e}^T K_\alpha (\delta - 1)\operatorname{sgn}(e) \\ & \le \sum_{i=1}^{n} |e_i| \left(-k_{f_i} + |\delta(t) - 1| k_{\alpha_i}\right)\end{aligned}$$

and taking into account that $0 \le \delta(t) \le \beta + 1$ for all $t \ge t_0$, we have

$$\begin{aligned}\dot{V}(\dot{q},e,x,\delta) \;\leq\; & -\dot{e}^T K_b \dot{e} - \lambda_{\min}(Q)\|x\|^2 \\ & -\alpha \log(1+\delta)(\log(1+\delta)+1) \;\leq 0\end{aligned} \tag{17}$$

Since $V(\dot{q},e,x,\delta)$ is positive definite and $\dot{V}(\dot{q},e,x,\delta)$ is a negative semi-definite decreasing function, it follows that the equilibrium point $(\dot{q},e,x,\delta) = (\dot{e},e,x,\delta) = 0$ of the closed-loop system (1)-(4) and (8)-(10) is uniformly stable, *i.e.*, $x(t), e(t), \dot{e}(t), \delta(t) \in L_\infty$. From (17), we can easily show that the squares of $x, \dot{e}, \delta$ are integrable with respect to time t; *i.e.*, $x(t), \dot{e}(t), \delta(t) \in L_2$. Next, Barbalat's lemma implies that $\dot{e}(t) \to 0, x(t) \to 0$ and $\delta(t) \to 0$. If $x(t) \to 0$ then $\dot{x}(t) \to 0$, and from (9), it follows that $e(t) \to 0$. This concludes our proof. ■

4 Application to an Industrial Robot Manipulator

4.1 Experimental Setup

The experimental setup designed in the research laboratory of CITEDI-IPN involves a three degrees-of-freedom (3-DOF) industrial robot manipulator manufactured be Amatrol, it is shown in Figure 1. This mechanical system presents Coulomb friction [8]. The base of the mechanical robot has a horizontal revolute joint, q_1, whereas two links have vertical revolute joints q_2 and q_3. The nominal parameter values of the mechanical manipulator are summarized in Table 1. A worm gear set, a helicon gear set and a roller chain are used for torque transmission to joints q_1, q_2 and q_3, respectively; there is a DC gear motor for each joint with a reduction ratio of 19.7:1 for q_1 and q_2 and 127.8:1 for q_3. The ISA Bus servo I/O card from the company *Servo To Go* is employed for the real time control system and it mainly consists of eight channels of 16-bit D/A outputs, 32 bits of I/O, and an interval timer capable of interrupting the PC. The controller is implemented using C++ programming language running on a 486 PC. Position measurements of each articulation of the robot are obtained using the quadrature encoder channel available on each DC gear-motor, connected to the I/O card, and programmed to provide the encoder signal processing every millisecond; the resolution of the encoders is 52 x 10^{-3} rad, 62 x 10^{-3} rad and 34 x 10^{-3} rad for q_1, q_2 and q_3, respectively. Along with this, a digital oscilloscope is used to store the control signal. Linear power amplifiers are installed en each servomotor which apply a variable torque to each joint. These amplifiers accept control inputs from the D/A converter in the range of ±10 volts. See Figure 2 for the hardware setup configuration. The dynamic model of the robot in the form of (1) is given in [8]. However, for control implementation, we only require $G(q)$.

Table 1. Nominal parameter values fo the mechanical manipulator

Description	Notation	Value	Units
Lenght of link 1	l_1	0.297	m
Lenght of link 2	l_2	0.297	m
Mass of link 1	m_1	0.38	Kg
Mass of link 2	m_2	0.34	Kg
Gravity acceleration	g	9.8	m/s^2

4.2. Experimental Results

The regulator performance was studied experimentally. The experiment was performed with the 3-DOF robot manipulator required to move in space from the origin $q_1(0) = q_2(0) = q_3(0) = 0$ to the desired position $q_{d1} = q_{d2} = q_{d3} = \pi/2$ [rad]. The initial velocities $\dot{q}(0) \in R^3$ and $\delta(0)$ were set to zero, respectively.

The control goal was achieved by implementing the control (8)-(10) where ([8])

$$G(q) = g\begin{bmatrix} 0 \\ m_1 l_1 \cos q_2 + m_2 l_1 \cos q_2 + m_2 l_2 \cos(q_1 + q_2) \\ m_2 l_2 \cos(q_2 + q_3) \end{bmatrix},$$

and the controller gains selected as follows:

$$K_p = diag\{15,40,40\}, \quad K_d = diag\{5,5,5\},$$
$$K_\alpha = diag\{3,3,3\}, \quad L = diag\{10,10,10\},$$

and $K_r = 2$ and $\alpha = 8$. The physical constant parameters, $m_i, l_i, i = 1,2$ are given in table 1.

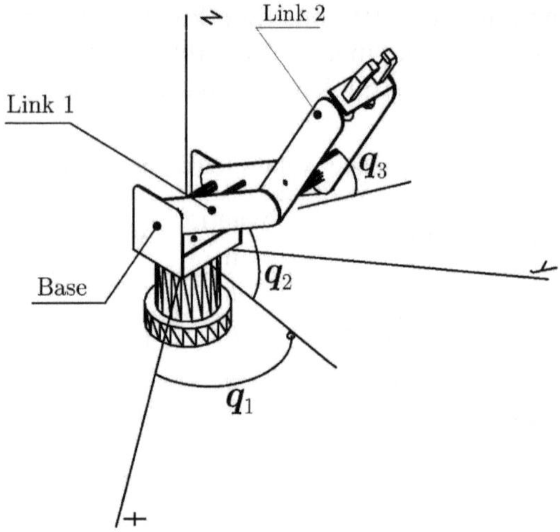

Fig. 1. Schematic diagram of the robot.

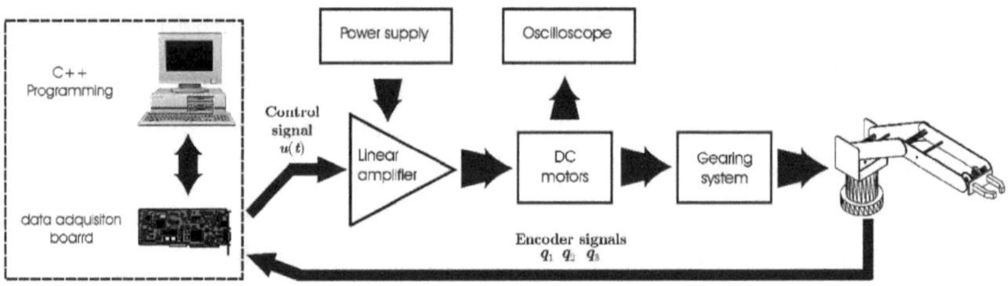

Fig. 2. Hardware setup.

The resulting joint positions and input torques are depicted in Figures 3 and 4, respectively. Figure 3 shows that joint positions converge to the desired position for the closed loop system [(1), (8)-(10)], whereas the fast switching due to LIP terms vanishes as t tends to ∞ (see Figure 4). Also, from Figure 3, a finite time convergence of the articulated positions to their desired positions is appreciated in about 3.2 seconds. The applied control inputs present chattering that is attenuated in about 8 seconds (see Figure 4). This chattering attenuation is good in mechanical systems, and was the main objective of the present paper. Finally, Figure 5 presents the time evolution of $\delta(t)$. It should be noted that the dynamic of $\delta(t)$ (10) is slower (smooth and slow variation) than [(1)-(4), (7), (8)].

5 Conclusions

We have developed a variable structure controller with chattering attenuation for robot manipulators in the presence of friction. The manipulator is governed by a second order differential equation with a right-hand discontinuous side admitting discontinuous terms to account for friction phenomena. The proposed controller uses Linear-in-the-Parameter Neural nets to attenuate the chattering signal inherent to variable structure systems without losing the robustness of the function framework. Effectiveness of the design is supported by the experiments made for a three degrees-of-freedom robot manipulator with frictional joints.

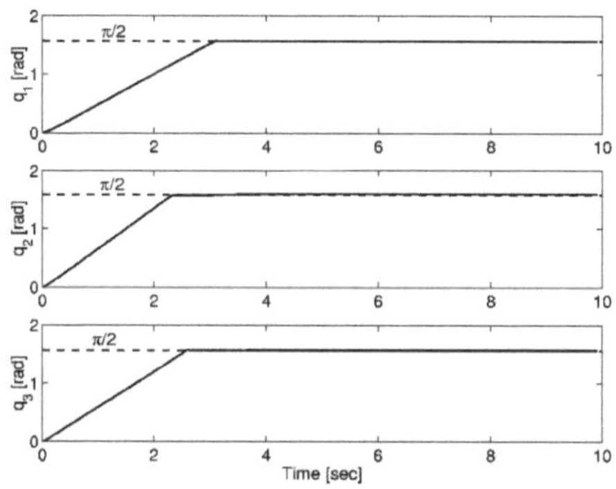

Fig. 3. Joint Positions.

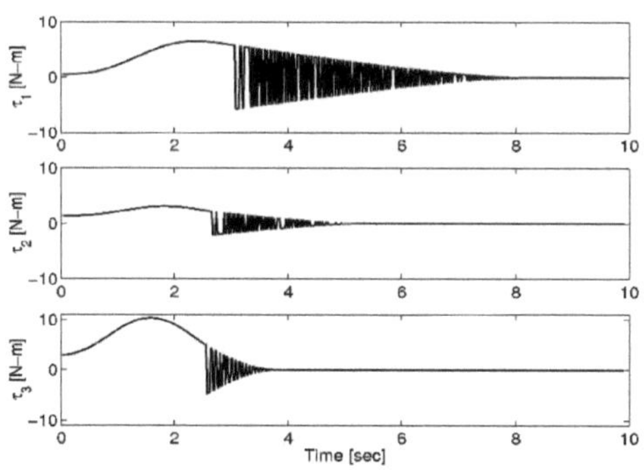

Fig. 4. Input Torques.

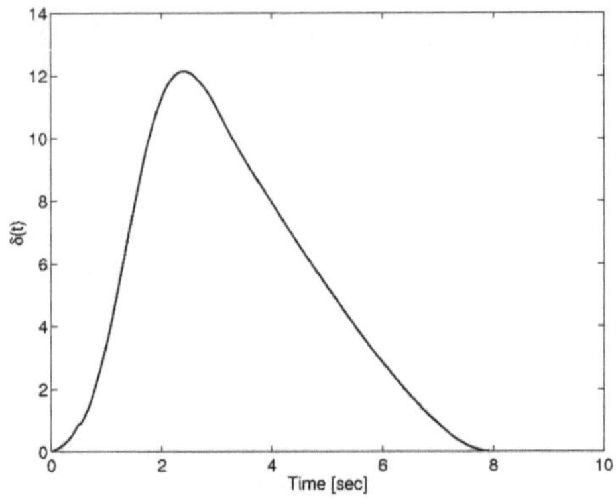

Fig. 5. Time evolution of $\delta(t)$.

References

1. Canudas de Wit C., Olsson H., Åström K.J., and Lischinsky P., A new model for control of systems with friction. *IEEE Trans. Aut. Ctrl.*, 40(3):419-425, 1995.
2. Friedland B. and Park Y., On adaptive friction compensation. *IEEE Trans. Aut. Ctrl.*, 37(10):1609-1612, 1992.
3. Huang J.-T., An adaptive compensator for a class of linearly parameterized systems. *IEEE Trans. Aut. Ctrl.*, 47(3):483-486, 2002.
4. Swevers J., Al-Bender F., Ganseman C. G., and Prajogo T., An integrated friction model structure with improved presliding behavior for accurate friction compensation. *IEEE Trans. Aut. Ctrl.*, 45(4):675-686, 2000.
5. Kelly R., Santibañez V., and González E., Adaptive friction compensation in mechanisms using Dahl model. In *Proc. Instn. Mech. Engrs.*, 218 Part I: J. Systems and Control Engineering, 53-57, 2004.
6. Cho S. and Ha I., A learning approach to tracking in mechanical systems with friction. *IEEE Trans. Aut. Ctrl.*, 45(1):111-116, 2000.
7. Dupont P. E. and Dunlap E. P., Friction modeling and PD compensation at very low velocities. *Trans. of the ASME*, 117(3): 8-14, 1995.
8. Orlov Y., Alvarez J., Acho L., and Aguilar L., Global position regulation of friction manipulators via switched chattering control. *Int. J. of Control,* 76(14):1446-1452, 2003.
9. Fridman L. and Levant A. Higher order sliding modes as a natural phenomenon in control theory. In Garafalo and Glielmo (Eds.) Robust Control via Variable Structure and Lyapunov Techniques, Lectures

notes in control and information science, 217, (Berlin: Springer, 1996), pp. 107-133.
10. Orlov Y., Aguilar L., and Cadiou J.C. Switched chattering control vs. back-lash/friction phenomena in electrical servomotors. Int. J. of Control, 76(9/10): 959-967, 2003.
11. Utkin V. I., *Sliding modes in control optimization.* Springer-Verlag, Berlin, Germany, 1992.
12. Rastko R. and Lewis F., Neural-network approximation of piewise continuous functions: Application to friction compensation. *IEEE Trans. Neural Networks,* 13(3):745-751, 2002.
13. Lewis F., Jagannathan S., and Yesildirek A., *Neural network control of robot manipulators and nonlinear systems.* Taylor and Francis, UK, 1999.
14. Polycarpou M. M., Stable adaptive neural control scheme for nonlinear systems. *IEEE Trans. Aut. Ctrl.,* 41(3):447-451, 1996.
15. Chen F. and Liu C., Adaptively controlling nonlinear continuous-time systems using multilayer neural networks, *IEEE Trans. Aut. Ctrl.,* 39(6):1306-1310, 1994.
16. Spooner J. T., Maggiore M., Ordóñez R., and Passino K. Stable adaptive control and estimation for nonlinear systems. Wiley Interscience, NY, 2002.
17. Parra-Vega V., Liu Y-H., and Arimoto S., Variable structure robot control undergoing chattering attenuation: adaptive and nonadaptive cases, in *Proc. Int. Conf. in Robotics and Automation,* pp. 1824-1829, 1994.
18. Bartolini G., Ferrara A., and Usai E., Chattering avoidance by second-order sliding mode control, *IEEE Trans. Automat. Contr.,* vol. 43, no. 2, pp. 241-246, 1998.
19. Berghuis H. and Nijmeijer H., Global regulation of robots using only position measurements. *Systems and Control Letters,* 21:289-293, 1993.
20. Hench J. J., On a class of adaptive suboptimal Riccati-based controllers. In *Proc. American Control Conference,* San Diego, CA, USA, 1999.

Tracking Control for a Unicycle Mobile Robot Using a Fuzzy Logic Controller

Selene L. Cárdenas, Oscar Castillo, Luis T. Aguilar, Nohé Cázarez

[1]Instituto Tecnológico de Tijuana, Clzd. Tecnológico S/N, C. P. 22379, Tijuana B. C. México, lilettecardenas@starmedia.com , {nohe, ocastillo}@tectijuana.mx.,

[2]CITEDI-IPN 2498 Roll Dr. # 757 Otay Mesa, San Diego, CA, USA, 92154 laguilar@citedi.mx

Abstract. We develop a tracking controller for the dynamic model of unicycle mobile robot by integrating a kinematic controller and a torque controller based on Fuzzy Logic Theory. Computer simulations are presented confirming the performance of the tracking controller and its application to different navigation problems.

1 Introduction

Mobile robots are nonholonomic systems due to the constraints imposed on their kinematics. The equations describing the constraints cannot be integrated simbolically to obtain explicit relationships between robot positions in local and global coordinate frames. Hence, control problems involve them have attracted attention in the control community in the last years [11].

Different methods have been applied to solve motion control problems. Kanayama et al. [10] propose a stable tracking control method for a nonholonomic vehicle using a Lyapunov function. Lee et al. [12] solved tracking control using backstepping and in [13] with saturation constraints. Furthermore, most reported designs rely on intelligent control approaches such as Fuzzy Logic Control [1][8][14][17][18][20] and Neural Networks [6][19].

However, the majority of the publications mentioned above, has concentrated on kinematics models of mobile robots, which are controlled by the velocity input, while less attention has been paid to the control problems of nonholonomic dynamic systems, where forces and torques are the true inputs: Bloch and Drakunov [2] and Chwa [4], used a sliding mode control

S. Cárdenas et al.: *Tracking Control for a Unicycle Mobile Robot Using a Fuzzy Logic Controller*, StudFuzz **208**, 243–253 (2007)
www.springerlink.com

to the tracking control problem. Fierro and Lewis [5] propose a dynamical extension that makes possible the integration of kinematic and torque controller for a nonholonomic mobile robot. Fukao et al. [7], introduced an adaptive tracking controller for the dynamic model of mobile robot with unknown parameters using backstepping.

In this paper we present a tracking controller for the dynamic model of a unicycle mobile robot, using a control law such that the mobile robot velocities reach the given velocity inputs, and a fuzzy logic controller such that provided the required torques for the actual mobile robot. The rest of this paper is organized as follows. Section II describes the formulation problem, which include: the kinematic and dynamic model of the unicycle mobile robot and introduces the tracking controller. Section III illustrates simulations results using the tracking controller. The section IV gives the conclusions.

2 Problem Formulation

2.1 The Mobile Robot

The model considered is a unicycle mobile robot (see Fig. 1), it consist of two driving wheels mounted on the same axis and a front free wheel [3].

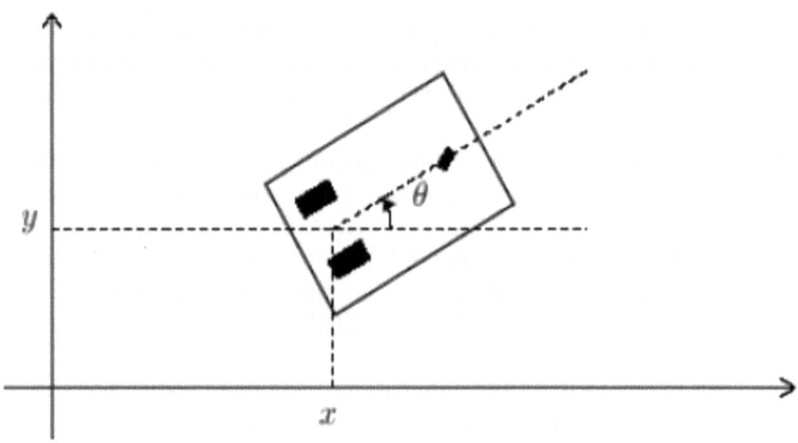

Fig. 1. Wheeled mobile robot.

The motion can be described with equation (1) of movement in a plane [5]:

$$\dot{q} = \begin{vmatrix} \cos\theta & 0 \\ \sin\theta & 0 \\ 0 & 1 \end{vmatrix} \begin{vmatrix} v \\ w \end{vmatrix} \tag{1a}$$

$$M(q)\dot{v} + V(q,\dot{q})v + G(q) = \tau \tag{1b}$$

Where $q = [x, y, \theta]^T$ is the vector of generalized coordinates which describes the robot position, *(x,y)* are the cartesian coordinates, which denote the mobile center of mass and θ is the angle between the heading direction and the x-axis (wich is taken counterclockwise form); $v = [v, w]^T$ is the vector of velocities, v and w are the linear and angular velocities respectively; $\tau \in R^r$ is the input vector, $M(q) \in R^{nxn}$ is a symetric and positive-definite inertia matrix, $V(q,\dot{q}) \in R^{nxn}$ is the centripetal and coriolis matrix, $G(q) \in R^n$ is the gravitational vector. Equation (1.a) represents the kinematics or steering system of a mobile robot. Notice that the no-slip condition imposed a nonholonomic constraint described by (2), that it means that the mobile robot can only move in the direction normal to the axis of the driving wheels.

$$\dot{y}\cos\theta - \dot{x}\sin\theta = 0 \tag{2}$$

2.2 Tracking Controller of Mobile Robot

Our control objective is established as follow: Given a desired trajectory $q_d(t)$ and orientation of mobile robot we must desing a controller that apply adequate torque τ such that the measured positions $q(t)$ achieve the desired reference $q_d(t)$ represented as (3):

$$\lim_{t\to\infty} \left\| q_d(t) - q(t) \right\| = 0 \tag{3}$$

To reach the control objective, we are based in the procedure of [5], we deriving a $\tau(t)$ of a specific $v_c(t)$ that controls the steering system (1.a) using a Fuzzy Logic Controller (FLC). A general structure of tracking control system is presented in the Fig. 2.

2.2.1 Control of the Kinematic Model

We are based on the procedure proposed by Kanayama et al. [10] and Nelson et al. [15] to solve the tracking problem for the kinematic model, this is denoted as vc(t). Suppose the desired trajectory qd satisfies (4):

$$\dot{q}_d = \begin{vmatrix} \cos\theta_d & 0 \\ \sin\theta_d & 0 \\ 0 & 1 \end{vmatrix} \begin{vmatrix} v_d \\ w_d \end{vmatrix} \tag{4}$$

Using the robot local frame (the moving coordinate system x-y in figure 1), the error coordinates can be defined as (5):

$$e = T_e(q_d - q), \begin{vmatrix} e_x \\ e_y \\ e_\theta \end{vmatrix} = \begin{vmatrix} \cos\theta & \sin\theta & 0 \\ -\sin\theta & \cos\theta & 0 \\ 0 & 0 & 1 \end{vmatrix} \begin{vmatrix} x_d - x \\ y_d - y \\ \theta_d - \theta \end{vmatrix} \tag{5}$$

And the auxiliary velocity control input that achieves tracking for (1.a) is given by (6):

$$v_c = f_c(e, v_d), \begin{vmatrix} v_c \\ w_c \end{vmatrix} = \begin{vmatrix} v_d + \cos e_\theta + k_1 e_x \\ w_d + v_d k_2 e_y + v_d k_3 \sin e_\theta \end{vmatrix} \tag{6}$$

Where k_1, k_2 and k_3 are positive constants.

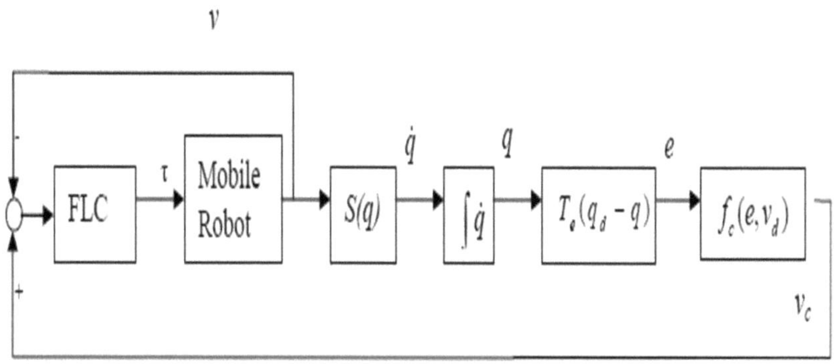

Fig. 2. Tracking control structure

2.2.2 Fuzzy Logic Controller (FLC)

The purpose of the FLC is to find a control input τ such that the current velocity vector v to reach the velocity vector v_c this is denoted as (7):

$$\lim_{t\to\infty}\left\|v_c - v\right\| = 0 \tag{7}$$

As is shown in Fig. 2, basically the FLC have 2 inputs variables corresponding the velocity errors obtained of (7) (denoted as e_v and e_w: linear and angular velocity errors respectively), and 2 outputs variables, the driving and rotational input torques τ (denoted by F and N respectively). The membership functions (MF)[9] are defined by 1 triangular and 2 trapezoidal functions for each variable involved due to the fact are easy to implement computationally.

Figure 3 and Fig. 4 depicts the MFs in which N, C, P represent the fuzzy sets[9] (Negative, Zero and Positive respectively) associated to each input and output variable, where the universe of discourse is normalized into [-1,1] range.

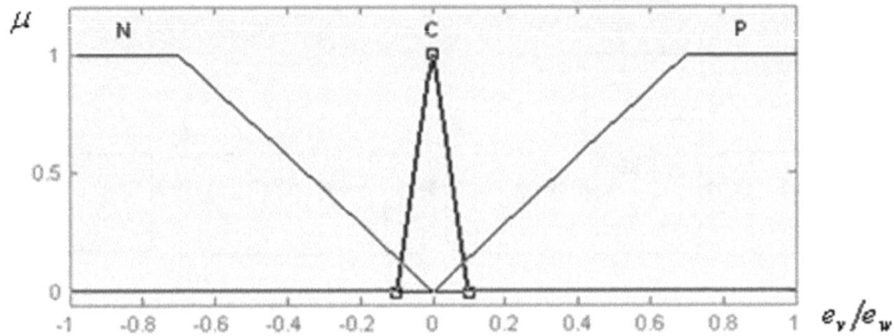

Fig. 3. Membership function of the input variables e_v and e_w

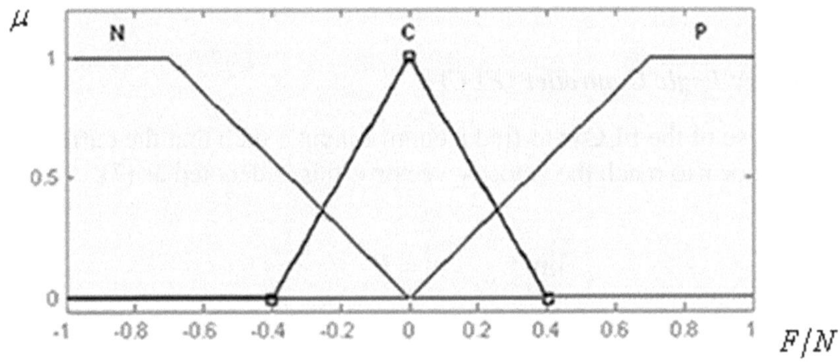

Fig. 4. Membership function of the output variables F and N.

The rule set of FLC contain 9 rules which governing the input-output relationship of the FLC and this adopts the Mamdani-style inference engine[16], and we use the center of gravity method to realize defuzzification procedure. In Table 1, we present the rule set whose format are established as follow:

Rule *i*: If e_v is G_1 and e_w is G_2 then F is G_3 and N is G_4

Where G_1..G_4 are the fuzzy set asociated to each variable and *i*= 1 ... 9.

Table 1. Rule set

e_v / e_w	N	C	P
N	*N/N*	*N/C*	*N/P*
C	*C/N*	*C/C*	*C/P*
P	*P/N*	*P/C*	*P/P*

3 Simulations Results

Simulations have been done in Matlab® to test the tracking controller of the mobile robot defined in (1). We consider the initial position $q(0) = (0, 0, 0)$ and initial velocity $v(0) = (0,0)$, in Table 2 we show the results of five simulations. The columns 2 to 6 indicate the reference position q_d and velocity v_d, the leftovers indicate the position and orientation errors obtained when finalizing the simulation.

Table 2. Simulation results of five experiments.

No.	$x_d(m)$	$y_d(m)$	θ_d (rad)	v_d (m/s)	w_d (rad/s)	e_x	e_y	e
1	0.15	0.15	1	1	0	0.0222	-0.0185	0.0321
2	0.15	0.15	1	0.5	0	-0.0083	-0.0022	-0.0022
3	0.25	0.25	0.45	1	0	0.0141	0.0005	0.0012
4	0.40	0.25	0.45	1	0	-0.0118	0.0160	0.0141
5	0.50	0.25	0.45	1	0	-0.0118	0.0079	0.0070

From Fig. 5 to Fig. 8 we show the results of the simulation for the case 1. Position and orientation errors are depicted in the Fig. 5 and Fig. 6 respectively, as can be observed the errors are sufficient close to zero, the trajectory tracked (see Fig. 7) is very close to the desired, and the velocity errors shown in Fig. 8 decrease to zero, achieving the control objective in less than 1 second of the whole simulation.

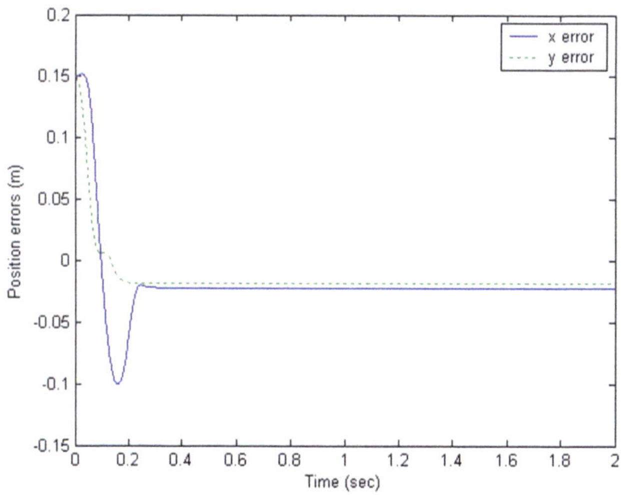

Fig. 5. Positions error with respect to the reference values. Solid: error in x, dotted: error in y.

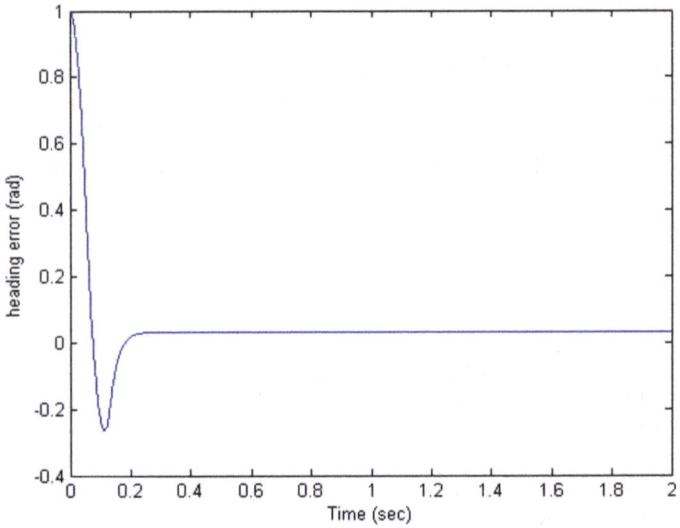

Fig. 6. Orientation error with respect to the reference values.

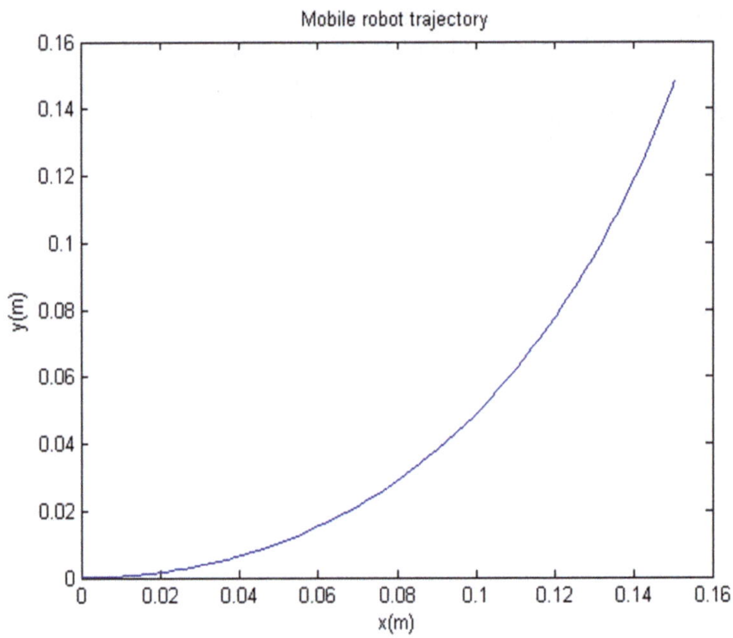

Fig. 7. Mobile Robot Trajectory

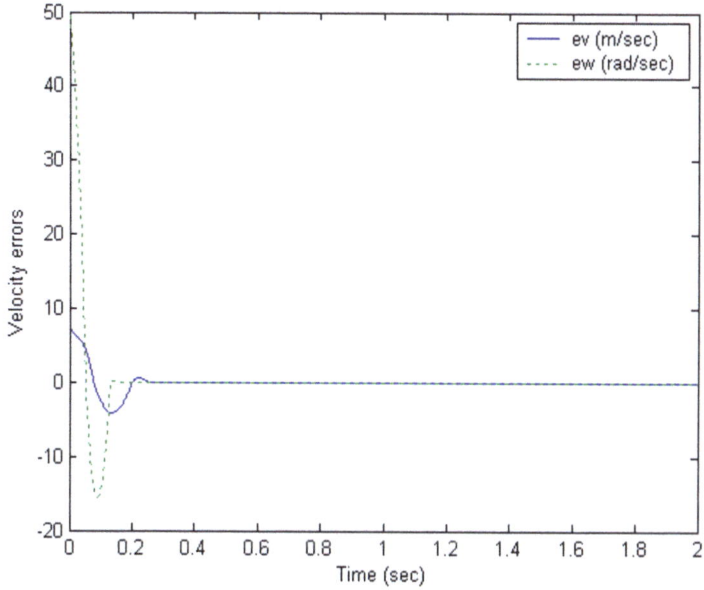

Fig. 8. Velocity errors. Solid: error in e_v, dotted: error in e_w

4 Conclusions

We described the development of a tracking controller integrating a fuzzy logic controller for a unicycle mobile robot with known dynamics, which can be applied for both, point stabilization and trajectory tracking. Computer simulation results confirm that the controller can achieve our objective. As future work, several extensions can be made to the control structure of Fig. 2, such as to increase the tracking accuracy and the performance level.

Acknowledgement

The authors would like to thanks CONACYT, Tijuana Institute of Technology and CITEDI-IPN for the support in our research work.

References

1. S. Bentalba, A. El Hajjaji, A. Rachid, Fuzzy Control of a Mobile Robot: A New Approach, Proc. IEEE Int. Conf. On Control Applications, Hartford, CT, pp 69-72, October 1997.
2. A. M. Bloch, S. Drakunov, Tracking in NonHolonomic Dynamic System Via Sliding Modes, *Proc. IEEE Conf. On Decision & Control*, Brighton, UK, pp 1127-1132, 1991.
3. G. Campion, G. Bastin, B. D'Andrea-Novel, Structural Properties and Classification of Kinematic and Dynamic Models of Wheeled Mobile Robots, IEEE Trans. On Robotics and Automation, Vol. 12, No. 1, February 1996.
4. D. Chwa., Sliding-Mode Tracking Control of Nonholonomic Wheeled Mobile Robots in Polar coordinates, IEEE Trans. On Control Syst. Tech. Vol. 12, No. 4, pp 633-644, July 2004.
5. R. Fierro and F.L. Lewis, Control of a Nonholonomic Mobile Robot: Backstepping Kinematics into Dynamics. *Proc. 34th Conf. on Decision & Control*, New Orleans, LA, 1995.
6. R. Fierro, F.L. Lewis, Control of a Nonholonomic Mobile Robot Using Neural Networks, *IEEE Trans. On Neural Networks*, Vol. 9, No. 4, pp 589 – 600, July 1998.
7. T. Fukao, H. Nakagawa, N. Adachi, Adaptive Tracking Control of a NonHolonomic Mobile Robot, *IEEE Trans. On Robotics and Automation*, Vol. 16, No. 5, pp. 609-615, October 2000.
8. S. Ishikawa, A Method of Indoor Mobile Robot Navigation by Fuzzy Control, *Proc. Int. Conf. Intell. Robot. Syst.*, Osaka, Japan, pp 1013-1018, 1991.
9. J. S. R. Jang, C.T. Sun, E. Mizutani, *Neuro Fuzzy and Soft Computing: A Computational Approach to Learning and Machine Intelligence*, Prentice Hall, Upper Sadle River, NJ, 1997.
10. Y. Kanayama, Y. Kimura, F. Miyazaki T. Noguchi, A Stable Tracking Control Method For a Non-Holonomic Mobile Robot, *Proc. IEEE/RSJ Int. Workshop on Intelligent Robots and Systems*, Osaka, Japan, pp 1236- 1241, 1991.
11. I. Kolmanovsky, N. H. McClamroch., Developments in Nonholonomic Nontrol Problems, *IEEE Control Syst. Mag.*, Vol. 15, pp. 20–36, December. 1995.
12. T-C Lee, C. H. Lee, C-C Teng, Tracking Control of Mobile Robots Using the Backsteeping Technique, *Proc. 5th. Int. Conf. Contr., Automat., Robot. Vision*, Singapore, pp 1715-1719, December 1998.
13. T-C Lee, K. Tai, Tracking Control of Unicycle-Modeled Mobile robots Using a Saturation Feedback Controller, *IEEE Trans. On Control Systems Technology*, Vol. 9, No. 2, pp 305-318, March 2001.
14. T. H. Lee, F. H. F. Leung, P. K. S. Tam, Position Control for Wheeled Mobile Robot Using a Fuzzy Controller, IEEE pp 525-528, 1999.
15. W. Nelson, I. Cox, Local Path Control for an Autonomous Vehicle, *Proc. IEEE Conf. On Robotics and Automation*, pp. 1504-1510, 1988.
16. K. M. Passino, S. Yurkovich, *"Fuzzy Control"*, Addison Wesley Longman, USA 1998.

17. S. Pawlowski, P. Dutkiewicz, K. Kozlowski, W. Wroblewski, Fuzzy Logic Implementation in Mobile Robot Control, *2nd Workshop On Robot Motion and Control,* pp 65-70, October 2001.
18. C-C Tsai, H-H Lin, C-C Lin, Trajectory Tracking Control of a Laser-Guided Wheeled Mobile Robot, *Proc. IEEE Int. Conf. On Control Applications*, Taipei, Taiwan, pp 1055-1059, September 2004.
19. K. T. Song, L. H. Sheen, Heuristic fuzzy-neuro Network and its application to reactive navigation of a mobile robot, *Fuzzy Sets Systems*, Vol. 110, No. 3, pp 331-340, 2000.
20. S. V. Ulyanov, S. Watanabe, V. S. Ulyanov, K. Yamafuji, L. V. Litvintseva, G. G. Rizzotto, Soft Computing for the Intelligent Robust Control of a Robotic Unicycle with a New Physical Measure for Mechanical Controllability, Soft Computing 2 pp 73 – 88, Springer- Verlag, 1998.

Intelligent Control and Planning of Autonomous Mobile Robots Using Fuzzy Logic and Genetic Algorithms

Julian Garibaldi, Azucena Barreras and Oscar Castillo

Dept. of Computer Science, Tijuana Institute of Technology, Mexico

Abstract. This paper describes the use of a Genetic Algorithm (GA) for the problem of Offline Point-to-Point Autonomous Mobile Robot Path Planning. The problem consist of generating "valid" paths or trajectories, for an Holonomic Robot to use to move from a starting position to a destination across a flat map of a terrain, represented by a two dimensional grid, with obstacles and dangerous ground that the Robot must evade. This means that the GA optimizes possible paths based on two criteria: length and difficulty.

1 Introduction

The problem of Mobile Robot Path Planning is one that has intrigued and has received much attention thru out the history of Robotics, since it's at the essence of what a mobile robot needs to be considered truly "autonomous". A Mobile Robot must be able to generate collision free paths to move from one location to another, and in order to truly show a level of intelligence these paths must be optimized under some criteria most important to the robot, the terrain and the problem given. GA's and evolutionary methods have extensively been used to solve the path planning problem, such as in (Xiao and Michalewicz, 2000) where a *CoEvolutionary* method is used to solve the path planning problem for two articulated robot arms, and in (Ajmal Deen Ali et. al., 2002) where they use a GA to solve the path planning problem in non-structured terrains for the particular application of planet exploration. In (Farritor and Dubowsky, 2002) an *Evolutionary Algorithm* is used for both off-line and on-line path planning using a linked list representation of paths, and (Sauter et. al., 2002) uses a *Particle swarm optimization* (PSO) method based on *Ant Colony Optimization* (ACO). However, the research work presented in this paper used as a basis for comparison and development the work done in (Sugihara, 1999). In this work, a grid representation of the terrain is used and different values

J. Garibaldi et al.: *Intelligent Control and Planning of Autonomous Mobile Robots Using Fuzzy Logic and Genetic Algorithms*, StudFuzz **208**, 255–265 (2007)
www.springerlink.com

are assigned to the cells in a grid, to represent different levels of difficulty that a robot would have to traverse a particular cell. Also they present a codification of all monotone paths for the solution of the path-planning problem.

2 Basic Theory

This section is intended to present some basic theory used to develop the GA's in this paper for use in the path planning problem, covering topics like basic Genetic Algorithm theory, Multi Objective optimization, Triggered Hypermutation and Autonomous Mobile Robot Point-to Point Path Planning.

2.1 Genetic Algorithms

A Genetic Algorithm is an evolutionary optimization method used to solve, in theory "any" possible optimization problem. A GA (Man et. al., 1999) is based on the idea that a solution to a particular optimization problem can be viewed as an *individual* and that these individual characteristics can be coded into a finite set of parameters. These parameters are the *genes* or the *genetic information* that makes up the *chromosome* that represents the real world structure of the individual, which in this case is a solution to a particular optimization problem. Because the GA is an evolutionary method, this means that a repetitive loop or a series of *generations* are used in order to evolve a *population* S of p individuals to find the *fittest* individual to solve a particular problem. The *fitness* of each *individual* is determined bye a given *fitness function* that evaluates the level of aptitude that a particular *individual* has to solve the given optimization problem. Each *generation* in the genetic search process produces a new set of individuals through *genetic operations* or *genetic operators: Crossover* and *Mutation,* operations that are governed by the *crossover rate* and the *mutation rate* μ respectively. These operators produce new *child chromosomes* with the intention of bettering the overall fitness of the population while maintaining a global search space. Individuals are selected for *genetic operations* using a *Selection method* that is intended to select the fittest individuals for the role of *parent chromosomes* in the *Crossover* and *Mutation* operations. Finally these newly generated child chromosomes are reinserted into the population using a *Replacement method.* This process is repeated a k number of *generations.*

2.2 Multiple Objective Genetic Algorithms

Real-world problem solving will commonly involve (Oliveira et. al., 2002) the optimization of two or more objectives at once, a consequence of this is that it's not always possible to reach an optimal solution with respect to all of the objectives evaluated individually. Historically a common method used to solve multi objective problems is by a linear combination of the objectives, in this way creating a single objective function to optimize (Sugihara, 1997) or by converting the objectives into restrictions imposed on the optimization problem. In regards to evolutionary computation, (Shaffer, 1985) proposed the first implementation for a multi objective evolutionary search. The proposed methods in (Fonseca and Fleming, 1993), (Srinivas, 1994) and (Goldberg, 1989), all center around the concept of *Pareto optimality* and the *Pareto optimal set*. Using these concepts of optimality of *individuals* evaluated under a multi objective problem, they each propose a *fitness* assignment to each individual in a current population during an evolutionary search based upon the concepts of *dominance* and *non-dominance* of *Pareto optimality*. Where the definition of *dominance* is stated as follows:

Definition 1: For an optimization (minimization) problem with n-objectives, solution u is said to be dominated by a solution v if:

$$\forall i = 1,2,....,n. \qquad f_i(u) \geq f_i(v) \, , \tag{1}$$

$$\exists j = 1,2,....,n, \qquad \because \; f_i(u) > f_i(v) \tag{2}$$

2.3 Triggered Hypermutation

In order to improve on the convergence of a GA, there are several techniques available such as (Man et. al. 1999) expanding the memory of the GA in order to create a repertoire to respond to unexpected changes in the environment. Another technique used to improve the overall speed of convergence for a GA is the use of a Triggered Hypermutation Mechanism (Cobb, 1990), which consists of using *mutation* as a control parameter in order to improve performance in a dynamic environment. The GA is modified by adding a mechanism by which the value of μ is changed as a result of a dip in the fitness produced by the best solution in each generation in the genetic search. This way μ is increased to a high *Hypermutation* value each time the top fitness value of the population at generation k dips below some lower limit set beforehand.

2.4 Autonomous Mobile Robots

An Autonomous Mobile Robot as defined in (Xiao and Michalewicz, 2000) can be seen as a vehicle that needs the capability of generating collision free paths that take the robot from a starting position *s* to a final destination *d*, and needs to avoid obstacles present in the environment. The robot must be able to have enough relevant information of his current position relative to *s* and *d,* and of the state of the environment or terrain that surrounds it. One advantage about generating paths or trajectories for these kinds of robots, compared to the more traditional robot arms, is that in general there are far less restrictions in regards to the precision with which the paths must be generated. The basic systems that operate in an Autonomous Mobile robot are:

1. Vehicle Control.
2. Sensor and Vision.
3. Navigation
4. Path Planning

2.5 Point-to-Point Path Planning Problem

The path planning problem when analyzed with the point-to-point technique, (Choset et. al., 1999) comes down to finding a path from one point to another (start and destination). Obviously, one of the most important reasons to generate an appropriate path for a robot to follow, is to help it avoid possible danger or obstacles along the way, for this reason an appropriate representation of the terrain is needed generating a sufficiently complete map of the given surroundings that the robot will encounter along its route. The general path-planning problem, that all autonomous mobile robots will face, has been solved (to some level of satisfaction) with various techniques, besides the evolutionary or genetic search, such as, using the *Voroni Generalized Graph* (Choset et. al., 1999), or using a *Fuzzy Controller* (Kim et. al., 1999), yet another is by the use of *Artificial Potential Fields* (Planas et. al., 2002).

3 Proposed Method

The first step before we can continue and give the details of the GA implementation used to solve the path-planning problem, is to explicitly define the problem and what is it that we are expecting out of the subsequent genetic search. To this end, we propose what will be the *input/output* pair that we are expecting from our GA as follows:

Input: 1) An ***n*** x ***n*** grid, where the starting cell ***s*** for the robot is in one corner and the destination cell ***d*** is diagonally across from it.
2) Each cell with a corresponding *difficulty weight* ***wd*** assigned to it ranging from [0, 1].
Output: A path, defined as a sequence of adjacent cells joining ***s*** and ***d***, and that complies with the following restrictions and optimization criteria:
1) The path most not contain cells with ***wd*** = 0 (solid obstacles).
2) The path must stay inside of the grid boundaries.
3) Minimize the path length (number of cells).
4) Minimize the total difficulty for the path, that means, the combined values of ***wd*** for all the cells in a given path.

We must also establish a set of ground rules or assumptions that our GA will be operating under.
1) The ***n*** x ***n*** grid isn't limited to all cells in the grid having to represent a uniform or constant size in the terrain, each cell is merely a conceptual representation of spaces in a particular terrain.
2) Each cell in a terrain has a given *difficulty weight* ***wd*** between the values of [0,1], that represents the level of difficulty that a robot would have to pass through it, where the lower bounds 0 represents a completely free space and the higher bounds 1 represents a solid impenetrable obstacle.
3) The terrain is considered to be static in nature.
4) It is assumed that there is a sufficiently complete knowledge in regards to the state of the terrain in which the robot will operate.
5) The paths produced by the GA are all monotone paths.

4 Architecture of the Genetic Algorithm

We now turn to the actual implementation of our GA, used to solve the path-planning problem for one and two optimization objectives. So we describe each of the parts of our GA and give a brief description of each, clearly stating any differences between the one and two optimization objectives implementations.

4.1 Individual Representation

Basically, the chromosome structure was taken from the work done in (Sugihara, 1999) where a binary string representation of monotone paths is used. The binary string chromosome is made up of ***n***-1 (where ***n*** is the number of columns and rows in the grid representing the map of a given terrain) pairs of *direction/distance* of length *3 + log[2]****n*****,** and an extra bit *a* which determines if the path is *x-monotone* (*a=0*) or *y-monotone* (*a=1*).

And each pair of *direction/distance* codes the direction in which a robot moves inside the grid and the number of cells it moves thru in that direction. The coding used greatly facilitates its use in a GA, because of its constant length no special or revamped genetic operators are needed, a problem that would be very cumbersome to solve if using a linked list chromosome representation of the path as done in (Xiao and Michalewicz, 2000).

4.2 Initial Population

The population *S* used in the genetic search is initialized with *p* total individuals. Of the *p* individuals in *S, p-2* of them are generated randomly while the remaining two represent straight line paths from *s* to *d,* one of this paths is *x-monotone* and the other is *y-monotone.*

So we can clearly define the population *S* as being made up by:

$$S = \{ x_0, x_1, x_2 \ldots\ldots\ldots\ldots x_{p-2}, a, b \} \quad (3)$$

Where x_i are randomly generated individuals, and by *a* and *b* that are *x-monotone* and *y-monotone* paths respectively that take a straight-line route from *s* to *d.*

4.3 Path Repair Mechanism

Each path inside of the population *S* is said to be either *valid* or *non-valid.* Where criteria for *non-validity* are:

- Path contains a cell with a solid obstacle (***wd*** = 1).
- Path contains cells out of bounds.
- The paths final cell isn't *d.*

Using this set of three rules to determine the state of validity of a given path for a particular genetic search, we can define a *subpopulation S',* which is made up by entirely *non-valid* paths in *S.*

The Path Repair Mechanism used with the GA is a *Lamarckian* process designed to take *non-valid x',* where *x' S',* and determine if they can be salvaged and return to a *valid* state, so as to be productive in the genetic search, because just because a particular path is determined to be *non-valid* this does not preclude it from having possible information coded in its chromosome that could prove to be crucial and effective in the genetic search process, this is way *non-valid* paths are given low fitness values with the penalty scheme used in the fitness evaluation, only after it has been determined that its *non-valid* state cant be reversed.

4.4 Fitness Evaluation

As was mentioned earlier, we introduce here both single and two objective optimization of the path planning problem, taking into account the length a given path and the difficulty of the same as the two criteria for optimization for paths in the population hence, the way in which each implementation of the GA assigns fitness values differs for obvious reasons.

4.4.1 Single Objective

Considering our Conventional GA, we can say that for paths inside *S* we optimize for only one objective, which is the path length, therefore we define fitness $f_1(x)$ as given by:

$$f_1(x) = (n^2) - (c) \tag{4}$$

Where *c* is the number of cells in a given path *x*.

4.4.2 Multiple Objective

Besides the fitness $f_1(x)$ used in Section 4.4.1 given for path length, a second fitness assignment $f_2(x)$ is given for path difficulty is given, and is calculated by,

$$f_2(x) = (n^2) - \Sigma wd_i \tag{5}$$

Where the second term in (5) is the sum of ***wd*** for each cell in a given path *x*. With this we are forced to use Pareto optimality for a rank-based system for individuals in population *S*. So for a path *x* where *x*   *S* its final fitness values is given by their *rank* value inside of *S* determined by,

$$\text{rank}(x) = p - t \tag{6}$$

Where *p* is the size of population *S* and *t* is the number of individuals that dominate *x* in *S*.

5 Simulation Results

We use the benchmark test presented in Figure 1, which was used in (Sugihara, 1997) due to its capability of projecting an accurate general performance score for the GA, and the performance measure of *probability optimality* $L_{opt}(k)$, which is a representation of the probability that a GA has of finding an optimal solution to a given problem. In this case, is the probability of finding a solution on the Pareto optimal front. Using

$L_{opt}(k)$ as the performance measure we present a set of optimal operating parameters for our MOGA using both a Generational and Elitist replacement scheme, Figures 2 to 3 show the simulation results that support this values. We also compare the two methods along with the GA proposed in (Sugihara, 1999) and the comparison is made under a normalized value for *kp=30,000* keeping the overall computational cost equal for each GA.

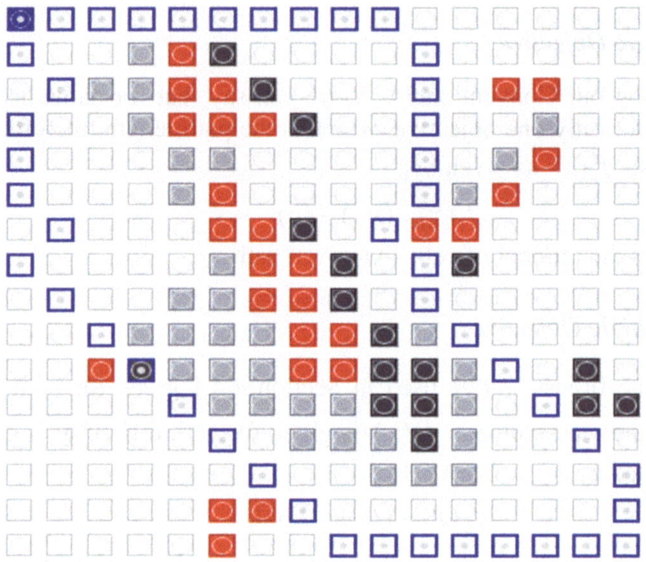

Fig. 1. Benchmark Test, with two paths on the Pareto Optimal Front.

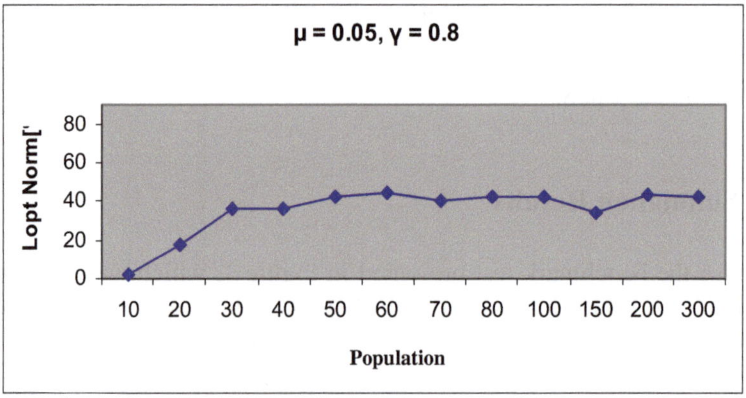

Fig. 2. Normalized $L_{opt}(k)$ and population size with Generational Replacement

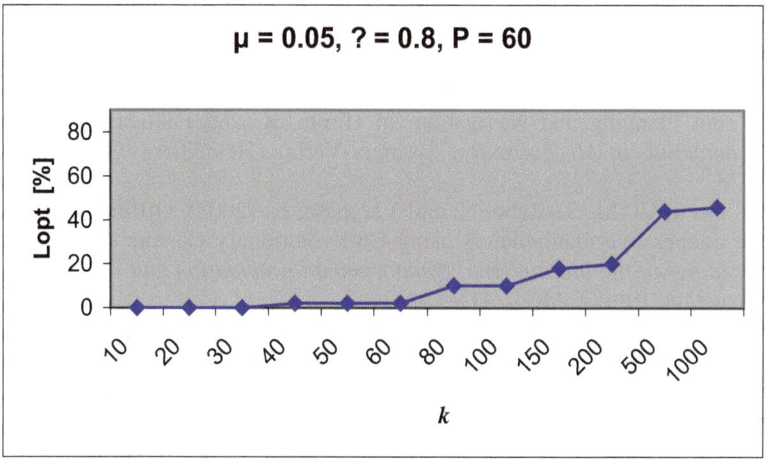

Fig. 3. $L_{opt}(k)$ and number of generations with Generational Replacement.

6 Conclusions

This paper presented a GA designed to solve the Mobile Robot Path Planning Problem. We showed with simulation results that both a Conventional GA and a MOGA, based on Pareto optimality, equipped with a basic repair mechanism for *non-valid* paths, can solve the point-to-point path planning problem when applied to grid representations of binary and continuous simulation of terrains respectively. From the simulation results gathered from experimental testing the Conventional GA with a Generational Replacement scheme and Triggered Hypermutation (which is commonly referred to as a conversion mechanism for dynamic environments) gave consistent performance to varying degrees of granularity in the representation of terrains with out a significant increase in population size or number of generations needed in order to complete the search in a satisfactory manner, while the MOGA based on Pareto Optimality combined with a Elitist replacement scheme clearly improves upon previous (Sugihara, 1999) work done with multiple objective path planning problem based on linear combination, with the added advantage of providing more than one equally usable solution.

References

Xiao, J. and Michalewicz, Z. (2000), An Evolutionary Computation Approach to Robot Planning and Navigation, in Hirota, K. and Fukuda, T. (eds.), *Soft Computing in Mechatronics*, Springer-Verlag, Heidelberg, Germany, 117 – 128.

Ajmal Deen Ali, M. S., Babu, N. and Varghese, K. (2002), Offline Path Planning of cooperative manipulators using Co-Evolutionary Genetic Algorithm, *Proceedings of the International Symposium on Automation and Robotics in Construction*, 19th (ISARC), 415-124.

Farritor, S. and Dubowsky, S. (2002), A Genetic Planning Method and its Application to Planetary Exploration, *ASME Journal of Dynamic Systems, Measurement and Control*, **124**(4), 698-701.

Sauter, J. A., Matthews, R., Parunak, H. V. D. and Brueckner, S. (2002), Evolving Adaptive Pheromone Path Planning Mechanisms, *First International Conference on Autonomous Agents and Multi-Agent Systems*, Bologna, Italy, 434-440.

Sugihara, K. (1999), Genetic Algorithms for Adaptive Planning of Path and Trajectory of a Mobile Robot in 2D Terrains, *IEICE Trans. Inf. & Syst.*, Vol. E82-D, 309 – 313.

Sugihara, K. (1997), *A Case Study on Tuning of Genetic Algorithms by Using Performance Evaluation Based on Experimental Design*, Tech. Rep. ICS-TR-97-01, Dept. of Information and Computer Sciences, Univ. of Hawaii at Manoa.

Sugihara, K. (1997), Measures for Performance Evaluation of Genetic Algorithms, *Proc. 3rd. joint Conference on Information Sciences*, Research Triangle Park, NC, vol. I, 172-175.

Cobb, H. G. (1990), An Investigation into the Use of Hypermutation as an Adaptive Operator in Genetic Algorithms Having Continuous, Time-Dependent Nonstationary Environments, Technical Report AIC-90-001, Naval Research Laboratory, Washington, D. C.

Fonseca, C. M. and Fleming, C. J. (1993), Genetic algorithms for multiobjective optimization: formulation, discussion and generalization, *5th Int. Conf. Genetic Algorithms*, 416-423.

Srinivas, M. and Deb, K. (1994), Multiobjective Optimization using Nondominated Sorting in Genetic Algorithms, *Evolutionary Computation*, **2**(3), 221-248.

Man, K. F., Tang, K. S. and Kwong, S. (1999), *Genetic Algorithms*, Ed. Springer, 1st Edition, London, UK.

Oliveira, G. M. B., Bortot, J. C. and De Oliveira, P. P. B. (2002), Multiobjective Evolutionary Search for One-Dimensional Cellular Automata in the Density Classification Task, in Proceedings of *Artificial Life VIII*, MIT Press, 202-206.

Schaffer, J. D. (1985), Multiple Objective Optimization with Vector Evaluated Genetic Algorithms, Genetic Algorithms and their Applications: Proceedings of the First International Conference on Genetic Algorithms, 93-100.

Goldberg, D. E. (1989), Genetic Algorithms in Search, Optimization and Machine Learning, Addison-Wesley, Reading, MA.

Spandl, H. (1992), *Lernverfahren zur Unterstützung der Routenplanung fur eine Mobilen Roboter*, Diss. Universität Karlsruhe, auch VDI-Forsch-Heft Reihe 10 Nr. 212, Düsseldorf, Germany, VDI-Verlag.

Choset, H., La Civita, M. L. and Park, J. C. (1999), Path Planning between Two Points for a Robot Experiencing Localization Error in Known and Unknown Environments, *Proceedings of the Conference on Field and Service Robotics (FSR'99)*, Pittsburgh, PA.

Kim, B. N., Kwon, O. S., Kim, K. J., Lee, E. H. and Hong, S. H. (1999), "A Study on Path Planning for Mobile Robot Based on Fuzzy Logic Controller", *Proceedings of IEEE TENCON'99*, 1-6.

Planas, R. M., Fuertes, J. M. and Martinez, A. B. (2002), Qualitative Approach for Mobile Robot Path Planning based on Potential Field Methods, *Sixteenth International Workshop on Qualitative Reasoning (QR'02)*, 1 -7.

Part IV Pattern Recognition Applications

The Role of Neural Networks in the Interpretation of Antique Handwritten Documents

Pilar Gómez-Gil, Guillermo De los Santos-Torres, Jorge Navarrete-García, Manuel Ramírez-Cortés

Department of Computer Science, (2) Department of Electrical Engineering. Universidad de las Américas, Puebla. MEXICO

Abstract. The need for accessing information through the web and other kind of distributed media makes it mandatory to convert almost every kind of document to a digital representation. However, there are many documents that were created long time ago and currently, in the best cases, only scanned images of them are available, when a digital transcription of their content is needed. For such reason, libraries across the world are looking for automatic OCR systems able to transcript that kind of documents. In this chapter we describe how Artificial Neural Networks can be useful in the design of an Optical Character Recognizer able to transcript handwritten and printed old documents. The properties of Neural Networks allow this OCR to have the ability to adapt to the styles of handwritten or antique fonts. Advances with two prototype parts of such OCR are presented.

1 The Problem of Antique Handwritten

Currently, web distribution of old documents is limited to a scanned image of the document because most of the commercial Optical Character Recognizers (OCR) do not obtain good recognition rates with old handwritten documents or with documents using old styles of fonts.

The recognition of old handwritten and printed documents is a challenge in pattern recognition, due to special characteristics that this recognition problem presents. Figure 1 shows an example of an old telegram, written by Gral. Porfirio Díaz, president of Mexico at the beginning of XX Century. Even for a non-expert person, who does not have some previous knowledge of this kind of writing, is very difficult to interpret the content of this document.

P. Gómez-Gil et al.: *The Role of Neural Networks in the Interpretation of Antique Handwritten Documents*, StudFuzz **208**, 269–281 (2007)
www.springerlink.com

Digital processing of old documents faces, among others, the following conditions:

- Old documents have been damaged with the pass of time. In most cases they present spots, color of paper have changed, or their texture is deteriorated.
- Digitalization process requires special cares to protect the documents. The production of a digital image that will feed the OCR is, by itself, a delicate process. It requires a special kind of scanner, which would not touch the document.
- The recognition process of old documents is off-line. There is no information about the dynamics of the writing or the pressure used by the writer.

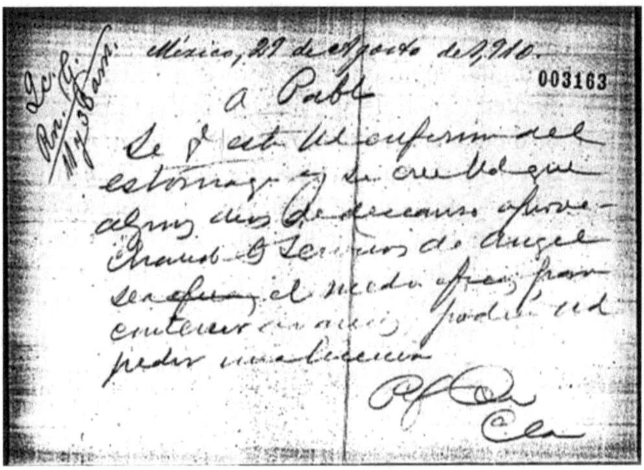
México, 27 de Agosto de 1910.
003163
a Pablo

Fig. 1. An example of a telegram written by Porfirio Díaz at the beginning of XX century [1]

Added to these conditions, there are also special complications during the recognition of old handwriting. Some of them are [2]:

- Old styles of handwriting have a lot of ornaments.
- Fonts are not uniform. For example, same character may look different in different places of a word, in different words or in different documents. Notice that this situation is presented in any kind of handwriting, and is much stronger if documents came from different writers.
- The shape and style of writing may be different even for the same person depending on environmental factors, mood, type of pens, age, etc.

- Character segmentation requires extra procedures, besides the common ones as identification of valleys and hills, due to the styles of different letters.
- In some patterns it is noticed that different classes of characters are very similar in shape.

Figure 2 shows some examples of handwritten words written by the same writer at different moments and documents. Notice that some letters have different shapes depending on their positions in the word, and when presented in different words. Some letters may be confused with a connection and some letters may be "embedded", looking two of them as one character. Therefore, in terms of a pattern recognition problem we have that:

- There are no evident prototypes to define each class
- The variance among members of the same class is greater than expected values
- Common similarity metrics, as Euclidian distance, are sometimes useless because it may be greater for patterns belonging to same class than for patterns belonging to different classes.

2 An OCR for Antique Handwritten Documents

The research group of Neural Networks and Pattern Recognition at Universidad de las Américas, Puebla, is currently working with the construction of an OCR able to recognize antique handwritten and printed documents. This OCR will be useful to our library, which posses a huge amount of such historical documents [3].

We propose the construction of an adaptive OCR, called *Priscus* (latin word meaning "antique") that have the following components (see figure 3):

- Digitization. Creation of a color or gray level image of the document to be recognized.
- Pre-processing. Cleaning of image, noise reduction and black and white conversion of the image.
- Segmentation of words. Given a binary map, this process obtains the words that are presented in the image.
- Segmentation training. Adaptive system that learns to identify segmentation points in a word, based on the handwriting or font presented to the OCR.

- Character segmentation. This process obtains possible segments that may contain characters, based on the knowledge obtained from the segmentation training.
- Recognizer training. Adaptive system that learns to identify characters from segments obtained from the binary image of the document.
- Recognition of characters. It receives segments of words, extract features of them and decide the most likely characters.
- Identification of words. Based on possible words obtained by the recognizer and a dictionary, this process decides the most likely words.
- Correction of style. Based on the identified most likely words and grammar rules, this process creates well formed sentences, obtained a transcription of the document.

At this point, we have focused our research in the segmentation and character recognition components, using artificial neural networks.

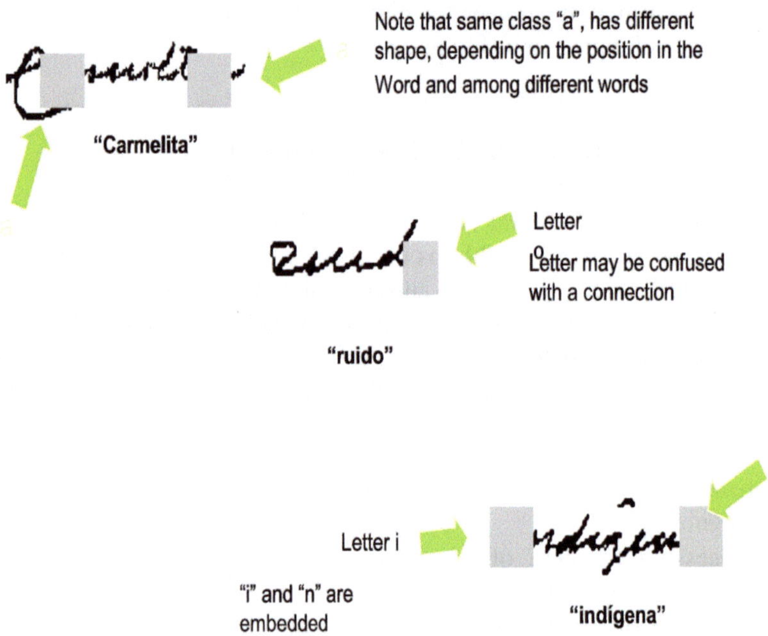

Fig. 2. Examples of old handwritten words [2]

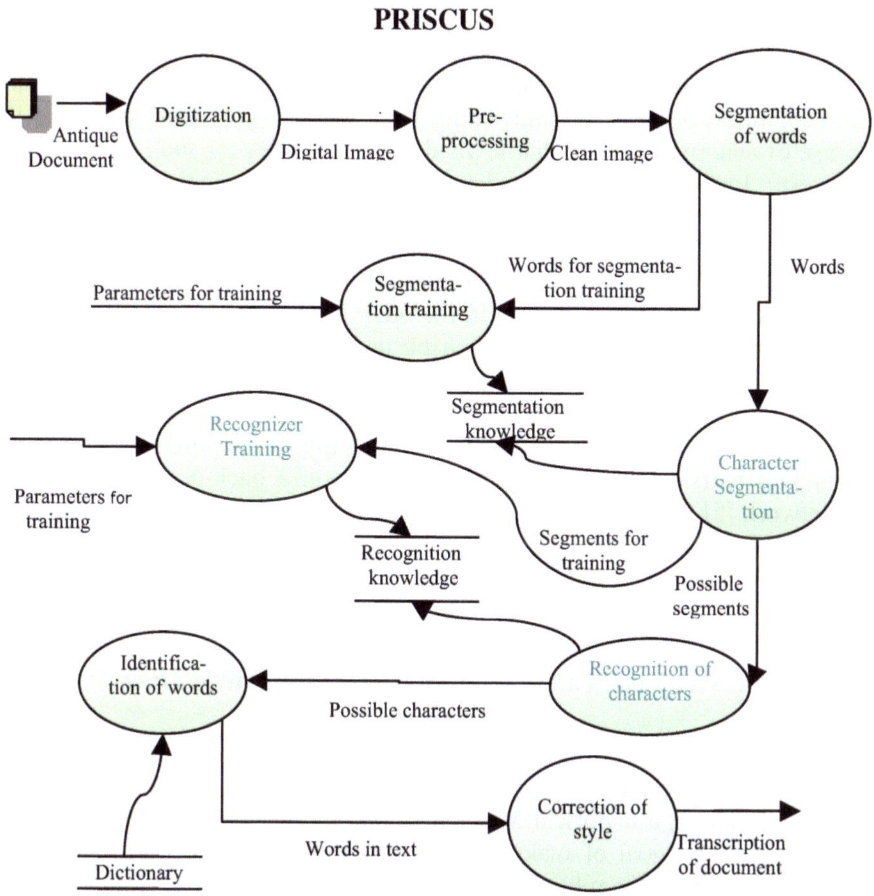

Fig. 3. The proposed OCR for transcription of old documents

3 Why Neural Networks?

Artificial Neural Networks (ANN) are mathematical models inspired in biological systems, able to learn the behavior of a system getting their knowledge from data. For our purposes we use ANN simulated in digital computers, but they are available also as hardware components. There are many types of ANN, and most of them present three important characteristics: abstraction, generalization and learning [4]. ANN are useful for problems were the functions describing a system are not explicit, but there is enough data that likely can be used to obtain some numerical representation of the knowledge ruling the behavior of the system.

It is evident that the definition of explicit rules for the segmentation and recognition process in handwriting is a very complex task, therefore the use of adaptive systems that learned from examples for such processes are useful. In the other hand, it is known that some models of neural networks are able to obtain better generalization rates than others adaptive systems. This generalization ability is an advantage in the case of handwriting recognition, given the spread presented in the patterns of the clusters of classes presented in this case. In the other hand, having trainable systems, segmentation and recognition can be tailored to the style of the writing found in specific applications or for fonts found in specific periods of the history.

Up to date, we have been working with the development of prototypes of two parts of the OCR: a system based on a back-propagation neural network [5] for segmentation of words and a character recognizer based on a SOFM neural network [6]. Both sub-systems will be explained in the next sections.

4 Test Case: Telegrams Written by Gral. Porfirio Diaz

The library of the Universidad de las Américas Puebla, contains a rich collection of old documents. Among them, there are about 70,000 telegrams written during a historical Mexican epoch known as "*Porfiriato.*" The library has the goal of making them available across the web. Up today, around 2,000 of them have been digitized and their contents transcribed by experts and they are available for consulting [1].

In order to test our segmentation and recognition systems, 25 of such telegrams were scanned, and their images were manually cleaned from noise and printed lines using commercial software, and manually a set of black and white isolated word images were cutting.

5 Segmentation Using a Back Propagation ANN

We build a segmentation application, called HOWOST, based on 3 components: the "white hole algorithm" proposed by Nicchiotti et al. [7], a vertical density algorithm proposed by Kussul & Kasaktina [8] and a Back propagation neural net trained to reduce the over segmentation generated by both algorithms.

The white hole algorithm detects white pixel areas rounded by black runs, in order to find caves or circumferences corresponding to letters as *a,b,c,d,e,g,h,n,o,p,q*. This method forces segmentation points before and after the white area. The second algorithm builds a black pixel density histogram for each word column so that valleys in the histogram indicate the presence of a ligature between letters.

The BP neural network is trained to learn which of the segmentation points generated by both algorithms are right and which are wrong. The supervised classification to train the network is given by a human expert that marked the correct and incorrect segmentation points in some examples automatically generated for the algorithms. After the network is trained with examples of the handwritten or font documents, the system may be used on-line to segment words. Figure 4 shows the interface of the segmentation subsystem.

Different configurations of networks with different data sets have been tested, but three major experiments were carried out to test the performance of the system. The first experiment uses very hard-to-read words, like words with short ligatures, overlapping, and bad quality, requiring an expert for their interpretation. The second experiment uses an ANN trained with "easy" words, like words with prominent ligatures easy to read for common people. The third experiment uses a training file with easy and hard words combined. All experiments use the same network topology (270-300-200-100-1), and a learning rate of 0.12. Training was stopped after 200 epochs. Table 1 shows the results of these experiments. As expected, training the network with difficult words improves de number of correct segmentation points when difficult data is tested. The low level of over segmentation obtained in the three cases demonstrates the success of the hybrid technique in this type of writing.

In general, using a set of 898 mixed patterns, we got 83% of accuracy in segmentation combining the two algorithms with the neural network. When tested independently, the white hole algorithm obtained 47% of success in the best case, and the vertical density algorithm obtained 53% .

6 Self-Organized Maps for Character Recognition

For the recognizer we chose a neural network able to create topological maps trained with a non supervised algorithm [10]. We decided to use topological maps because, given the special characteristics of these problems, it was mandatory to have several prototypes of each class, and to relate them in a way that similar prototypes were near in a way that their relative relation were shown. The decision of use non supervised algorithm was based on the idea that the learning showed by humans when reading old handwriting is no supervised.

Inspired in the organization by maps of human brain T. Kohonen developed the self organizing feature mapping algorithm (SOFM) [6]. The goal of SOFM algorithm is to store a set of input patterns $\mathbf{x} \in X$ by finding a set of prototypes $\{\mathbf{w}j \mid j = 1, 2...N\}$ that represent the best feature map $\mathbf{\Phi}$, following some topological fashion. The map is formed by the weights connection $\mathbf{w}_j$ of a one or two-dimensional lattice of neurons, where the neurons are also related each other in a competitive way.

This learning process is stochastic and off-line; that is, two possible stages are distinguished for the net: learning and evaluation. It is important to notice that the success of map forming is highly dependent on the learning parameters and the neighborhood function defined in the model. The map is defined by the weights connecting the output neurons to the input neurons.

Following is a description of the SOFM algorithm as applied in the construction of the recognizer [11].

1. Initialize the weights with random values:

$$\mathbf{w}_j(0) = \text{random}() \quad , j = 1..N \text{ (number of neurons)} \tag{1}$$

2. Chose randomly a pattern $\mathbf{x}(t)$ from the training set X at iteration t.
3. For each neuron i in the map feature map $\mathbf{\Phi}$ calculate the similarity among its corresponding weight set $\mathbf{w}_i$ and $\mathbf{x}$. The Euclidian distance may be used:

$$d^2(\mathbf{w}_i, \mathbf{x}) = \sum_{k=1}^{n} \left(w_{ik} - x_k\right)^2 \quad i = 1..N \tag{2}$$

4. Find a wining neuron i^* which is the one with maximum similarity (minimum distance).

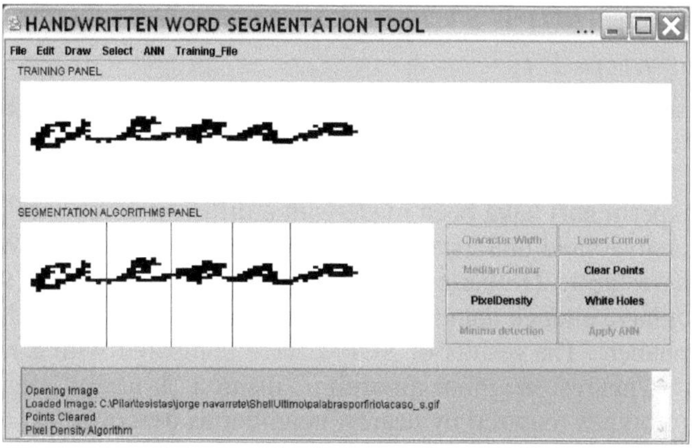

Fig. 4. Our segmentation software [9]

Table 1. Results obtained by the segmentation subsystem

Training input	Ideal number of segmentation points in the test set	Correct segmentation points found by HAWOST	Incorrect segmentation points found by HOWOST
Hard-to-read words	82	67	22
Easy-to-read words	82	43	23
Mixed words	82	55	18

5. Update the weights of winning neuron i^* and their neighbors as:

$$\mathbf{w}_j(t+1) = \mathbf{w}_j(t) + \alpha(t)(\mathbf{x}(t) - \mathbf{w}_i(t)) \quad \text{for j} \in \Lambda_{i^*}(t) \tag{3}$$

6. Where $\Lambda_{i^*}(t)$ corresponds to a neighborhood function centered on the winning neuron. For this problem, we choose a neighborhood distance

of 0 neurons. $\alpha(t)$ is a learning rate function depending on time. We choose: $\alpha(t) = 1/t$.

7. Go to step 2 until no more changes in the feature map are observed or a maximum number of iterations is reached.

Several experiments have been made with a different number of classes in order to analyze and understand the behavior of this network. We started with 3 classes up to 21 classes. Unfortunately, at the moment of this work, we did not have enough data to test the whole alphabet with 27 classes (the whole alphabet). The results of SOFM were compared with a recognizer based on a "nearest neighbor algorithm" using a "k-means" algorithm to get the prototypes required by nearest neighbor as described at [12]. Table 2 shows the results obtained by both the SOFM network and the nearest neighbor classifier. Notice that in all cases the SOFM network gets better results that the nearest neighbor algorithm.

Figure 5 shows some topological maps generated by the SOFM using patterns of the five vowels. Notice that the maps result as expected. Similar prototypes are generated near each other. Figure 6 shows the topological maps for 21 classes generated by the network.

Table 2. Results of recognizer with different number of classes [10]

Number of classes	Number of training patterns	Type of Recognizer	Recognition rate on Training set
3	13	Nearest neighbor	84%
		SOFM (3x3)	92%
5	56	Nearest neighbor	58%
		SOFM (5x1)	58%
		SOFM (5x2)	71%
		SOFM (5x5)	73%
21	86	Nearest neighbor	6%
		SOFM (5x12)	63%
		SOFM (2x30)	70%

7 Conclusions and Future Work

We presented the overall issues associated to the construction of a OCR able to process antique handwritten and printed documents. The special characteristics associated to this problem were discussed, as well as the role of artificial neural networks in the implementation of useful segmentation and recognition systems. The advances obtained in these two subsystems were also presented.

It is clear that there is still a lot of work to be done, because each component of this recognizer is by itself a complete system. At this moment we are working with the integration of segmentation and recognition subsystems, as well with the identification of a systematic way to look for the best SOFM topology. It can be noticed that we did not present results from the application of the recognizer or the segmentation system in old printed documents; we are also working with this part.

Acknowledgments

Authors would like to thank Mr. Alberto García-García, for the advisements given during the development of this work, as well as for providing us with the images of the telegrams.

Fig. 5. Topological maps generated for vowel using different topologies [10].

Fig. 6. Topological maps generated for 21 classes using 2 different topologies [10].

References

1. Universidad de las Américas, Puebla. Digitalización, Codificación y el Acceso Vía Internet de los Telegramas del ex presidente de México Porfirio Díaz. In: Colecciones Digitales Biblioteca (2002) http://biblio.udlap.mx/telegramas
2. Gomez-Gil, P.; Navarrete-García, J.: Analysis of a Neural-net based Algorithm for the Segmentation of Difficult-to-read Handwritten Letters." In: WSEAS Transactions on Systems. Issue 4, Vol. 3 (2004) 1426 – 1429
3. García-García, A.: Digitalización y Divulgación Digital de Acervos Antiguos. In: Servicios Digitales. Bibliotecas de la Universidad de las Américas Puebla. http://ict.udlap.mx/projects/cudi/buap/ (2004)
4. Haykin S.: Neural Networks: a Comprehensive Foundation. Macmillan College Publishing Company. New York. (1994)
5. Rumelhart, D.E. G. E. Hinton and R.J. Williams.: Learning Internal Representation by error propagation. In: Parallel Distributed Processing: Explorations in the Microstructure of Cognition D.E. Rumelhart and J.L. McClelland, eds. Vol. 1, Chapter 8. Cambridge, MA: MIT Press. (1986)
6. Kohonen, T.: Self-Organized formation of topologically correct feature maps. Biological Cybernetics, 43, (1982) 59-69.
7. Nicchiotti G., Scagliola, C., Rimassa. S.: A Simple and Effective Cursive Word Segmentation Method. Proceedings of the Seventh International Workshop on Frontiers in Handwriting Recognition, Amsterdam, (2000) 499-504.
8. Kussul Mikhailovich, E. and Kasaktina, L.M : Neural Network System for continuous handwritten Words Recognition. Book of Summaries of International Joint Conference on Neural Networks. Washington, D.C., (1999) 22.
9. Navarrete-García, J.: Mejora en el algoritmo de segmentación para el reconocimiento de caracteres de telegramas escritos por el Gral. Porfirio Díaz.

Tesis para obtener el grado de Licenciatura. Departamento de Ingeniería en Sistemas Computacionales. Universidad de las Américas, Puebla. (2002).
10. De-los-Santos-Torres, G.: Reconocedor de Caracteres Manuscritos. Master thesis. Departamento de Ingeniería en Sistemas Computacionales. Universidad de las Américas, Puebla. (2003).
11. Gómez-Gil, Pilar, De los Santos-Torres, M., Ramírez-Cortés, Manuel: Feature Maps for Non-supervised Classification of Low-uniform Patterns of Handwritten Letters. Progress in Pattern Recognition, Image Analysis and Applications, Lecture Notes in Computer Science Vol. 3287 (2004) 203-207.
12. Tao, J.T. and Gonzalez, R.C. Pattern Recognition Principles. Addison-Wesley (1974)

Object Recognition Using Fuzzy Inferential Reasoning

Thompson Sarkodie-Gyan

Department of Electrical and Computer Engineering, University of Texas, El Paso TX 79968, Email: tsarkodi@utep.edu

Abstract. This paper introduces a vision-based pattern recognition scheme for the identification of very high tolerances of manufactured industrial objects. An image-forming device is developed for the generation and the capture of images/silhouettes of the components. A simple but effective feature extraction algorithm is employed to produce distinguishable features of the components in question. Radial basis function (RBF) based membership functions are used as classifiers for the pattern classification. For the decision making process, a fuzzy logic based inferential reasoning algorithm is implemented for the approximate reasoning scheme.

1 Introduction

In recent years, the concept of *fuzzy logic* has become a favourite technique for inferential reasoning processes in many control and decision-making applications ranging from manufacturing processes to medical diagnosis. Fuzzy logic, developed by Zadeh [1,2,3], has provided a more natural and human-like interface between human and machine capable of making many rational decisions in environments with llevels of uncertainty and imprecision. In fact, many Japanese consumer products such as washing machines, air conditioning systems and camcorders, to mention just a few, have already successfully applied fuzzy logic concepts for control purposes. There are two main reasons for the current use of fuzzy logic. First, the basic design principle of fuzzy logic systems is very similar to the way of human reasoning, based on simple conditional rules. Second, expert knowledge (normally in way of natural language) can be built into the systems in advance and can easily be adapted and improved.

In this paper, the concept underlying fuzzy inferential reasoning for the identification of high tolerances in manufactured components is proposed

T. Sarkodie-Gyan: *Object Recognition Using Fuzzy Inferential Reasoning*, StudFuzz **208**, 283–297 (2007)
www.springerlink.com

and explained. Several different test objects, compressor impellers, which serve as significant components of turbochargers, are used for the identification.

A turbocharger is a device fitted to vehicle engines to increase engine brake power and torque over parts of the engine speed range. They are also used for the avoidance of higher peak pressures beyond those of the normally aspirated vehicle engines [4]. Since different classes of turbochargers have different shapes, different characteristics and, therefore, different performances, the use of the correct components to assemble the turbocharger is highly significant in terms of performance and working-life of both the turbocharger and the engine.

Incorrect impellers fitted into the turbocharger may result in the degradation of the engine performance and even damage to the engine itself. The difference between different classes of impellers are so small such that it is almost impossible to identify them by human visual perception. The use of human inspectors to check every impeller is impractical especially at high rates of production. Therefore, there is a need for an easy-to-operate vision-based checking system at the assembly unit or parts dispatch stage to verify the identities of parts.

In this paper, the design configuration, the development and testing of the system are all described. Both the advantages and the shortcomings of the system are discussed as well.

2 Image-Forming Device

A high quality input image is always the fundamental requirement of an accurate and high performance image analysis process. The device used for forming and storing the images/silhouettes of the impeller blade(s) is the *mechano-optical arrangement* [5,6] as in Fig. 1. This image-forming device consists of an incandescent light source, two special purposed reflecting mirrors, a stepper motor-driven rotary table, and a CCD camera. The operating principle of this image-forming device is described as below.

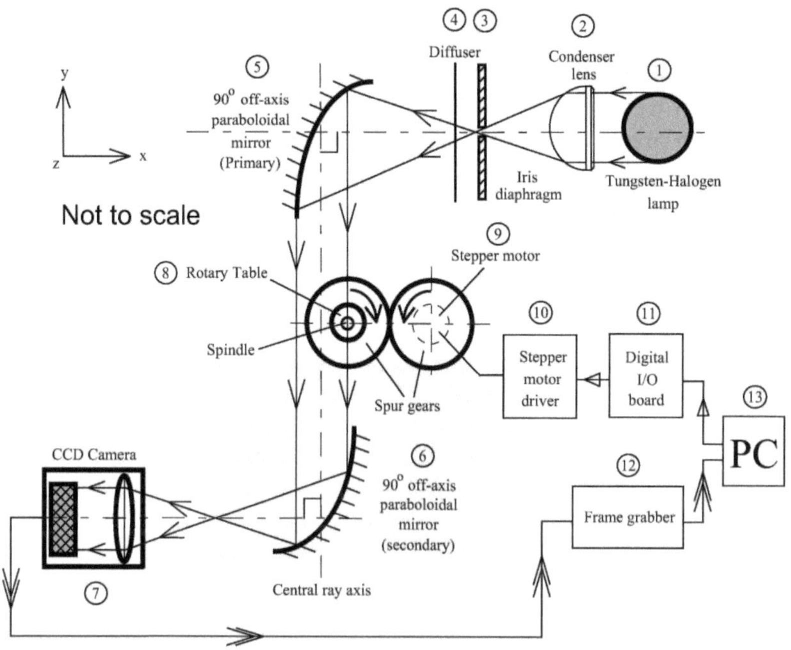

Fig. 1. Design of the mechano-optical arrangement

The device consists of a *tungsten-halogen lamp* ① for the generation of a white beam as the light source of the system. The beam is then focused onto an *iris diaphragm* ③ by passing it through an *aspheric condenser lens* ②. After the beam has been diffused by a *circular holographic diffuser* ④, a picture of the silhouette of the impeller blade(s) will be reflected to a *CCD camera* ⑦ by a pair of *90° off-axis paraboloidal mirrors* ⑤&⑥.

In this experimental investigation, impellers with different tolerances are screwed down separately to a spindle vertically mounted on a *rotary table* ⑧. 25 sequential images of the blade silhouettes of the impeller were recorded by rotating a *stepper motor* ⑨ through 45° in pre-programmed steps (i.e., 1.8°/step) and images were captured at each interval. The images were subsequently digitised by a *frame grabber* ⑫ into 25 imaging files.

The stepper motor is driven by a *stepper motor driver* ⑩. A *digital input/output board* ⑪ together with a tailored software program are used to produce the control signal from the *PC* ⑬ to the stepper motor driver. The programs for the stepper motor movement control and the image acquisition process are both written in "C" and run on a 486 PC. The capture time for 25 images is approximately 13 seconds. Once the images are digitised and stored in the memory, feature extraction operations can be performed. It may be noted that, since the impeller is symmetrical in shape, only a half of it is used for the process. Fig. 2 illustrates the form of the input image.

Fig. 2. Input image

3 Feature Extraction

The main objective of feature extraction is to transform the output data acquired from the sensing device (pattern space) into a new lower dimensional space (feature space) for later pattern representation and/or class discrimination. In order to gain processing speed, a reduction of the data dimensionality becomes a main feature of extraction. However, information is always lost in the feature extraction stage, and it is possible that some of this information is valuable for classification. Therefore, selecting the optimal feature(s) is always important to the performance of the classification process.

A one-dimensional edge detector and a feature transformation plus indexing technique are used to extract the distinguishable features of the images. The flow diagram below (Fig. 3) depicts the principle underlying the feature extraction algorithm. A full description of this algorithm can be found in [7].

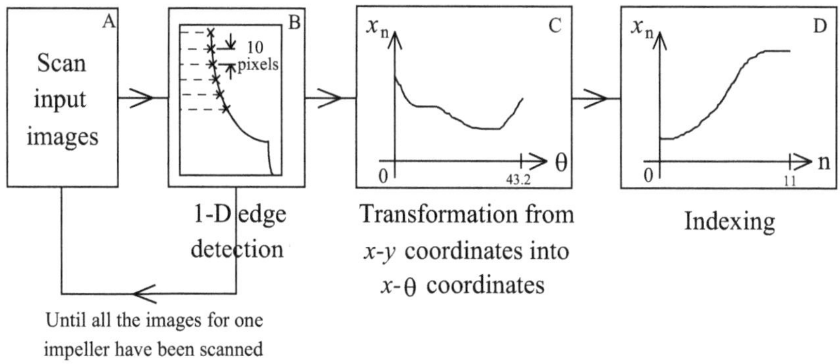

Fig. 3. Principle of the feature extraction algorithm

4 Design of the Classifier

In general, a classifier sorts out patterns into mutually exclusive regions or classes. Since classifiers are usually designed with labelled patterns, their design is sometimes unknown as the *supervised learning* [10].

Classifier design normally consists of two stages, *training* and *testing*. After the establishment of the basic decision rules, the boundaries that separate the pattern classes may be obtained. This is done by training the classifier on a group of known objects (training set). Once the training is finished, a classifier evaluation process (testing) has to be performed to ensure the accuracy of the classifier. One of the most common methods for classifier testing is by presenting another group of known objects to the classifier and counting the number of misclassifications.

In this study, a Radian Basis Function (RBF) based membership function is used as the classifier. A modified Gaussian-like RBF is expressed as

$$\mu(x_i) = h \cdot \exp\left[\frac{(x_i - \overline{x})^2}{2\sigma^2}\right] \qquad i = 1,2,3,\ldots\ldots$$

where $\overline{x}$ is the mean of x (the central position), σ is the standard deviation of x (the width) and,

$$h = \frac{1}{\sigma\sqrt{2\Pi}}$$

is a real constant which represents the height of the function. A graphical representation of the membership function is illustrated in the Figure 4.

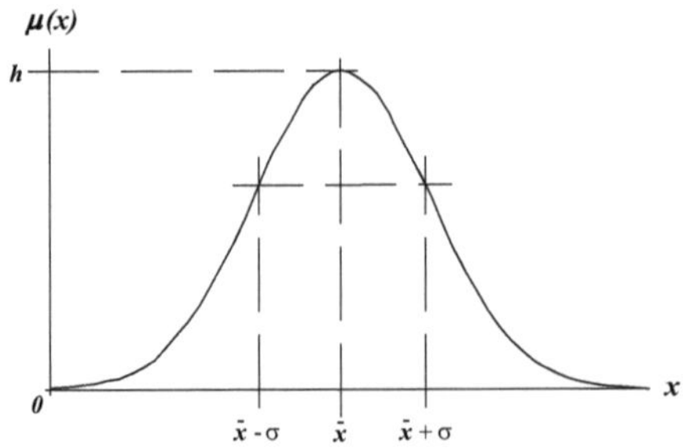

Fig. 4. Graphical explanation of the membership function

Since the classes of the impellers are known to possess subtle differences by virtue of manufacturing tolerances, it may be possible to recognise these impeller classifications by applying three characteristic attributes as the membership function parameters ($\overline{x}$, σ and h) as inputs to a decision-making algorithm.

Table 1 shows the knowledge-base of the membership parameters for the 1700 and 5300 impeller series. In view of the manufacturing tolerances of the impellers provided by the manufacturer, it was necessary to define not only three parameters for each class of impellers, but also two limits for each parameter to allow for all the possibilities of impeller acceptance within one class.

Table 1. Illustration of the knowledge-base of the membership parameters for the 1700 and 5300 series impellers.

Class	1710	1711	1712	5358	5398
$\overline{x}$	> = 514 < = 523	> = 499 < = 508	> = 484 < = 493	> = 560 < = 569	> = 519 < = 528
σ	> = 99.3 < = 103.8	> = 104.5 < = 109.0	> = 110.1 < = 114.6	> = 76.0 < = 79.5	> = 92.7 < = 96.6
h	> = 0.0038 < = .00405	> = 0.00362 < = 0.00385	> = 0.00346 < = 0.00366	> = 0.00500 < = 0.00526	> = 0.00410 < = 0.00434

5 Classification : Fuzzy Logic Based Inferential Reasoning

Intelligent reasoning and clever inference of conclusions — especially in the presence of ambiguity, confusion, missing information, deception, and so forth — are major attributes of human problem solving capabilities [11]. The main task of this system is to perform intelligent decision-making through inference and extraction of conclusions from available relative data — an inferential reasoning system that can more closely emulate human performance.

To cope with this style of reasoning, a suitable modelling technique is developed using fuzzy logic based on fuzzy set theory since this system deals with inexact, uncertain and incomplete data.

Fuzzy logic based on fuzzy set theory provides at least two advantages in the field of pattern recognition [12], first, it serves as an interface between the linguistic variables which seem to be preferred by humans, and also the quantitative characterisations appropriate for machines. Secondly, it emphasises the possibility distribution interpretation of the concept of fuzziness. Fuzzy logic legitimises and provides a meaningful interpretation for some distributions that human beings believe are useful, but might have difficulty for their justification on the basis of objective probabilities.

For this particular project, a cross-correlation algorithm based on the theory of fuzzy sets, the *fuzzy cross-correlation algorithm* [7,9], has been developed to measure the fuzzy similarity relationship between two feature vectors obtained from two similar objects.

5.1. Principle of the Fuzzy Cross-Correlation

Let A be a fuzzy subset of a classical set X, then we can write,

$$A = \{(x_i, \mu_A(x_i)) : x_i \in X; \mu_A(x_i) \in [0,1]\}$$

In the case of finite universes, it is more convenient to use vector notation,

$$\begin{aligned} a^T &= [a_1, \ldots, a_n] \\ &= [\mu_A(x_1), \ldots, \mu_A(x_n)] \end{aligned}$$

The Cartesian product $X \times Y$ of finite universes is expressed as,

$$X = \{x_1, \ldots, x_n\} \text{ and } Y = \{y_1, \ldots, y_m\}$$

and may be written in a corresponding matrix notation as,

$$A = \begin{bmatrix} a_{1,1} & \cdots & a_{1,m} \\ \vdots & & \vdots \\ a_{n,1} & \cdots & a_{n,m} \end{bmatrix} = \begin{bmatrix} \mu_A(\bar{x}_1, \bar{y}_1) & \cdots & \mu_A(\bar{x}_1, \bar{y}_m) \\ \vdots & & \vdots \\ \mu_A(\bar{x}_n, \bar{y}_1) & \cdots & \mu_A(\bar{x}_n, \bar{y}_m) \end{bmatrix} \tag{1}$$

Fuzzy sets of the type given by Eqn (1) play an important role in fuzzy relations in $X \times Y$ indicating the strength of the relations between, $x_i \in X$ and $y_j \in Y$.

The composition $U \circ V$ of the fuzzy relations U in $X \times Y$ and V in $Y \times Z$, where $Z = \{z_1, \ldots, z_p\}$ is an additional universe, is the fuzzy relation in $X \times Z$,

$$R = U \circ V = \begin{bmatrix} r_{1,1} & \cdots & r_{1,p} \\ \vdots & & \vdots \\ r_{n,1} & \cdots & r_{n,p} \end{bmatrix}$$

where the element $r_{i.k}$ $(i = 1,..,n; \quad k = 1,\ldots,p)$,

$$r_{i,k} = u_i^T \circ v_k = \max_j \left[\min\left(u_{i,j}, v_{j,k}\right)\right] \tag{2}$$

are commonly interpreted as the strength of the chains linking x_i to z_k.

Equation (2) can be represented by an equivalent notion,

$$r_{i,k} = hgt\left(u_i \cap v_k\right) \tag{3}$$

where $hgt(\cdot)$ is the height of the fuzzy set which results from the intersection $u_i \cap v_k$.

When both u_i and v_k are non-normalised fuzzy sets, with height less than 1, it may be useful to write,

$$r_{i,k} = \frac{hgt\left(u_i \cap v_k\right)}{hgt\left(u_i \cup v_k\right)}$$

instead of Eqn (3).

The disadvantage of the max-min composition as a measure of composition between two fuzzy sets is that $r_{i.k}$ would become equal to 1 even though $u_i \neq v_k$.

Therefore, an improved version of a measure is obtained by the relationship,

$$r_{i,k}^{*} = \frac{|u_i \cap v_k|}{|u_i \cup v_k|}$$

where $|\cdot|$ indicates the fuzzy cardinality of the corresponding fuzzy set. In this case, the measure of $r_{i,k}^{*}$ is equal to 1, if and only if $u_i = v_k$.

By means of $*$–composition, the equation,

$$\left(r_i^{*}\right)^T = u_i^T * v \tag{4}$$

may be constructed as a set of comparative operations between a fuzzy sample set u_i and a reference set of fuzzy sets $V = \{v_1,\ldots,v_p\}$, where u_i as well as $v_1,\ldots,v_p$ are fuzzy subsets of Y. Thus the fuzzy set r_i^{*} of Z indicates the grade of conformity of the pairs (u_i, v_k) for a fixed i and $k = 1,\ldots,p$.

Obviously, the composition in Eqn (4), is the cross-correlation of discrete sequences, and it is, therefore, called the fuzzy cross-correlation which may complement the notion of fuzzy convolution.

5.2 Fuzzy Similarity Measure

Equation (5) illustrates the mathematical expression to compute the fuzzy cross-correlation algorithm,

$$\mu_{ref*test} = \omega_{ref}^{n} * \omega_{test}^{n}$$

or,

$$\mu_{ref*test} = \frac{\sum_{i=1}^{n} \min[R_i, T_i]}{\sum_{i=1}^{n} \max[R_i, T_i]} \qquad i = 1,2,3,\ldots \tag{5}$$

$$= \frac{[R_1 \cap T_1] + [R_2 \cap T_2] + \ldots}{[R_1 \cup T_1] + [R_2 \cup T_2] + \ldots}$$

where $\omega_{ref}^{n} = [R_i]$ and $\omega_{test}^{n} = [T_i]$ represent the reference and the testing feature matrices respectively, n is the number of elements in the matrix, and the R_i and T_i are the elements in the matrix. The symbol "$*$" represents the fuzzy cross-correlation operator, the "min" represents the fuzzy logic intersection, the "max" represents the fuzzy logic union and the $\mu_{ref*test}$ is the grades of similarity of the testing component to the reference component ranging from 0 to 1.

Figure 5 shows the principle of a *Neuro-fuzzy algorithm* based on the radial basis function as membership functions of the objects (impellers) and the fuzzy cross-correlation algorithm.

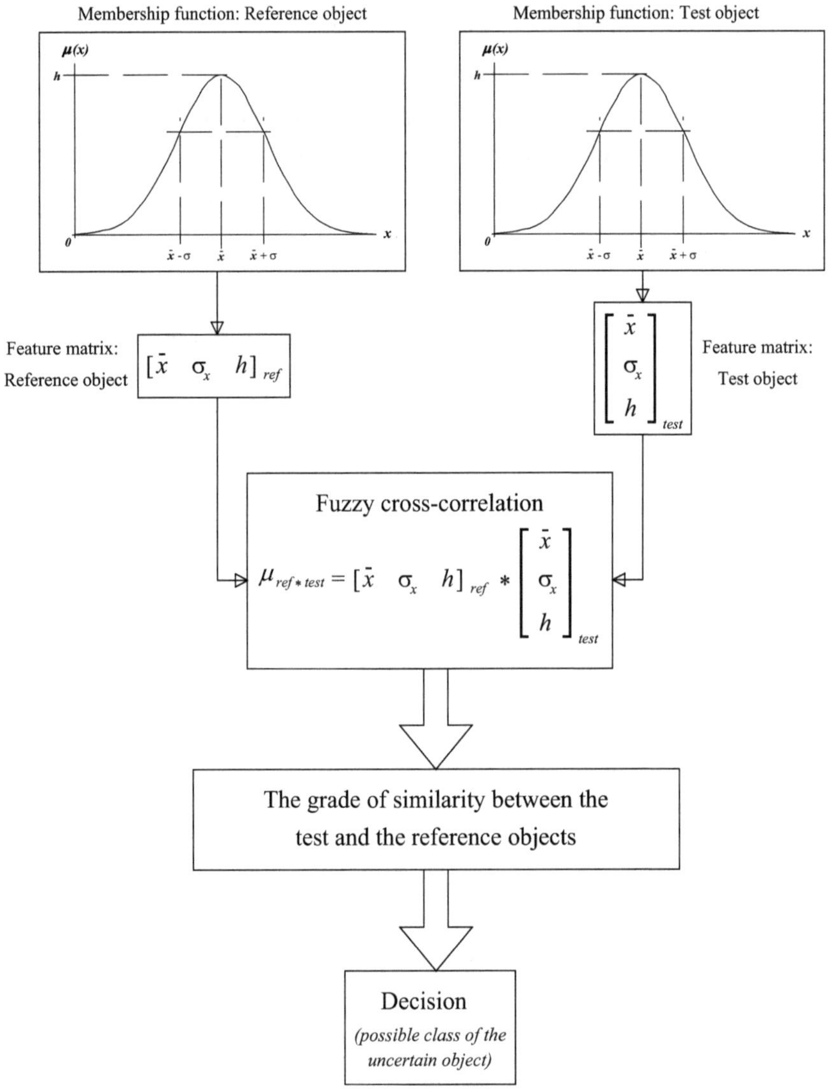

Fig. 5. Principle of the Neuro-fuzzy algorithm

The attributive parameters ($\bar{x}$ - the position; σ_x - the width; and h - the height) of the membership function of the reference object are first extracted and stored in a knowledge-base. The attributive parameters of the membership function of the test objects are then extracted and expressed in a matrix format for similarity measure. The following matrices represent the feature matrices of the reference object and the test object respectively.

$$\begin{bmatrix} \bar{x} & \sigma_x & h \end{bmatrix}_{ref} \qquad \begin{bmatrix} \bar{x} \\ \sigma_x \\ h \end{bmatrix}_{test}$$

Feature matrix: Reference object. Feature matrix: Test object.

For mathematical reasons, the number of columns in the matrix of the reference object should be equal to the number of rows in the matrix of the test object.

From Eqn (5), the grade of similarity between the reference object and the test objects may be computed. "1" can be obtained if the two objects are identical or a number between 0 and 1 is given to show the similarity between the two objects.

$$\begin{aligned} \mu_{ref*test} &= \begin{bmatrix} \bar{x} & \sigma_x & h \end{bmatrix}_{ref} * \begin{bmatrix} \bar{x} \\ \sigma_x \\ h \end{bmatrix}_{test} \\ &= \frac{\left[\bar{x}_{ref} \cap \bar{x}_{test}\right] + \left[\sigma_{x,ref} \cap \sigma_{x,test}\right] + \left[h_{ref} \cap h_{test}\right]}{\left[\bar{x}_{ref} \cup \bar{x}_{test}\right] + \left[\sigma_{x,ref} \cup \sigma_{x,test}\right] + \left[h_{ref} \cup h_{test}\right]} \\ &= Decision \end{aligned}$$

In this study, impeller 5358 is used as the reference model. Using Table 1 as the input values to Eqn (5), the grade of similarity of all five classes of impeller can be defined as in Table 2.

Table 2. Summary of the $\mu_{ref*test}$ for each impeller.

$\mu_{5358T*5358}$	$\mu_{5358T*5398}$	$\mu_{5358T*1710}$	$\mu_{5358T*1711}$	$\mu_{5358T*1712}$
0.9903	0.9078	0.8907	0.8616	0.8325

6 System Evaluation

In this section, the performance of the impeller recognition system will be evaluated. Three impellers with various tolerances are used to test the efficiency and sensitivity of the system. It may be pointed out that these three impellers were not used for the development of the knowledge-base. In order to show the effectiveness of the system, the mechano-optical arrangement was subjected to the influence of ambient lighting during the data acquisition process. Tables 3 to 5 show the results of the evaluation.

Table 3. System evaluation - sensitivity of the impeller recognition system (*1710*).

Test no.	**1**	**2**	**3**	**4**	**5**
$\overline{X}$	516.167	514.583	515.417	515.250	515.917
σ	100.396	102.080	104.024	103.282	100.922
h	0.00397	0.00391	0.00384	0.00386	0.00395
Acceptable	✓	✓		✓	✓
Unclassified			✓		
Grade of similarity			0.8827		
Similar to			1710		

Table 4. System evaluation - sensitivity of the impeller recognition system (*1711*).

Test no.	**1**	**2**	**3**	**4**	**5**
$\overline{X}$	500.333	503.250	501.667	500.750	500.917
σ	107.287	105.535	105.572	105.913	105.772
h	0.00372	0.00378	0.00378	0.00377	0.00377
Acceptable	✓	✓	✓	✓	✓
Unclassified					
Grade of similarity					
Similar to					

Table 5. System evaluation - sensitivity of the impeller recognition system (*1712*).

Test no.	1	2	3	4	5
$\overline{X}$	485.667	484.750	484.083	483.917	487.750
σ	111.092	112.980	113.752	112.935	110.772
h	0.00359	0.00353	0.00351	0.00353	0.00360
Acceptable	✓	✓	✓		✓
Unclassified				✓	
Grade of similarity				0.8250	
Similar to				1712	

Tables 3 to 5 show that most of the classifications are correctly done even though the system suffers from interference by unexpected factors, e.g., ambient lighting. In Table 3 (Test no. 3) and Table 5 (Test no. 4), the classification results using the direct comparison scheme as depicted in [7] (i.e., the direct comparison of the three characteristic parameters of the membership function) were unsuccessful. However, the Neuro-fuzzy system demonstrates its effectiveness with imprecise data. The uncertain impeller ***X*** can be assigned as a member of the reference class that shows the maximum similarity with impeller ***X***.

7 Conclusions

A pattern recognition scheme based on machine vision technology and Neuro-fuzzy based decision-making algorithm is used to identify high tolerance manufactured components for identity verification. An overall flow diagram of this pattern recognition scheme is illustrated in Fig. 6.

In this study, a neuro-fuzzy algorithm for the impeller recognition scheme, as shown in Fig. 7, has been developed and tested. In this design, we demonstrate the ease with which good performance with respect to high tolerance recognition is achieved and could be improved by applying the neuro-fuzzy algorithm to the classification of manufactured components. In the proposed algorithm, characteristics of the application task can be built into the neeural network model in advance by employing a logical structure, in the form of fuzzy inference rules. Therefore, it is easier to improve the performance of the proposed algorithm in which the internal state can be observed because of its structure, than in an ordinary neural network model, which is like a *black box* (non-algorithmic classification).

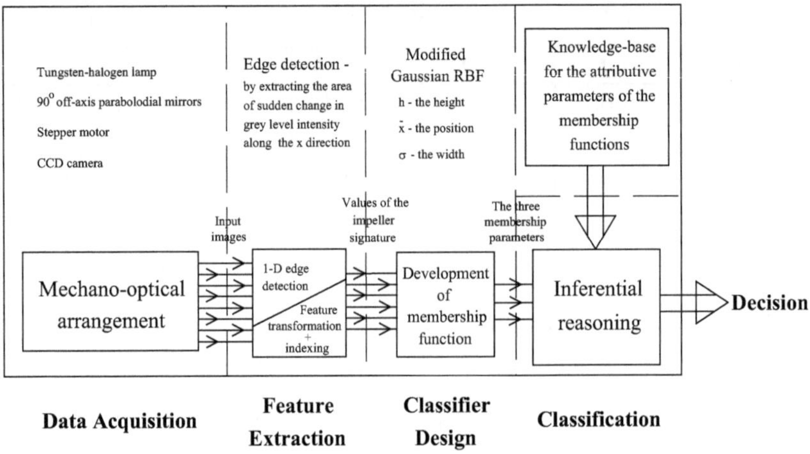

Fig. 6. Impeller recognition scheme.

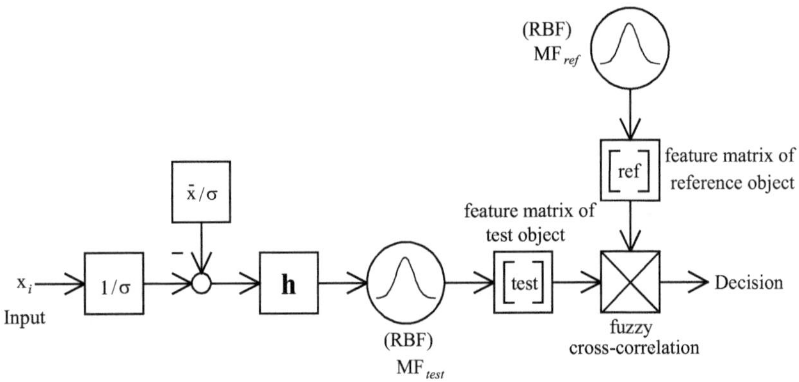

Fig. 7. Neuro-fuzzy classification system for impellers.

References

1. L.A. Zadeh, "Fuzzy Algorithms," *Information and Control*, vol. 12, pp. 94-102, 1968.
2. L.A. Zadeh, "A Rationale for Fuzzy Control," *Journal Dynamic Systems, Measurement and Control*, vol. 94, Series G, pp. 3-4, 1972.

3. L.A. Zadeh, "Outline of a New Approach to the Analysis of Complex Systems and Decision Processes," *IEEE Trans. on Systems, Man. and Cybernetics*, vol. SMC-3, pp. 28-44, 1973.
4. L.J.K. Setright, *Turbocharging and Supercharging for Maximum Power and Torque*. J.H. Haynes and Company Limited, Sparkford Yeovil Somerset, 1976.
5. A.W. Campbell; T. Sarkodie-Gyan and C.W. Lam, "Identification of Turbocharger Components which are Similar but not Identical," *Proc. of IMC-11 on Lean Production*, Queen's University of Belfast, Northern Ireland, pp. 471-474, 1994.
6. T. Sarkodie-Gyan; C.W. Lam; D. Hong and A.W. Campbell, "A Fuzzy Clustering Method for Efficient 2-D Object Recognition," *Proc. of 5th IEEE Int. Conf. on Fuzzy Systems*, New Orleans, USA, vol. 2, pp. 1400-1406, 1996.
7. Sarkodie-Gyan, T. et al " (1997) "An Efficient Object Recognition Scheme for a Prototype Component Inspection", *Int. Journal of Mechatronics.* , Elsevier Science Ltd., vol.7, No.2, pp.185-197.
8. T. Sarkodie-Gyan, *Lecture Notes on Intelligent Control Systems - Fuzzy Logic and Applications*. CAD Lab. for Systems/Robotics, Dept. of Electrical and Computer Engineering, University of New Mexico, Albuquerque, USA, (1989-90).
9. T. Sarkodie-Gyan, C.W. Lam, and A.W. Campbell, "Development of a Novel Image Sensor and Its Application to Analysis of Automobile Components," IEEE/ASME Transactions on Mechatronics, June 1977, vol.2, No.2, pp. 144-150.
10. J.C. Bezdek, "A Review of Probabilistic, Fuzzy, and Neural Models for Pattern Recognition," *Jour. of Intelligent and Fuzzy Systems*, vol. 1, no. 1, pp. 1-25, 1993.
11. M.A. Eshera and S.C. Barash, "Parallel Rule-Based Fuzzy Inference on Mesh-Connected Systolic Arrays," IEEE Expert, vol. , no. , pp. 27-35, 1989.
12. Y.-H. Pao, *Adaptive Pattern Recognition and Neural Networks*. Addison-Wesley, Reading, MA, 1989.

The Fuzzy Sugeno Integral as a Decision Operator in the Recognition of Images with Modular Neural Networks

Olivia Mendoza Duarte, Patricia Melin

[1]Tijuana Institute of Technology, México, Unidad Tomás Aquino, Calzada Tecnológico s/n, Fracc. Tomás Aquino, Apartado Postal 1166, C.P. 22000 Tijuana, B.C., omendozad@yahoo.com, pmelin@tectijuana.mx, http://fca.tij.uabc.mx/docentes/omendoza/hmr

Abstract. In a previous paper we presented the implementation of the Fuzzy Sugeno Integral formulas developed with Matlab 6.5™. The programs are now included in a System called "Herramientas Multired" ("hmr"). In this paper we will review an example of modular neural network for image recognition, using images divided in four parts. The Fuzzy Sugeno Integral was used to make a final decision for pattern recognition.

1 The Fuzzy Sugeno Integral

The Fuzzy Integral is an operator introduced in 1974 by Sugeno [1]. This operator is used to resolve problems of multicriteria decision making, where the information that is combined is based in fuzzy measures determined by an expert.

The goal is the simulation of the human process for the integration of different source of information [2].

Fuzzy measures are functions applied to fuzzy sets and they consist of different coefficients call fuzzy densities. Each fuzzy density rate the relevance of the different sets and their combinations, in order to satisfy certain hypothesis.

There are two types of Fuzzy Integral: Choquet Fuzzy Integral (1) and Sugeno Fuzzy Integral (2). [3]

$$h(\sigma_1,\cdots,\sigma_n)=\max_{i=1}^{n}(\min(\sigma_i,\mu(x_i,\cdots,x_n))) \tag{1}$$

$$h(\sigma_1,\cdots,\sigma_n)=\sum_{i=1}^{n}(\sigma_i-\sigma_{i-1})\mu(\{x_i,\cdots,x_n\}) \tag{2}$$

Where $\sigma i=\sigma(xi)$ and $0\le\sigma 1\le\ldots\le\sigma n\le 1$

O.M. Duarte and P. Melin: *The Fuzzy Sugeno Integral as a Decision Operator in the Recognition of Images with Modular Neural Networks*, StudFuzz **208**, 299–310 (2007)
www.springerlink.com

1.1 Fuzzy Measures

A fuzzy measure μ, respect to the data set X, it must satisfy the following conditions: [4] [5].

1) $\mu(X)=1, \mu(\emptyset)=0$
2) If $S \subseteq T$, then $\mu(S) \leq \mu(T)$

Where S y T are subsets of X.

One fuzzy measure is a Sugeno Measure or λ-fuzzy, if it satisfies the following condition of addition for some $\lambda > -1$.

$$\mu(S \cup T) = \mu(S) + \mu(T) + \lambda \mu(S)\mu(T) \tag{3}$$

λ can be calculated of the following by (4) or (5):

$$\mu(S) = \left[\prod_{x \in S} (1 + \lambda \mu(\{x\}))\right] \Big/ \lambda \tag{4}$$

$$\lambda + 1 = \prod_{i=1}^{n} (1 + \lambda \mu(\{x_i\})) \tag{5}$$

The method used to calculate Sugeno measures, it is carrying out the calculation of recursive way, [6][7] using (6),(7).

$$\mu(A1) = \mu(x1) \tag{6}$$

$$\mu(Ai) = \mu(xi) + \mu(Ai-1) + \lambda \mu(xi)\mu(Ai-1) \tag{7}$$

Where $1 < i \leq n$, and the values to μ(xi) corresponds to the fuzzy densities determined by an expert.

A fundamental restriction to use the recursive formulas (6) and (7) is the reordering of the fuzzy densities. The fuzzy densities must be ordered respect the descendent order of the respective values to combine.

1.2 Example for Calculation of Sugeno Measures

Consider the set X={x1,x2,...xn}, the fuzzy density values are given as follows:

Fuzzy densities: $\mu(x1)=0.3$, $\mu(x2)=0.4$, $\mu(x3)=0.1$
Values to combine: $\sigma(x1)=0.9$, $\sigma(x2)=0.6$, $\sigma(x3)=0.3$

The value of λ can be calculated by (5), solving the following equation, using some numeric method to found the root of $f(\lambda)$ [8]:

$$1+\lambda=(1+0.3\lambda)(1+0.4\lambda)(1+0.1\lambda)$$

The solutions are $\lambda= -16.8$ y $\lambda=0.9906$, if $\lambda>-1$, then $\lambda=0.9906$
The Sugeno measures (3) can be constructed as follows:

$$\mu(x1)=0.3,\ \ \mu(x2)=0.4,\ \ \mu(x3)=0.1$$
$$\mu(x1,x2)=\mu(x1)+\mu(x2)+\lambda(\mu(x1)\mu(x2))=0.8189$$
$$\mu(x1,x3)=\mu(x1)+\mu(x3)+\lambda(\mu(x1)\mu(x3))=0.4297$$
$$\mu(x2,x3)=\mu(x2)+\mu(x3)+\lambda(\mu(x2)\mu(x3))=0.5396$$
$$\mu(x1,x2,x3)=1$$

In this example the values to combine are already ordered descendent, then the reordering of fuzzy densities is not necessary to do the calculation using the recursive formulas (6) and (7).

$$\mu(A1)=0.3$$
$$\mu(A2)=0.4+0.3+(0.9906)(0.4)(0.3)= 0.8189$$
$$\mu(A3)=0.1+0.8189+(0.9906)(0.1)(0.8189)= 1$$

1.3 Example for the Calculation of Sugeno Integral

The Fuzzy Sugeno Integral, can now calculate using (1)

$$h(0.9,0.6,0.3)=\max(\min(0.9,0.3),\min(0.6,0.8189),\min(0.3,1))$$
$$h(0.9,0.6,0.3)=\max(0.3,0.6,0.3)=0.6$$

2 The Modular Neural System "Fotos 4 Partes"

The model "Fotos 4 partes" shown in figure 1, contains one neural network node to each part of the face of 5 people. Each neural network was trained with one of 4 parts of 5 different people, with 7 samples.

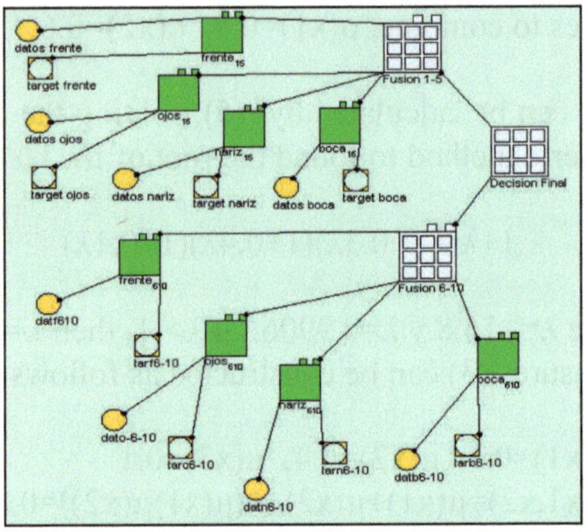

Fig. 1. Model "Fotos 4 partes"

The steps to create the modular multi-net system are:

- Divide *n* images in *p* parts with *s* samples.
- Train one monolithic neural network for each part.
- Simulate each neural net trained, using a complete image like input data.
- Use the result of simulation of many neural nets to build a matrix contains one row for each person, and one column for each part of the image.

For example, the module "frente" was trained with 7 samples of the superior part of the face of 5 different people, as shown in figure 2

To explain the recognition process, we use the module "Fusion 1-5", that combines four modules (neural nets) trained to recognize one of the parts of the face of five people.

All the neural networks in this model, was trained with samples 1,2,3,4,5,6 and 10, for each person, the samples 7,8 and 9, are not included in the train data set, then this three samples are the most difficult images to recognize, and are used to test the precision of the calculation.

In this example, we use as input data for the module "Fusion 1-5", a complete image of the sample number 8 of the person number 3.

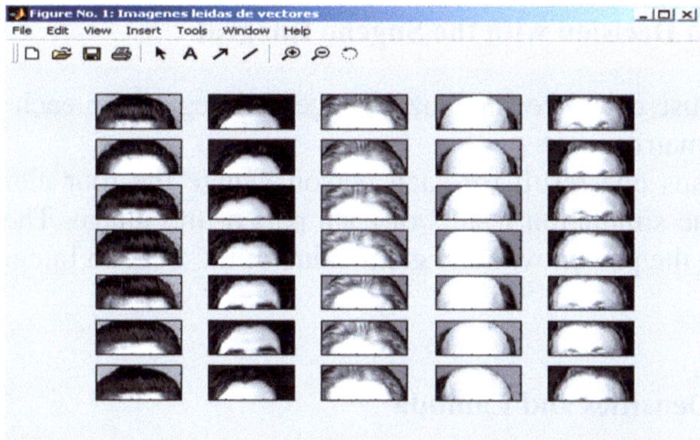

Fig. 2. Images assigned to the input data node: "datos frente".

The matrix shown in figure 3, have the simulation results of the people 1,2,3,4 and 5, one row for each person. Each column contains the result of simulate: 1(forehead), 2(eyes), 3(nose), 4(mouth) of each person.

	1	2	3	4
1	0.43849	-0.29364	-0.38769	-0.69856
2	0.50203	-0.28418	2.1399	0.64638
3	0.89923	1.1218	-0.13577	0.63817
4	-0.20142	-0.73076	-0.17258	0.009603
5	-0.8161	0.64992	0.65338	0.36658

Fig. 3. Results of the simulation, of the person number 3, with the sample number 8 in the cooperative module "Fusion 1-5".

The figure 4 shows what person is selected for each part of the image, using the maximum value method. If we use vote method to make a final decision, we have 2 votes for the person 3 and 2 votes to the person 2.

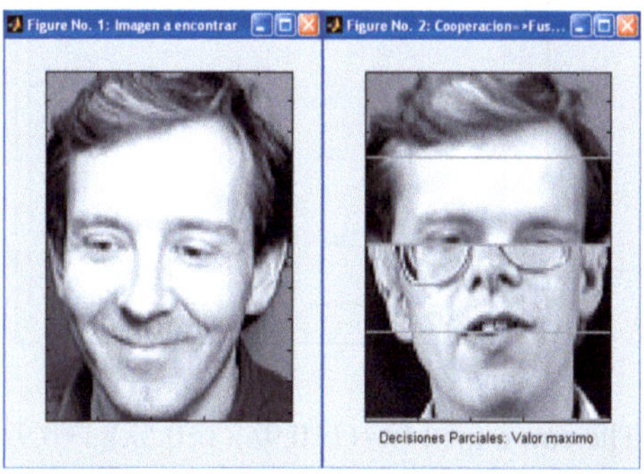

Fig. 4. Results of search of the person number 3, with the sample number 8 in the cooperative module "Fusion 1-5".

3. Making a Decision with the Sugeno Integral

First we must calculate the Fuzzy Sugeno Integral for each row in the simulation matrix.

That means one result for each person, where the four elements combined are the simulation results of each part or the image. The final decision will be the person with the great value of the Sugeno Integral.

3.1 Fuzzy Densities and Lambda

The fuzzy density values are given as follows:

μ(forehead)	μ(eyes)	μ(nose)	μ(mouth)
0.9	0.5	0.9	0.9

That values means, all the parts have the same relevance to the recognition, except the module "eyes", if we suppose is the part with minor relevance because the person can ware eyeglasses, or maybe eye make up, but in a specific system this values must be determined by an expert.

The value of λ can be calculated by (5), solving the following equation, using some numeric method to found the root of $f(\lambda)$ [8]:

$$1+\lambda=(1+0.9\lambda)(1+0.5\lambda)(1+0.9\lambda)(1+0.9\lambda)$$

The curve to the previous function is shown in the figure 5

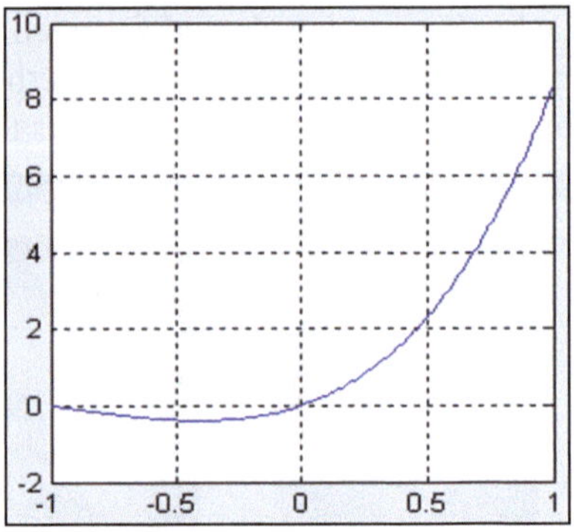

Fig. 5. The curve for $1+\lambda=(1+0.9\lambda)(1+0.5\lambda)(1+0.9\lambda)(1+0.9\lambda)$

The program calc_lambda21.m implemented with matlab, use the function fzero, that realize a scalar nonlinear zero finding.

The solution using the program calc_lambda21.m is:

```
>>L=calc_lambda21([.9 .5 .9 .9])
L =

     -0.9995
```

The solution is λ= -0.9995.

3.2 Fuzzify the Simulation Matrix

Once we have the matrix with the simulation results of the modular system, the next step is the fuzzification of the results of simulation

The following is the simulation matrix for person 3 and sample number 8:

Person	forehead	eyes	nose	mouth
1	0.43849	-0.29364	-0.38769	-0.69856
2	0.50203	-0.28418	2.1399	0.64638
3	0.89923	1.1218	-0.13577	0.63817
4	-0.20142	-0.73076	-0.17258	0.009603
5	-0.8161	0.64992	0.65338	0.36658

Then, to fuzzify the simulation matrix, we use the gaussmf.m program with parameters:[.4 1 .9], as shown in figure 6.[9]

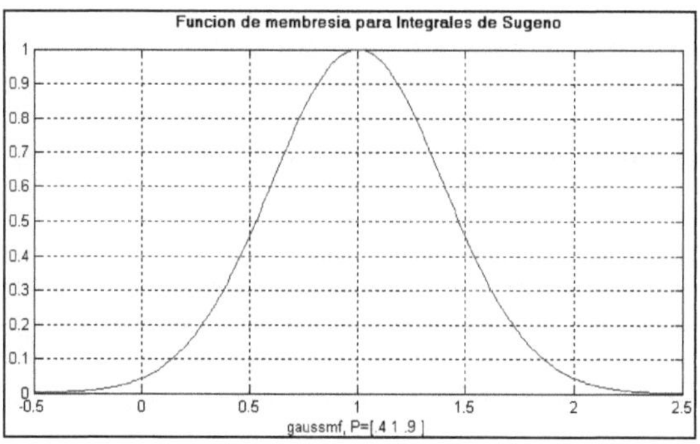

Fig. 6. Membership function to fuzzify the values of simulation.

Then, the simulation matrix is now fuzzified and the fuzzy values are:

Person	forehead	eyes	nose	mouth
1	0.37333	0.0053553	0.002435	0.000121
2	0.46074	0.005779	0.017237	0.67653
3	0.96876	0.95468	0.017754	0.66423
4	0.010991	8.6008e-005	0.013613	0.046641
5	3.3401e-005	0.68183	0.68698	0.28541

3.3 Recursive Formula for Sugeno Measures

To use the recursive formula to calculate the Sugeno Measures, we must order the values in a descendent fashion. These values are in the fuzzy simulation matrix.

Example for Person 2:

	forehead	eyes	nose	mouth
Simulation results	0.46074	0.005779	0.017237	0.67653
μ	0.9	0.5	0.9	0.9

Now must order the table in a descendent way with respect to the simulation results.

	mouth	forehead	nose	eyes
Simulation results	0.67653	0.46074	0.017237	0.005779
μ	0.9	0.9	0.9	0.5

The Sugeno measures (3) can be constructed using (6) and (7)

$$\mu(A1)=0.9$$
$$\mu(A2)=0.9+0.9+(-0.9905)(0.9)(0.9)= 0.9904$$
$$\mu(A3)=0.9+0.9904+(-0.9905)(0.9)(0.9004)= 0.9995$$
$$\mu(A4)=0.5+0.9995+(-0.9905)(0.5)(0.9995)= 1$$

Then, the Sugeno measures for each person are:

Person	μ(A1)	μ(A2)	μ(A3)	μ(A4)
1	0.9000	0.9502	0.9955	1.0000
2	0.9000	0.9904	0.9995	1.0000
3	0.9000	0.9502	0.9955	1.0000
4	0.9000	0.9904	0.9995	1.0000
5	0.9000	0.9502	0.9955	.0000

3.4 The Final Decision

Once we have the Sugeno Measures for each person, we can calculate the respective Sugeno Integral, with (1).

IS1=max(min(0.37333,0.9),min(0.0053551,0.95023),
min(0.0024351,0.99546),min(0.00012146,1))

IS2=max(min(0.67653,0.9),min(0.46074,0.99041),
min(0.01724,0.99949),min(0.0057792,1))

IS3=max(min(0.96877,0.9),min(0.9547,0.95023),
min(0.66423,0.99546),min(0.017754,1))

IS4=max(min(0.046641,0.9),min(0.013613,0.99041),
min(0.010991,0.99949),min(8.6011e-005,1))

IS5=max(min(0.68698,0.9),min(0.68182,0.95023),
min(0.28541,0.99546),min(3.3401e-005,1))

Person	Fuzzy Sugeno Integral
1	0.37333
2	0.67653
3	0.95023
4	0.046641
5	0.68698

IS1=max(0.37333,0.0053551,0.0024351,0.00012146)
IS2= max(0.67653,0.46074,0.01724,0.0057792)
IS3= max(0.9,0.95023,0.66423,0.017754)
IS4= max(0.046641,0.013613,0.010991,8.6011e-005)
IS5= max(0.68698,0.68182,0.28541,3.3401e-005)

The person selected is the number 3, because his Fuzzy Sugeno Integral is greater than the obtained for each person 1,2,4 and 5.

All the previous calculations are implemented in the computer program final_sugeno_individual.m, included in the System "hmr", we can execute the function for each row of the simulation matrix, or execute the image recognition within the graphic interface.

The input parameters for the function final_sugeno_individual are:

- row_individuo: The simulation data for one person.
- row_g: The fuzzy densities for each part of the image.
- mfparams: The parameters for the membership function.
- mftype: The type of membership function.

In the following example we use the function final_sugeno_individual to calculate the Fuzzy Sugeno Integral for the Person number 3:

```
>>row_individuo=[0.89923 1.1218 -0.13577
0.63817]
>>row_g=[0.9,0.5,0.9,0.9]
>>mfparams=[.4 1 .9]
>>mftype='gaussmf'
>>[inte-
gral_sugeno,simulacion_fuzzy_row]=finalsugeno_in
dividual(row_individuo,row_g,mfparams,mftype)

L =
   -0.9995
simulacion_fuzzy_row =
     0.9688     0.9547     0.0178     0.6642
medidas =
     0.9000     0.9502     0.9955     1.0000
minimos_texto =
min(0.96877,0.9),min(0.9547,0.95023),
min(0.66423,0.99546),min(0.017754,1)
integral_sugeno =
     0.95023
```

In the System “hmr” the results can be shown like in figure 7, the first image is the input data to simulation, the second are the partial result if we use the maximum value for each part of the image, and the third is the person selected with the Fuzzy Sugeno Integral.

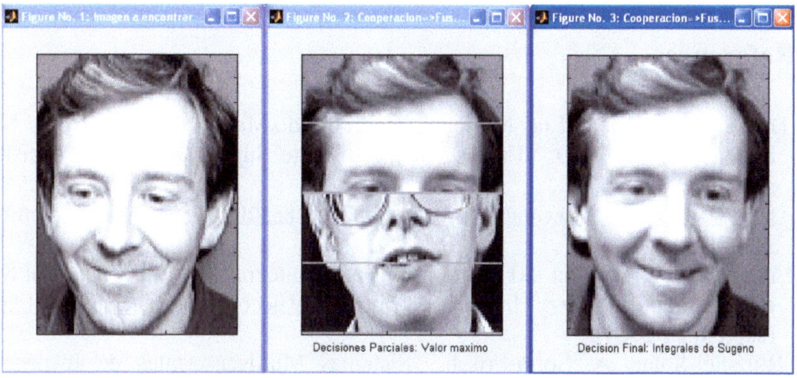

Fig. 7. Results as shown in System “hmr”

4 Conclusions

The “Fuzzy Sugeno Integral” allows us to test many combinations of input values, including a fuzzy membership function. These features help us to set the best combination of parameters for many kinds of applications. That parameters can be change recurrently, until found the best results.

The model “Photos in 4 Parts” contains one cooperative module for each five people to recognize, then, if we want to complete the model to recognize the 40 people of the OCR database [10], our model should contain 8 cooperative modules. Then, to make a final decision, the input data to the competitive module “Decision Final” will be a simulation matrix with 40 rows and 4 columns, one column for each part of the image, and one row for each person.

In this particular example, the values to combine are always 4, because the images are divided in 4 parts, and the Fuzzy Sugeno Integral combines the values of simulation of each part for each person, but the programs implemented in the system “hmr” allow to combine any number of elements, so we can build another models with images divided in different number of parts, and the system can calculate it with out problems about the input values.

As shown in the results of this particular example, the Fuzzy Sugeno Integral is a very useful operator to make decisions that are very difficult to obtain with another method such voting or maximum value.

5 References

1. Didier Dubois, Jean-Luc Marichal, Henri Prado,Marc Roubens, R´egis Sabbadin, Francia, 2000. "The use of the discrete Sugeno integral in decision-making: a survey", http://www.worldscinet.com/ijufks/09/0905/S0218488501001058.html, (september, 2004).
2. Erkan Duman, Turquía, 2003. "A New Fuzzy Integral Model For Control Systems: Adaptive Fuzzy Integral", http://www.ijci.org/product/tainn/E08012.pdf, (september 2004).
3. Ruiz-del-Solar, A. Soria-Frisch, "Sistemas Multisensoriales de Inspección Industrial: Procesamiento Conjunto de Imágenes de Color e Infrarrojas", (september, 2004).
4. H. R. Tizhoosh, Waterloo, Inglaterra, 1997. "Fuzzy Measure Theory", http://watfor.uwaterloo.ca/tizhoosh/measure.htm, (may, 2004).
5. Arunas Lipnickas, Lithuania, 2001. "Classifiers Fusion With Data Dependent Aggregation Schemes", http://www.elen.ktu.lt/~arunas/public/alincfddas.pdf, (september, 2004),
6. A.Verikas, A. Lipnickas, K. Malmqvist, Korea, 2000. "Fuzzy measures in neural networks fusion", http://www.elen.ktu.lt/~arunas/public/aviconip2000.pdf, (september, 2004).
7. A.Lipnickas, Bielorrusia, 2001. "Classifiers Fusion with Data Dependent Aggregation Schemess", http://www.elen.ktu.lt/~arunas/public/alincfddas.pdf, (september, 2004).
8. Antonio Nieves, México, 1998. "Métodos Numéricos Aplicados a la Ingeniería", Ed. CECSA, p.p 34-57.
9. The MathWorks, Inc. ©1994-2004. "Fuzzy Logic Tool Box, For use with Matlab" AT&T Laboratories, Cambridge, 2002, "The ORL Database of Faces", http://www.uk.research.att.com/facedatabase.html, (May, 2004).

Modular Neural Networks and Fuzzy Sugeno Integral for Pattern Recognition: The Case of Human Face and Fingerprint

Patricia Melin, Claudia Gonzalez, Diana Bravo, Felma Gonzalez and Gabriela Martinez

Dept. of Computer Science, Tijuana Institute of Technology, Tijuana, Mexico

Abstract. We describe in this paper a new approach for pattern recognition using modular neural networks with a fuzzy logic method for response integration. We proposed a new architecture for modular neural networks for achieving pattern recognition in the particular case of human faces and fingerprints. Also, the method for achieving response integration is based on the fuzzy Sugeno integral with some modifications. Response integration is required to combine the outputs of all the modules in the modular network. We have applied the new approach for fingerprint and face recognition with a real database from students of our institution.

1 Introduction

Response integration methods for modular neural networks that have been studied, to the moment, do not solve well real recognition problems with large sets of data or in other cases reduce the final output to the result of only one module. Also, in the particular case of face recognition, methods of weighted statistical average do not work well due to the nature of the face recognition problem. For these reasons, a new approach for face and fingerprint recognition using modular neural networks and fuzzy integration of responses was proposed in this paper.

The basic idea of the new approach is to divide a human face into three different regions: the eyes, the nose and the mouth, and the fingerprint also into three parts, top, middle and bottom. Each of these regions is assigned to one module of the neural network. In this way, the modular neural network has three different modules, one for each of the regions of the human face and the fingerprint. At the end, the final decision of face and fingerprint recognition is done by an integration module, which has to take into

P. Melin et al.: *Modular Neural Networks and Fuzzy Sugeno Integral for Pattern Recognition: The Case of Human Face and Fingerprint*, StudFuzz **208**, 311–326 (2007)
www.springerlink.com

account the results of each of the modules. In our approach, the integration module uses the fuzzy Sugeno integral to combine the outputs of the three modules. The fuzzy Sugeno integral allows the integration of responses from the three modules of the eyes, nose and mouth of a human specific face and the integration of the responses from the three modules of the fingerprint parts. Other approaches in the literature use other types of integration modules, like voting methods, majority methods, and neural networks.

The new approach for face and fingerprint recognition was tested with a database of students and professors from our institution. This database was collected at our institution using a digital camera for the faces and a special scanner for the fingerprints. The results with our new approach for face and fingerprint recognition on this database were excellent.

2 Modular Neural Networks

There exists a lot of neural network architectures in the literature that work well when the number of inputs is relatively small, but when the complexity of the problem grows or the number of inputs increases, their performance decreases very quickly. For this reason, there has also been research work in compensating in some way the problems in learning of a single neural network over high dimensional spaces.

In the work of Sharkey (Sharkey 1998), the use of multiple neural systems (Multi-Nets) is described. It is claimed that multi-nets have better performance or even solve problems that monolithic neural networks are not able to solve. It is also claimed that multi-nets or modular systems have also the advantage of being easier to understand or modify, if necessary.

In the literature there is also mention of the terms “ensemble” and “modular” for this type of neural network. The term “ensemble” is used when a redundant set of neural networks is utilized, as described in Hansen and Salomon (Hansen and Salomon 1990). In this case, each of the neural networks is redundant because it is providing a solution for the same task, as it is shown in Figure 1.

On the other hand, in the modular approach, one task or problem is decompose in subtasks, and the complete solution requires the contribution of all the modules, as it is shown in Figure 2.

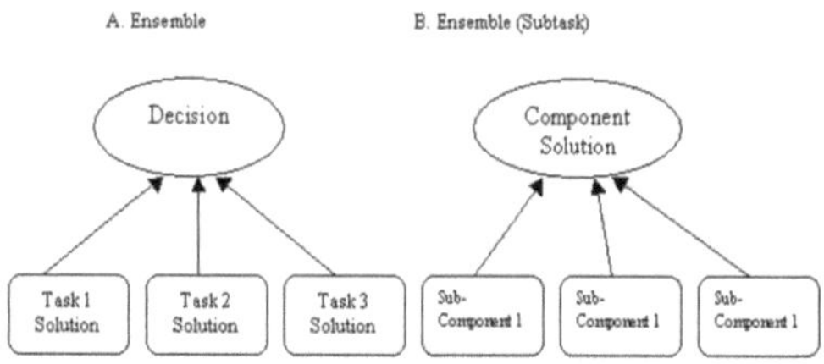

Fig. 1. Ensembles for one task and subtask.

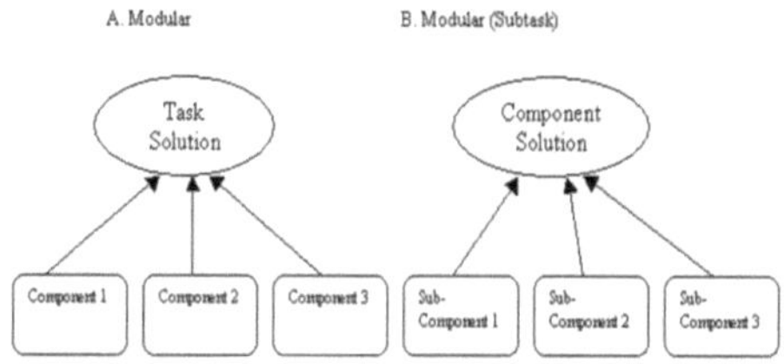

Fig. 2. Modular approach for task and subtask.

2.1 Multiple Neural Networks

In this approach we can find networks that use strongly separated architectures. Each neural network works independently in its own domain. Each of the neural networks is build and trained for a specific task. The final decision is based on the results of the individual networks, called agents or experts. One example of this decision is shown by (Albrecht 1996), as shown in Figure 3, where a multiple architecture is used, one module consists of a neural network trained for recognizing a person by the voice, while the other module is a neural network trained for recognizing a person by the image.

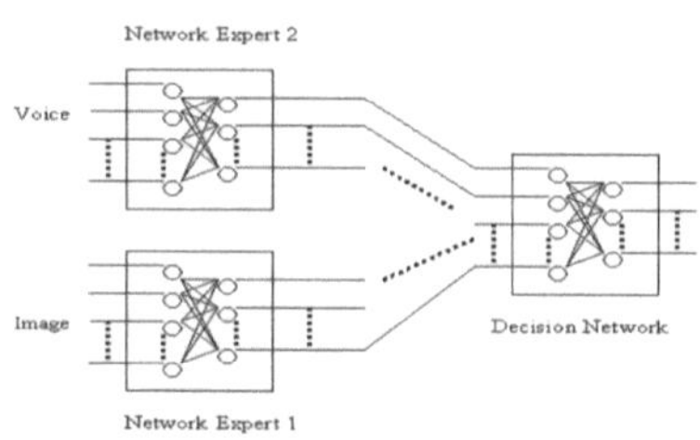

Fig. 3. Multiple networks for voice and image.

The outputs by the experts are the inputs to the decision network, which is the one making the decision based on the outputs of the expert networks.

2.2 Main Architectures with Multiple Networks

Within multiple neural networks we can find three main classes of this type of networks (Fu et al. 2001):

- Mixture of Experts (ME): The mixture of experts can be viewed as a modular version of the multi-layer networks with supervised training or the associative version of competitive learning. In this design, the local experts are trained with the data sets to mitigate weight interference from one expert to the other.
- Gate of Experts: In this case, an optimization algorithm is used for the gating network, to combine the outputs from the experts.
- Hierarchical Mixture of Experts: In this architecture, the individual outputs from the experts are combined with several gating networks in a hierarchical way.

2.3 "Modular" Neural Networks

The term "Modular Neural Networks" is very fuzzy. It is used in a lot of ways and with different structures. Everything that is not monolithic is said to be modular. In the research work by (Boers and Kuiper 1992), the concept of a modular architecture is introduced as the development of a large network using modules.

One of the main ideas of this approach is presented in (Albrecht 1996), where all the modules are neural networks. The architecture of a single module is simpler and smaller than the one of a monolithic network. The tasks are modified in such a way that training a subtask is easier than training the complete task. Once all modules are trained, they are connected in a network of modules, instead of using a network of neurons. The modules are independent to some extent, which allows working in parallel. Another idea about modular networks is presented by (Boers and Kuiper 1992), where they used an approach of networks not totally connected. In this model, the structure is more difficult to analyze, as shown in Figure 4. A clear separation between modules can't be made. Each module is viewed as a part of the network totally connected.

In this figure, we can appreciate two different sections from the monolithic neural network, namely A and B. Since there are no connections between both parts of the network, the dimensionality (number of weights) is reduced. As a consequence the required computations are decreased and speed of convergence is increased.

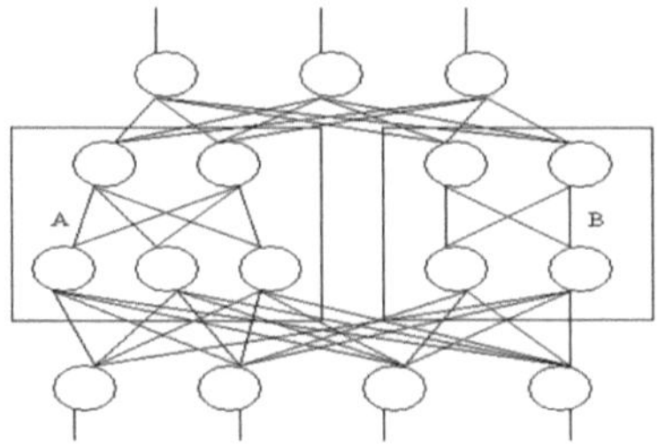

Fig. 4. One type of modular neural network.

2.4 Advantages of Modular Neural Networks

A list of advantages of modular networks is given below:

- They give a significant improvement in the learning capabilities, over monolithic neural networks, due to the constraints imposed on the modular topology.
- They allow complex behavior modeling, by using different types of knowledge, which is not possible without using modularity.
- Modularity may imply reduction of number of parameters, which will allow and increase in computing speed and better generalization capabilities.
- They avoid the interference that affects “global” neural networks.
- They help determine the activity that is being done in each part of the system, helping to understand the role that each network plays within the complete system.
- If there are changes in the environment, modular networks enable changes in an easier way, since there is no need to modify the whole system, only the modules that are affected by this change.

2.5 Elements of Modular Neural Networks

When considering modular networks to solve a problem, one has to take into account the following points (Ronco and Gawthhrop 1995):

- Decompose the main problem into subtasks.
- Organizing the modular architecture, taking into account the nature of each subtask.
- Communication between modules is important, not only in the input of the system but also in the response integration.

In the particular case of this paper, we will concentrate in more detail in the third point, the communication between modules, more specifically information fusion at the integrating module to generate the output of the complete modular system.

2.6 Main Task Decomposition into Subtasks

Task Decomposition can be performed in three different ways, as mentioned by (Lu and Ito 1998):

- Explicit Decomposition: In this case, decomposition is made before learning and requires that the designer has deep knowledge about the problem. Of course, this maybe a limitation if there isn't sufficient knowledge about the problem.
- Automatic Decomposition: In this case, decomposition is made as learning is progressing.
- Decomposition into Classes: This type of decomposition is made before learning, a problem is divided into a set of sub-problems according to the intrinsic relations between the training data. This method only requires knowledge about the relations between classes.

2.7 Communication Between Modules

In the research studies made by (Ronco and Gawthrop 1995), several ways of achieving communication between modules are proposed. We can summarize their work by mentioning the following critical points:

1. How to divide information, during the training phase, between the different modules of the system.
2. How to integrate the different outputs given by the different modules of the system to generate the final output of the complete system.

2.8 Response Integration

Response integration has been considered in several ways, as described by (Smith and Johansen 1997) and we can give the following list:

- Using Kohonen's self organizing maps, Gaussian mixtures, etc.
- The method of "Winner Takes All", for problems that require similar tasks.
- Models in series, the output of one module is the input to the following one.

- Voting methods, for example the use of the "Softmax" function.
- Linear combination of output results.
- Using discrete logic.
- Using finite state automata.
- Using statistical methods.
- Using fuzzy logic (Castillo and Melin 2003).

3 Methods of Response Integration

The importance of this part of the architecture for pattern recognition is due to the high dimensionality of this type of problems. As a consequence in pattern recognition is good alternative to consider a modular approach. This has the advantage of reducing the time required of learning and it also increases accuracy. In our case, we consider dividing the images of a human face in three different regions. We also divide the fingerprint into three parts, and applying a modular structure for achieving pattern recognition.

In the literature we can find several methods for response integration, that have been researched extensively, which in many cases are based on statistical decision methods. We will mention briefly some of these methods of response integration, in particular the ones based on fuzzy logic. The idea of using these types of methods, is that the final decision takes into account all of the different kinds of information available about the human face and fingerprint. In particular, we consider aggregation operators, and the fuzzy Sugeno integral (Sugeno 1974).

Yager (1999) mentions in his work, that fuzzy measures for the aggregation criteria of two important classes of problems. In the first type of problems, we have a set Z={z1,z2,…,zn} of objects, and it is desired to select one or more of these objects based on the satisfaction of certain criteria. In this case, for each zi∈Z, it is evaluated D(zi)=G(Ai(zi),…,Aj(zi)), and then an object or objects are selected based on the value of G. The problems that fall within this structure are the multi-criteria decision problems, search in databases and retrieving of documents.

In the second type of problems, we have a set G={G1,G2,…,Gq} of aggregation functions and object z. Here, each Gk corresponds to different possible identifications of object z, and our goal is to find out the correct identification of z. For achieving this, for each aggregation function G, we obtain a result for each z, Dk(z)=Gk(A1(z), A2(z), … ,An(z)). Then we associate to z the identification corresponding to the larger value of the aggregation function.

A typical example of this type of problems is pattern recognition. Where Aj corresponds to the attributes and Aj(z) measures the compatibility of z with the attribute. Medical applications and fault diagnosis fall into this type of problems. In diagnostic problems, the Aj corresponds to symptoms associated with a particular fault, and Gk captures the relations between these faults.

3.1 Fuzzy Integral and Fuzzy Measures

Fuzzy integrals can be viewed as non-linear functions defined with respect to fuzzy measures. In particular, the "gλ-fuzzy measure" introduced by (Sugeno 1974) can be used to define fuzzy integrals. The ability of fuzzy integrals to combine the results of multiple information sources has been mentioned in previous works.

Definition 1. A function of sets g:2x-(0.1) is called a fuzzy measure if:

1. 1) g(0)=0 g(x)=1
2. 2) g(A)≤ g(B) if A⊂B
3. 3) if {Ai}iα =1 is a sequence of increments of the measurable set then

$$\lim_{i \to \infty} g(Ai) = g\,(\lim_{i \to \infty} Ai) \tag{1}$$

From the above it can be deduced that g is not necessarily additive, this property is replaced by the additive property of the conventional measure.

From the general definition of the fuzzy measure, Sugeno introduced what is called "gλ-fuzzy measure", which satisfies the following additive property:

For every A, B ⊂ X and A ∩ B = θ,

$$g(A \cup B) = g(A) + g(B) + \lambda\, g(A)g(B), \tag{2}$$

for some value of λ>-1.

This property says that the measure of the union of two disjunct sets can be obtained directly from the individual measures. Using the concept of fuzzy measures, (Sugeno 1974) developed the concept of fuzzy integrals, which are non-linear functions defined with respect to fuzzy measures like the gλ-fuzzy measure.

Definition 2. let X be a finite set and h:X→[0,1] be a fuzzy subset of X, the fuzzy integral over X of function h with respect to the fuzzy measure g is defined in the following way,

$$h(x) \circ g\,(x) = \max_{E \subseteq X} [\,\min(\,\min_{x \in E} h(x), g(E))] \tag{3}$$

$$= \sup\,[\min(\,\alpha\,,\, g(h\alpha\,))]\ \alpha \in [0,1]$$

where hα is the level set α of h,

$$h\alpha = \{ x \mid h(x) \geq \alpha \}. \quad (4)$$

We will explain in more detail the above definition: h(x) measures the degree to which concept h is satisfied by x. The term min(hx) measures the degree to which concept h is satisfied by all the elements in E. The value g(E) is the degree to which the subset of objects E satifies the concept measure by g. As a consequence, the obtained value of comparing these two quantities in terms of operator min indicates the degree to which E satifies both criteria g and min(hx). Finally, operator max takes the greatest of these terms. One can interpret fuzzy integrals as finding the maximum degree of similarity between the objective and expected value.

4 Proposed Architecture and Results

In the experiments performed in this research work, we used 20 photographs that were taken with a digital camera and 20 fingerprints from students and professors of our Institution. The photographs were taken in such a way that they had 148 pixels wide and 90 pixels high, with a resolution of 300x300 ppi, and with a color representation of a gray scale, some of these photographs are shown in Figure 5. In addition to the training data (20 photos) we did use 10 photographs that were obtained by applying noise in a random fashion, which was increased from 10 to 100%.

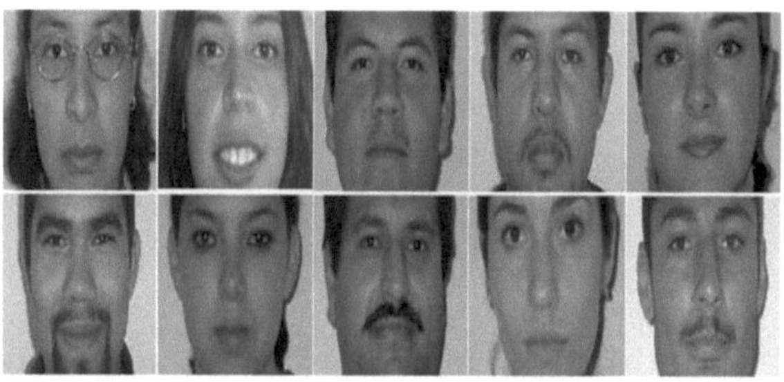

Fig. 5. Sample Faces Used for Training.

The images of fingerprints (Quezada 2004) were taken in such a way that they had 198 pixels wide and 200 pixels high, with a resolution of 300x300 ppi, and with a color representation of a gray scale, some of these images are shown in Figure 6. In addition to the training data (20 fingerprints) we did use 10 images that were obtained by applying noise in a random fashion, which was increased from 10 to 100%.

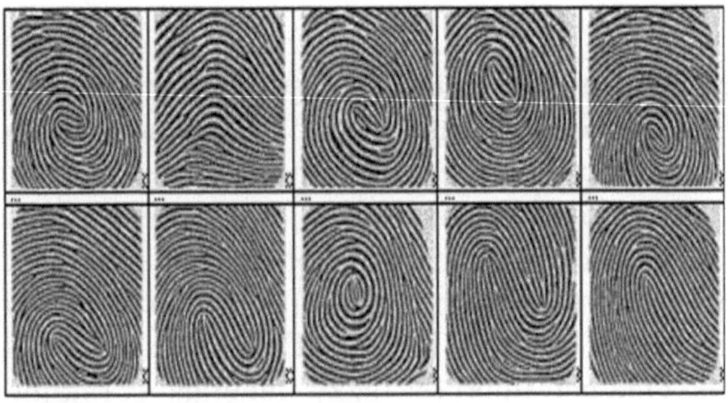

Fig. 6. Sample fingerprints used for training.

4.2 Proposed Architecture

The architecture proposed in this work consist of three main modules, in which each of them in turn consists of a set of neural networks trained with the same data, which provides the modular architecture shown in Figure 7.

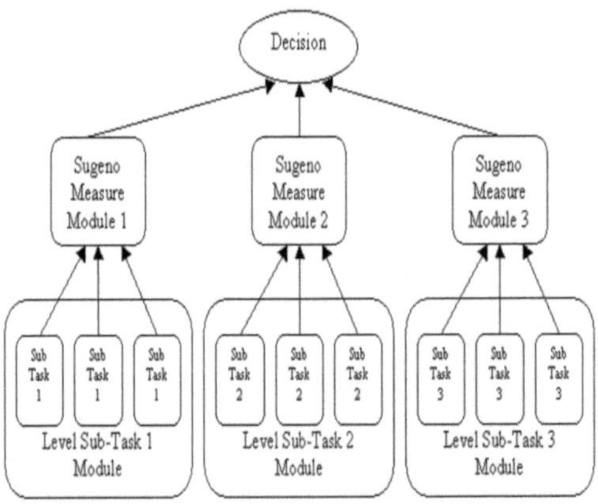

Fig. 7. Final proposed architecture.

The input to the modular system is a complete photograph. For performing the neural network training, the images of the human faces were divided in three different regions. The first region consists of the area around the eyes, which corresponds to Sub Task 1. The second region consists of the area around the nose, which corresponds to Sub Task 2. The third region consists of the area around the mouth, which corresponds to Sub Task 3. An example of this image division is shown in Figure 8.

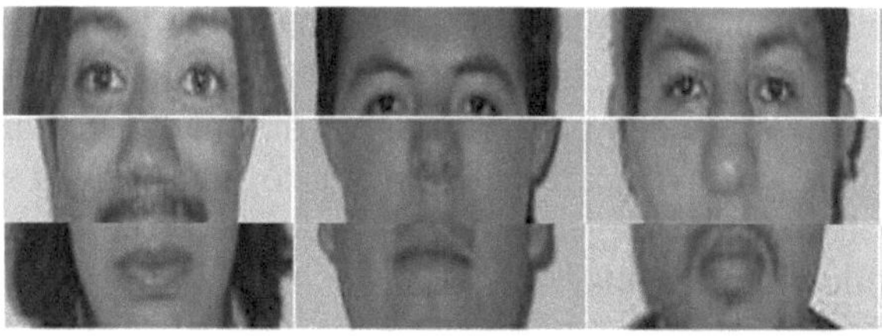

Fig. 8. Example of Image Division.

As output to the system we have an image that corresponds to the complete image that was originally given as input to the modular system, we show in Figure 9 an example of this for face recognition. In the same way the fingerprints are divided in three parts and given to the corresponding Sub task module. This is ilustrated in Figure 10.

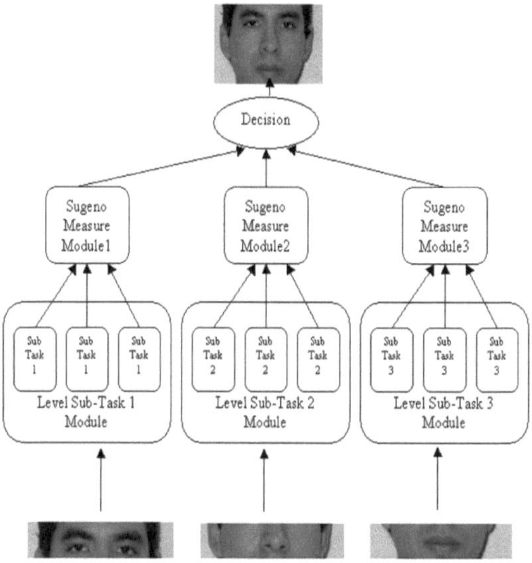

Fig. 9. Final architecture showing inputs and outputs.

4.2 Description of the Integration Module

The integration modules performs its task in two phases. In the first phase, it obtains two matrices. The first matrix, called h, of dimension 3x3, stores the larger index values resulting from the competition for each of the members of the modules. The second matrix , called I, also of dimension 3x3, stores the photograph number corresponding to the particular index.

Once the first phase is finished, the second phase is initiated, in which the decision is obtained. Before making a decision, if there is consensus in the three modules, we can proceed to give the final decision, if there isn't consensus then we have search in matrix g to find the larger index values and then calculate the Sugeno fuzzy measures for each of the modules, using the following formula,

$$g(Mi) = h(A) + h(B) + \lambda\, h(A)\, h(B) \tag{5}$$

Where λ is equal to 1. Once we have these measures, we select the largest one to show the corresponding photograph.

4.3 Summary of Results

We describe in this section the experimental results obtained with the proposed approach using the 20 photographs as training data. We show in Table 1 the relation between accuracy (measured as the percentage of correct results) and the percentage of noise in the figures.

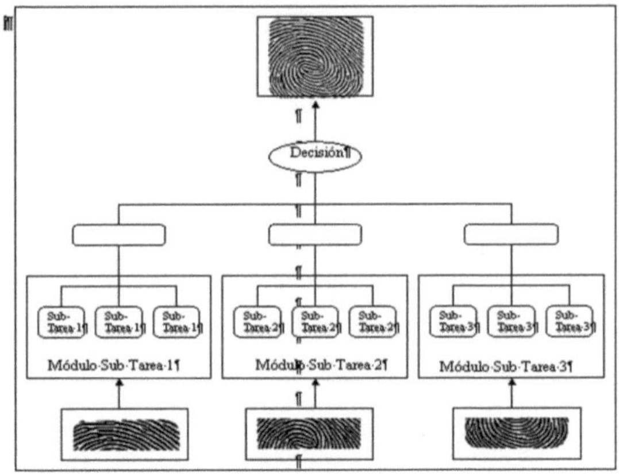

Fig. 10. Final architecture for the fingerprints.

In Table 1 we show the relation that exists between the % of noise that was added in a random fashion to the testing data set, that consisted of the 20 original photographs, plus 200 additional images. We show in Figure 11 sample images with noise.

In Table 2 we show the reliability results for the system. Reliability was calculated as shown in the following formula.

$$\text{Reliability} = \frac{\text{correct results-error}}{\text{correct results}}$$

Table 1. Relation between the % of noise and the % of correct results

% of noise	% accuracy
0	100
10	100
20	100
30	100
40	95
50	100
60	100
70	95
80	100
90	75
100	80

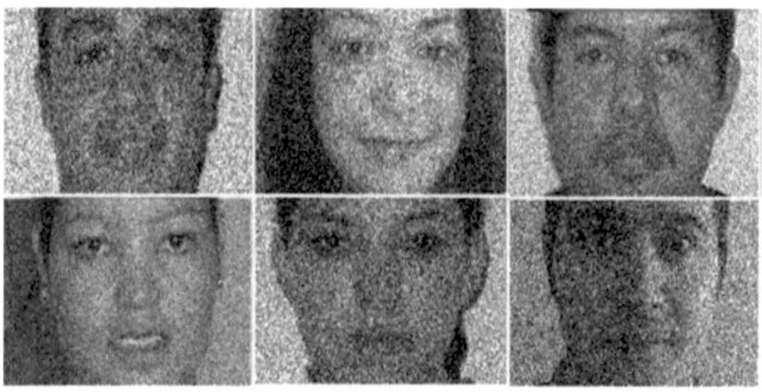

Fig. 11. Sample images with noise.

Table 2. Relation between reliability and accuracy.

% errors	%reliability	%correct results
0	100	100.00
0	100	100.00
0	100	100.00
0	100	100.00
5	94.74	95.00
0	100	100.00
0	100	100.00
5	94.74	95.00
0	100	100.00
25	66.67	75.00
20	75	80.00

We show in Figure 12 a plot relating the percentage of recognition against the number of examples used in the experiments.

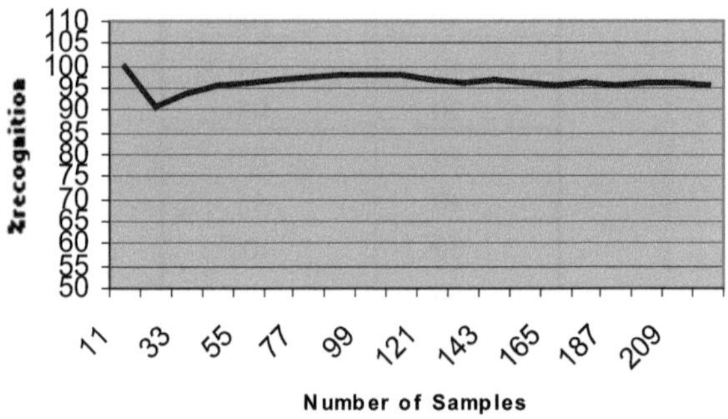

Fig. 12. Relation between recognition and number of examples.

In addition to the results presented before, we also compared the performance of the modular approach, against the performance of a monolithic neural network approach. The conclusion of this comparison was that for this type of input data, the monolithic approach is not feasible, since not only training time is larger, also the recognition is too small for real-world use. We show in Figure 13 a plot showing this comparison but now in a graphical fashion.

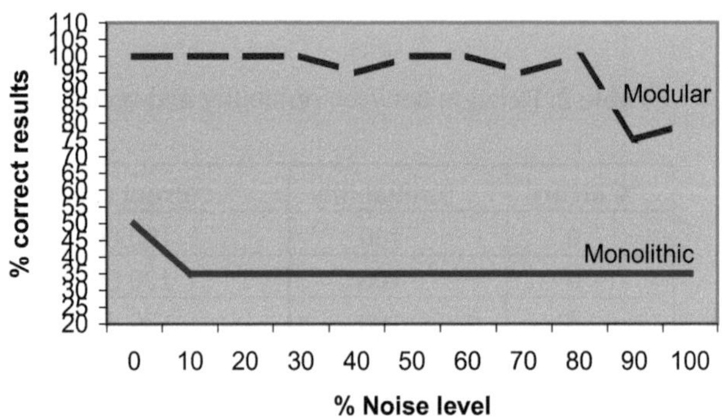

Fig. 13. Comparison between modular and monolithic approach.

5 Conclusions

We showed in this paper the experimental results obtained with the proposed modular approach. In fact, we did achieve a 98.9% recognition rate on the testing data, even with an 80% level of applied noise. For the case of 100% level of applied noise, we did achieve a 96.4 % recognition rate on the testing data. The testing data included 10 photographs for each image in the training data. These 10 photographs were obtained by applying noise in a random fashion, increasing the level of noise from 10 to 100 %, to the training data. We also have to notice that it was achieved a 96.7 % of average reliability with our modular approach. This percentage values was obtained by averaging

In light of the results of our proposed modular approach, we have to notice that using the modular approach for human face pattern recognition is a good alternative with respect to existing methods, in particular, monolithic, gating or voting methods. As future research work, we propose the study of methods for pre-processing the data, like principal components analysis, eigenfaces, or any other method that may improve the performance of the system. Other future work include considering different methods of fuzzy response integration, or considering evolving the number of layers and nodes of the neural network modules.

Acknowlegments

We would like to express our gratitude to the Research Grant Committee of COSNET, for the financial support given to this research work under grant 422.03-P. We would also like to thank CONACYT for the scholarships for the students working in this project (Claudia Gonzalez, Diana Bravo, Felma Gonzalez and Gabriela Martinez).

References

S. Albrecht. (1996), A Modular Neural Network Architecture with Additional Generalization Abilities for High Dimensional Input Vectors.

E.J.W. Boers and H. Kuiper. (1992), "Biological Metaphors and the Design of Modular Artificial Neural Networks". *Departments of Computer Science and Experimental and Theoretical Psychology at Leid University, the Netherlands.*

O. Castillo and P. Melin. (2003), "Soft Computing and Fractal Theory for Intelligent Manufacturing". *Springer-Verlag, Heidelberg, Germany.*

H-C. Fu, Y-P. Lee, C-C Chiang and H-T. Pao. (2001), "Divide-and-Conquer Learning and Modular Perceptron Networks", *IEEE Transaction. Neural Networks,* vol. 12, pp 250-263.

L. K. Hansen and P. Salomon. (1990), "Neural Network Ensembles", *IEEE Transactions on Pattern Analysis and Machine Intelligence. Neural Networks*, vol 12, pp 993-1001.

B. Lu and M. Ito. (1998), "Task Decomposition and module combination based on class relations: modular neural network for pattern classification", *Technical Report, Nagoya Japan.*

R. Murray-Smith and T.A. Johansen. (1997), Multiple Model Approaches to Modeling and Control. *Taylor and Francis.*

A. Quezada (2004), "Reconocimiento de Huellas Digitales Utilizando Redes Neuronales Modulares y Algoritmos Geneticos". *Thesis of Computer Science, Tijuana Institute of Technology.*

E. Ronco and P. Gawthrop. (1995), "Modular neural networks: A State of the Art. Technical Report", *Center for System and Control. University of Glasgow, Glasgow, UK.*

A.J.C. Sharkey. (1998), "Combining Artificial Neural Nets: Ensemble and Modular Multi-Nets Systems", *Ed. Springer-Verlag, New York.*

M. Sugeno. (1974), "Theory of fuzzy integrals and its application", *Doctoral Thesis, Tokyo Institute of Technology.*

R.R. Yager. (1999), "Criteria Aggregations Functions Using Fuzzy Measures and the Choquet Integral", *International Journal of Fuzzy Systems, Vol. 1, No. 2.*

Part V Time Series and Diagnosis

Optimal Training for Associative Memories: Application to Fault Diagnosis in Fossil Electric Power Plants

Jose A. Ruz-Hernandez[1], Edgar N. Sanchez[2] and Dionisio A. Suarez[3]

[1] Universidad Autonoma del Carmen, Av. 56 # 4 X Av. Concordia, CP 24180, Cd. del Carmen, Campeche, Mexico, on Ph. D studies at CINVESTAV, Unidad Guadalajara, jruz@pampano.unacar.mx,
[2] CINVESTAV, Unidad Guadalajara, Apartado Postal 31-430, Plaza La Luna, Guadalajara, Jalisco C.P. 45091, MEXICO, sanchez@gdl.cinvestav.mx, [3] Instituto de Investigaciones Electricas, Calle Reforma # 113, Col. Palmira, C.P. 62490, Cuernavaca, Morelos, MEXICO, suarez@iie.org.mx

Abstract. In this chapter, the authors discuss a new synthesis approach to train associative memories, based on recurrent neural networks. They propose to update the weight vector as the optimal solution of a linear combination of support patterns. The proposed training algorithm maximizes the margin between the training patterns and the decision boundary. This algorithm is applied to the synthesis of an associative memory, for fault diagnosis in fossil electric power plants. The scheme is evaluated via a full scale simulator to diagnose the main faults occurred in this kind of power plants.

1 Introduction

The implementation of associative memories via recurrent neural networks is discussed in [1]. The goal of associative memories is to store a set of desired patterns as stable memories such that a stored pattern can be retrieved when the input pattern (or the initial pattern) contains sufficient information about that stored pattern. In practice the desired memory patterns are usually represented by bipolar vectors (or binary vectors).

There are several well-known methods available in the literature, which solve the synthesis problem of recurrent neural networks for associative memories, including the *outer product method* [2], the *projection learning rule* [3], [4], and the *eigenstructure method* [5], [6], (for a review of this

J.A. Ruz-Hernandez et al.: *Optimal Training for Associative Memories: Application to Fault Diagnosis in Fossil Electric Power Plants*, StudFuzz **208**, 329–356 (2007)
www.springerlink.com

synthesis methods, see [7]). The eigenstructure method has also been generalized for the synthesis of recurrent neural networks with predetermined constraints on the interconnecting structure [8]-[11]. These synthesis methods use a set of linear equations for the design of recurrent neural networks. For the design method discussed in [12] and [13], a set of linear inequalities is formulated and solved via the optimal mean square-error (MSE), using the Hop-Kashyap method [14]. The design method presented in [15] utilizes a set of linear inequalities solved using linear programming.

An important contribution to recurrent neural networks for associative memories is presented in [1]; a new synthesis approach is developed based on the perceptron training algorithm. Due to their high generalization performance, support vector machines (SVMs) have attracted great attention for pattern recognition, machine learning, neural networks and so on [16], [17]. Learning of a SVM leads to a quadratic programming (QP) problem, which can be solved by many techniques [18].

This chapter proposes an optimal training algorithm to design associative memories implemented by recurrent neural networks. The algorithm uses a synthesis approach based on the SVMs, updating weight vector via a solution, which is expressed as a linear combination of support patterns. The training algorithm maximizes the margin between the training patterns and the decision boundary. The proposed algorithm is used for a neural network based scheme of fault diagnosis in fossil electric power plants.

2 Preliminaries

This section introduces useful preliminaries about associative memories implemented by recurrent neural networks, the perceptron training algorithm and a synthesis for recurrent neural networks based on this algorithm, which is proposed in [1]. The class of recurrent neural networks considered is described by equations of the form

$$\begin{aligned} \frac{dx}{dt} &= -Ax + T\text{sat}(x) + I \\ y &= \text{sat}(x) \end{aligned}, \tag{1}$$

where x is the state vector, $y \in D^n = \{x \in R^n \mid \leq x_i \leq 1\}$ is the output vector, $A = \text{diag}\,[a_1, a_2, \ldots, a_n]$ with $a_i > 0$ for $i = 1, 2, \ldots, n$, $T = [T_{ij}] \in R^{nxn}$ is the connection matrix, $I = [I_1, I_2, \ldots, I_n]^T$ is a bias vector, and $\text{sat}(x) = [\text{sat}(x_1), \ldots, \text{sat}(x_n)]^T$ represents the activation function, where

$$\text{sat}(x_i) = \begin{cases} 1 & x_i > 1 \\ x_i & -1 \le x_i \le 1 \\ -1 & x_i < 1 \end{cases}.$$

It is assumed that the initial states of (1) satisfy $|x_i(0)| \le 1$ for $i = 1, 2, ..., n$. System (1) is a variant of the analog Hopfield model [19] with activation function sat(•).

2.1 The Perceptron

A number of different types of perceptrons are described in [20] and [21]. The one which will be utilized in the present paper is described by:

$$Z = \text{sign}(Wu) \ , \tag{2}$$

where Z is the perceptron output, $u = [u_1, u_2, \ldots, u_n, 1]^T$ is the input vector, $W = [w_1, w_2, \ldots, w_n, \theta]^T$ is the weight vector, and

$$\text{sign}(\xi) = \begin{cases} 1 \text{ if } \xi \ge 0 \\ -1 \text{ if } \xi < 0 \end{cases}.$$

This simple perceptron can perform pattern classification (between two classes denoted by X_1 and X_2). The weight vector W can be obtained by the following training algorithm ([20], [21]).

Perceptron Training Algorithm: Given m training patterns α^k, $k = 1, 2, ..., m$ which are known to belong to class X_1 (corresponding to $Z = 1$) or X_2 (corresponding to $Z = -1$), the weight vector W can be obtained by the following algorithm.

1. Initialize the weight vector $W(l)$ for $l = 0$;
2. For $l = 0, 1, 2, ...$
   a) if $W(l)\, u(l) \ge 0$ and $u(l) \in X_2$, then update $W(l+1) = W(l) - \eta\, u(l)$.
   b) if $W(l)\, u(l) < 0$ and $u(l) \in X_1$, then update $W(l+1) = W(l) + \eta\, u(l)$.
   c) otherwise, $W(l+1) = W(l)$, where $u(l) = \alpha^k$ for some k, $1 \le k \le m$ and $\eta > 0$ is the perceptron learning rate;
3. Stop the training when no more updates for the weight vector W are needed, i. e., stop the training when all the training patterns can be correctly classifies by W.

2.2 Synthesis for Recurrent Neural Networks Based on the Perceptron Algorithm

For the sake of completeness, the following results are taken from [1] and included in this section. A vector α will be called a (*stable*) *memory vector* (or simply, *a memory*) of system (1) if $\alpha = \text{sat}(\beta)$ and if β is an *asymptotically stable* equilibrium point of system (1). In the following lemma, B^n is defined as a set of n-dimensional *bipolar vectors* $B^n = \{x \in R^n \mid x_i \vee x_j = -1, i = 1, 2, ..., n\}$. For $\alpha = [\alpha_1, \alpha_2, \ldots, \alpha_n]^T \in B^n$ define $C(\alpha) = \{x \in R^n \mid x_i\alpha_i > 1, i = 1, 2, ..., n\}$.

Lemma 2.1 If $\alpha \in B^n$ and if

$$\beta = A^{-1}(T\alpha + I), \tag{3}$$

then (α, β) is a pair of stable memory vector and an asymptotically stable equilibrium point of (1).

The following synthesis problem concerns the design of (1) for associative memories.

Synthesis Problem: Given m vectors $\alpha^1, \alpha^2, ..., \alpha^m$ in the set of n-dimensional bipolar vectors, B^n, choose $\{A, T, I\}$ in such a manner that:

1. $\alpha^1, \alpha^2, ..., \alpha^m$ are stable memory vectors of system (1);
2. the system has no oscillatory and chaotic solutions;
3. the total number of spurious memory vectors (i.e., memory vectors which are not desired) is as small as possible, and the domain (or basin) of attraction of each desired memory vectors is as large as possible.

Item 1) of the synthesis problem can be guaranteed by choosing the $\{A, T, I\}$ such that every α^i satisfies condition 3 of Lemma 2.1. Item 2 can be achieved by designing a neural network with symmetric connection matrix T. Item 3) can be partly ensured by constraining the diagonal elements of the connection matrix.

The previous synthesis problem can be solved by applying the perceptron training algorithm. To solve the synthesis problem, one needs to determine A, T and I from (3) with $\alpha = \alpha^k$, $k = 1, 2,..., m$.

Condition given in (3) can be equivalently written as

$$\begin{cases} T_i\alpha^k + I_i > a_i & \text{if } \alpha_i^k = 1 \\ T_i\alpha^k + I_i < -a_i & \text{if } \alpha_i^k = -1 \end{cases}, \tag{4}$$

for $k = 1,2,...,m$ and $i = 1,2,...,n$ where T_i represents el ith row of T, I_i denotes the ith element of I, a_i is the i-th diagonal element of A, and α_i^k is the

i-th entry of α^k. From (4), the following synthesis algorithm based on the perceptron training algorithm can now be obtained [1] .

Synthesis Algorithm 2.1: Using the perceptron algorithm to obtain n perceptrons

$$W^i = \left[w_1^i, w_2^i, \ldots, w_n^i, w_{n+1}^i \right],$$

i = 1, 2, ...,n, such that:

$$\begin{cases} W_i \alpha^{-k} \geq 0 \text{ if } \alpha_i^k = 1 \\ W_i \alpha^{-k} < 0 \text{ if } \alpha_i^k = -1 \end{cases}, \tag{5}$$

and for k = 1, 2, ..., m where

$$\alpha^{-k} = \begin{bmatrix} \alpha^k \\ \ldots \\ 1 \end{bmatrix},$$

choose A = diag $[a_1, a_2, \ldots, a_n]$ with $a_i > 0$. For i, j = 1,2,...,n choose $T_{ij} = w_j^i$ if $i \neq j$, $T_{ij} = w_j^i + a_i \mu_i$, with $\mu_i > 1$ e $I_i = w_{n+1}^i$.

We have used this algorithm to train associative memories based on a recurrent neural network and isolate faults in a neural networks-based scheme for fault diagnosis in fossil electric power plants [22]. As an extension of these results, in this chapter, we propose to use the support vector machine ideas to propose a new optimal hyperplane algorithm to train this class of associative memories.

3 Support Vector Machines and Optimal Hyperplane Algorithm

This section reviews the method of optimal hyperplane [23] for separation of training data. Two subsections are presented: the first one describes SVMs and the problem of finding an optimal hyperplane; and the second one describes the optimal hyperplane algorithm.

3.1 Support Vector Machines and Optimal Hyperplanes

The support vector machine is a learning mechanism for two-group classification problems. The machine implements the following idea: input vectors are nonlinearly mapped to a very high-dimension feature space. In this feature space, a linear decision surface is constructed. Special properties of the decision surface ensures high generalization ability [17].

The problem is how to find a separating hyperplane, which will generalize well. It was solved in 1965, for the case of *separable classes* [23]. An *optimal hyperplane* is defined as the linear function with maximal margin between the vectors of the two classes, (see Fig. 1).

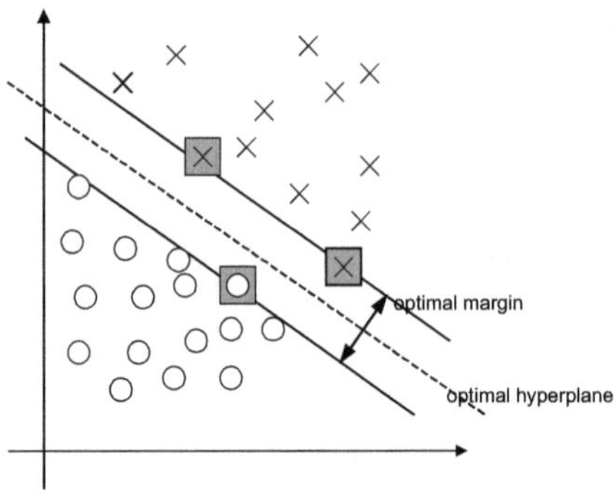

Fig. 1. An example of a separable problem in R^2. The support vectors, marked with gray squares, define the margin of largest separation between the two classes.

It was observed that to construct such optimal hyperplanes one only needs to take into account a small amount of the training data, the so called *support vectors*, which determine this margin. It was shown that if the training vectors are separated without errors by an optimal hyperplane the expectation value of probability of doing an error on a test example is bounded by the ratio between the expectation value of the number of support vectors and the number of training vectors:

$$E[\Pr(\text{error})] \leq \frac{E[\text{number of support vectors}]}{\text{number of training vectors}} . \tag{6}$$

Let

$$\mathrm{w}_0 \bullet \mathrm{z} + b_0 = 0 , \tag{7}$$

be the optimal hyperplane in the feature space, where $\mathrm{w}_0 \bullet \mathrm{z}$ is the dot-product between the weights and *vector* z in this space. The weights w_0 for the optimal hyperplane in feature space can be written as some linear combination of support vectors

$$\mathrm{w}_0 = \sum_{\text{support vectors}} \lambda_i \mathrm{z}_i . \tag{8}$$

The linear decision function $I(z)$, in the feature space will accordingly be of the form

$$I(\mathrm{z}) = \mathrm{sign}\left(\sum_{\text{support vectors}} \lambda_i \mathrm{z}_i \bullet \mathrm{z} + b_0 \right), \tag{9}$$

where $\mathrm{z}_i \bullet \mathrm{z}$ is the dot-product between support vectors z_i and *vector* z in feature space.

3.2 The Optimal Hyperplane Algorithm

This section closely follows [17]. The set of labelled training patterns

$$(y_1, x_1), \ldots, (y_l, x_l), \qquad y_i \in \{-1, 1\} \tag{10}$$

is said to be linearly separable if there exists a vector w and a scalar b such that the inequalities

$$\begin{aligned} \mathrm{w} \bullet \mathrm{x}_i + b &\geq 1 \ \text{ if } \ y_i = 1, \\ \mathrm{w} \bullet \mathrm{x}_i + b &\leq -1 \ \text{ if } \ y_i = -1, \end{aligned} \tag{11}$$

are valid for all elements of the training set (10). Below, we write these inequalities (11) as

$$\mathrm{y}_i (\mathrm{w} \bullet \mathrm{x}_i + b) \geq 1, \qquad i = 1, \ldots, l. \tag{12}$$

The optimal hyperplane

$$\mathrm{w}_0 \bullet \mathrm{x} + b_0 = 0\,, \tag{13}$$

is the unique one which separates the training data with a maximal margin: it determines the direction $\mathrm{w} / |\mathrm{w}|$ where the distance between the projections of the training vectors of two different classes is maximal, (recall Fig. 1). This distance $\rho(\mathrm{w}, b)$ is given by

$$\rho(\mathrm{w}, b) = \min_{\{x:y=1\}} \frac{\mathrm{x} \bullet \mathrm{w}}{|\mathrm{w}|} - \max_{\{x:y=-1\}} \frac{\mathrm{x} \bullet \mathrm{w}}{|\mathrm{w}|}\,. \tag{14}$$

The optimal hyperplane (w_0, b_0) is the arguments which maximize the distance (14). It follows form (14) and (12) that

$$\rho(\mathrm{w}_0, b_0) = \frac{2}{|\mathrm{w}_0|} = \frac{2}{\sqrt{\mathrm{w}_0 \bullet \mathrm{w}_0}}\,. \tag{15}$$

This means that the optimal hyperplane is the unique one which minimizes $\mathrm{w} \bullet \mathrm{w}$ under the constraints (12). Vectors x, for which $y_i(\mathrm{w}\bullet \mathrm{x}_i+b)=1$, will be support vectors. To solve the associate optimization problem, it is possible to construct the Lagrangian

$$L(\mathrm{w},b,\Lambda)=\frac{1}{2}\mathrm{w}\bullet\mathrm{w} - \sum_{i=1}^{l}\lambda_i[y_i(\mathrm{x}_i\bullet\mathrm{w}+b)-1] \ , \tag{16}$$

where $\Lambda^T = (\lambda_1, \ldots,\lambda_l)$ is the vector of non-negative Lagrange multipliers corresponding to the constraints (12). It is known that the solution for this optimization problem is determined by the saddle point of this Lagrangian in the $2l + 1$ dimensional space of w, Λ, and b, where the minimum should be taken with respect to the parameters w and b, and the maximum should be taken with respect to the Lagrange multipliers Λ [17]. At the minimum point (with respect to w and b), one obtains

$$\frac{\partial L(\mathrm{w},\Lambda,b)}{\partial \mathrm{w}}\bigg|_{\mathrm{w}=\mathrm{w}_0}=(\mathrm{w}_0-\sum_{i=1}^{l}\lambda_i y_i \mathrm{x}_i)=0 \, , \tag{17}$$

$$\frac{\partial L(\mathrm{w},\Lambda,b)}{\partial b}\bigg|_{b=b_0}=\sum_{\alpha_i} y_i\lambda_i=0 \, . \tag{18}$$

From equality (17), we derive

$$\mathrm{w}_0=\sum_{i=1}^{l}\lambda_i y_i \mathrm{x}_i \ , \tag{19}$$

which expresses that the optimal hyperplane solution can be written as a linear combination of training vectors. Note that only training vectors x_i with $\lambda_i > 0$ have an effective contribution to the sum (19).

Substituting (18) and (19) into (16), we obtain

$$W(\Lambda)=\sum_{i=1}^{l}\lambda_i-\frac{1}{2}\mathrm{w}_0\mathrm{w}_0 \, , \tag{20}$$

$$W(\Lambda)=\sum_{i=1}^{l}\lambda_i-\frac{1}{2}\sum_{i=1}^{l}\sum_{j=1}^{l}\lambda_i\lambda_j y_i y_j \mathrm{x}_i \mathrm{x}_j \ . \tag{21}$$

In vector notation, this can be rewritten as

$$W(\Lambda)=\Lambda^T\mathrm{Q}-\frac{1}{2}\Lambda^T D\Lambda \ , \tag{22}$$

where Q is an l-dimensional unit vector, and D is a symmetric l x l matrix with elements

$$D_{ij} = y_i y_j \mathrm{x}_i \mathrm{x}_j \ . \tag{23}$$

To find the desired saddle point it remains to locate the maximum of (21) under the constraints (17)

$$\Lambda^T \mathrm{Y} = 0 \ , \tag{24}$$

where $\mathrm{Y}^T = (y_{1,\ldots,} \ y_l)$, and

$$\Lambda \geq 0. \tag{25}$$

The Kuhn-Tucker theorem plays an important part in this optimization problem [24]. According to this theorem, at our saddle point in w_0, b_0, Λ_0, any Lagrange multiplier λ_i^0, and its corresponding constraints are connected by an equality

$$\lambda_i [y_i(\mathrm{x}_i \bullet \mathrm{w}_0) + b - 1] = 0, \quad i = 1,\ldots,l. \tag{26}$$

From this equality comes that non-zero values λ_i are only achieved in the cases where

$$y_i(\mathrm{x}_i \bullet \mathrm{w}_0) + b - 1 = 0 \ . \tag{27}$$

In other words, $\lambda_i \neq 0$ only for cases where the *inequality* (12) satisfies the *equality condition*. We call vectors x_i for which

$$y_i(\mathrm{x}_i \bullet \mathrm{w}_0) + b = 1 \ , \tag{28}$$

the support vectors. From (19), w_0 can be expanded on support vectors. Based on equation (18) and (19) for the optimal solution, the relationship between the maximal value $W(\Lambda_0)$ and the separation distance ρ_0:

$$\mathrm{w}_0 \mathrm{w}_0 = \sum_{i=1}^{l} \lambda_i^0 y_i \mathrm{x}_i \bullet \mathrm{w}_0 = \sum_{i=1}^{l} \lambda_i^0 (1 - y_i b_0) = \sum_{i=1}^{l} \lambda_i^0 \ . \tag{29}$$

Substituting this equality into the expression (20) for $W(\Lambda_0)$, we obtain

$$W(\Lambda_0) = \sum_{i=1}^{l} \lambda_i^0 - \frac{1}{2} \mathrm{w}_0 \mathrm{w}_0 = \frac{1}{2} \mathrm{w}_0 \mathrm{w}_0 \ . \tag{30}$$

Taking into account the expression (14), we obtain

$$W(\Lambda_0) = \frac{2}{\rho_0^2} \ , \tag{31}$$

where ρ_0 is the margin for the optimal hyperplane. If for same Λ_* and large constant W_0 the inequality

$$W(\Lambda_*) > W_0 \ , \tag{32}$$

is valid, then all hyperplanes which separate the training data have a margin

$$\rho \prec \sqrt{\frac{2}{W_0}} \ . \tag{33}$$

If the training set can not be separated by an hyperplane, the margin between the patterns of the two classes becomes arbitrary small, forcing the value of the functional $W(\Lambda)$ to be arbitrary large.

4 New Synthesis Approach

This section contains our principal contribution. Considering the determination of the optimal hyperplane explain in section 3.2 and the training discussed in [1], we propose the following innovative algorithm for training associative memories implemented by recurrent neural networks.

Synthesis Algorithm 4.1: Given m training patterns α^k, k = 1,2,…,m which are known to belong to class X_1 (corresponding to Z = 1) or X_2 (corresponding to Z = -1), the weight vector W can be determined by means of the following algorithm.

1. Start out by solving the quadratic programming problem given by (22), (24) and (25) to obtain $\Lambda^T = (\lambda_1, \ldots, \lambda_m)$.
2. Compute the weight vector

$$W^i = \sum_{j=1}^{m} \lambda_j \alpha_j^k \alpha^{-k} = \left[w_1^i, w_2^i, \ldots, w_n^i, w_{n+1}^i \right], \tag{34}$$

i = 1, 2, …, n, such that

$$\begin{aligned} &W^i \bullet \alpha^{-k} + b \geq 1 \ \text{ if } \ \alpha_i^k = 1, \\ &W^i \bullet \alpha^{-k} + b \leq -1 \ \text{ if } \ \alpha_i^k = -1, \end{aligned} \tag{35}$$

and for k = 1, 2, ..., m where

$$\alpha^{-k} = \begin{bmatrix} \alpha^k \\ \ldots \\ 1 \end{bmatrix} .$$

3. Choose A = diag $[a_1, a_2, \ldots, a_n]$ with $a_i > 0$. For $i, j = 1,2,\ldots,n$ choose $T_{ij} = w_j^i$ if $i \neq j$, $T_{ij} = w_j^i + a_i\mu_i$, with $\mu_i > 1$ e $I_i = w_{n+1}^i$.

5 Application to Fault Diagnosis in Fossil Electric Power Plants

In order to illustrate the applicability of the above proposed optimal procedure to train associative memories, based on recurrent neural networks, we discuss its application to fault diagnosis in fossil electric power plants.

Fault diagnosis can be performed by a three steps algorithm [25]. First, one or several signals are generated which reflect faults in the process behavior. These signals are called residuals. For the second step, the residual are evaluated. A decision has to be taken in order to determine time and location of possible faults from the residuals. Finally, the nature and the cause of the fault is analyzed by the relations between the symptoms and their physical causes.

In order to describe the fault free nominal behavior of the process under supervision, a model (mathematic or heuristic) is employed, giving to this concept the name of model-based fault diagnosis. Model-based approaches have dominated the fault diagnosis research [26], [27].

Employing measurements of the process under normal operation, if possible, or with the help of a simulator as realistic as possible, a suitable neural network can be trained to learn the process input-output behaviour [28].

This section presents a neural network scheme for fault diagnosis. It uses for residual generation a predictor which consists of a bank of recurrent multilayer perceptron neural network models. Fault diagnosis is carried out by an associative memory, which is based on a recurrent neural network, and trained with the proposed optimal learning algorithm.

5.1 Problem Description

Fault diagnosis in fossil electric power plants is a task carried out by an expert operator. This operator recognizes typical faults via supervision of key variables evolution. Adequate fault detection and diagnosis aids will help the human operator in order to take the right decisions to maintain the required electric energy production, avoiding failures and even accident risky to humans and the environment [29].

For this kind of plants, the main faults can be clustered as: faults related to temperature control of the superheated and reheated steam, faults related to combustion control and faults related to the steam generator drum water level. In order to understand the first group of faults, it is helpful to briefly describe the steam generator and superheated and reheated steam system. A simplified scheme is presented in Fig. 2, which illustrates the main components of a typical steam generator and superheating/ reheating system.

Feedwater from the economizer enters the steam drum, and by forced circulation, the drum water flows down the downcomers and rises through the furnace wall tubes to generate steam by means of the hot combustion gases in the furnace. The water and steam in the drum are separated by steam separators and the steam becomes superheated as it passes through various superheaters. The turbine exhaust steam is again superheated in the reheater before generating power in the intermediate and low pressure turbines. For this system, the main automatic control loops are the main steam temperature control and the reheated steam temperature control. The first one is controlled by the temperation spray, and the second one is controlled by the burners inclination angle, as well as by another temperation spray.

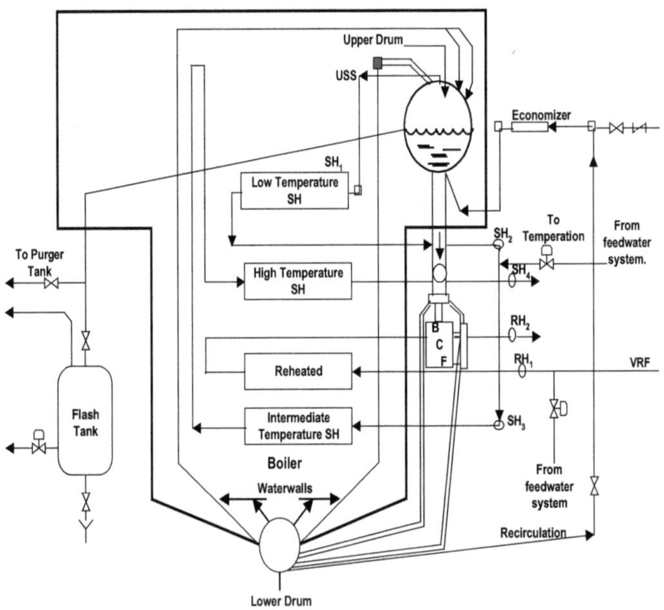

Fig 2. Steam Generator and Reheating / Superheating System.

As an example, we discuss a typical fault: *waterwall tubes breaking*, which is part of the above first fault group. It could be due to inadequate design, selection of materials, and/or unsuitable start-up operations [30]. In presence of this fault, the combustion gases do not circulate properly and the waterwall tubes are not suitable cold. Additionally, the water level on the steam generator drum goes down and the level control tries to keep it by means of varying the feedwater flow. However if the maximum value of this flow is reached, and the water level continues to decrease, the low level monitoring orders the steam generator out of operation. If this order takes a long time to be executed or if it is not performed, the waterwalls tubes operating normally will suffer strong damages.

This fault also diminishes the steam generator drum pressure, causing reductions on the superheated and reheated steam pressures. The combustion control tries to correct this situation by increasing the air flow and the fossil oil flow; these actions could increase the steam generator pressure beyond the allowed limit, and as a consequence the steam generator would be taken out of operation. If the human operator, in presence of this fault, does not take the adequate corrective actions, the healthy waterwalls tubes could be damaged due to thermal stress. The turbine will also suffer from thermal and mechanical stress.

5.2 Scheme for Fault Diagnosis

This scheme has two components: residual generation and fault diagnosis. The scheme is displayed in Fig. 3.

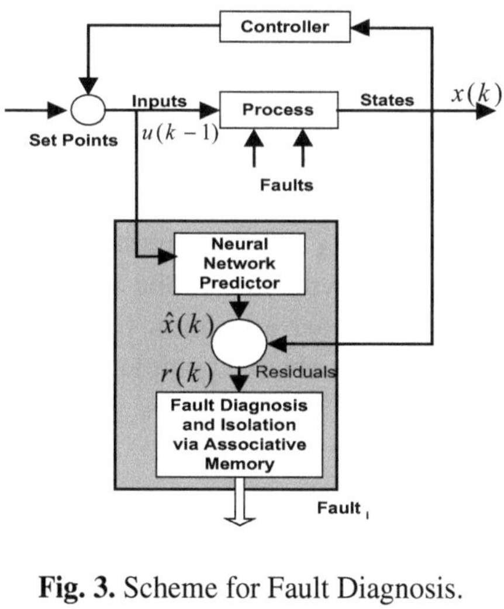

Fig. 3. Scheme for Fault Diagnosis.

The first component is based on comparison between the measurements coming from the plant and the predicted values generated by a neural network predictor. The predictor is based on neural network models, which are trained using healthy data from the plant. The differences between these two values, named as residuals, constitute a good indicator for fault detection. The residuals are calculated as

$$r_i(k) = x_i(k) - \hat{x}_i(k), \; i = 1, 2, \ldots, n. \tag{36}$$

where $x_i(k)$ are the plant measures and $\hat{x}_i(k)$ are the predictions. The residuals should be independent of the system operating state under nominal plant operating conditions. In absence of faults, the residuals are only due to noise and disturbance. When a fault occurs in the system, the residuals deviate from zero in characteristic ways.

For the second component, residuals are encoded in bipolar or binary vectors using thresholds to obtain fault patterns. These fault patterns are used to train an associative memory based on a recurrent neural network, which is employed to carry out the fault diagnosis. Our proposed optimal algorithm is used to train this associative memory.

5.2.1 Residual Generation

For residual generation purposes the neural network replaces the analytical model describing the process under normal operation. The neural networks training is done using the series-parallel scheme [31]. After finishing the training, the neural networks are applied for residual generation (Fig. 4); its weights are fixed and used as a parallel scheme to carry out predictions. The neural network predictor is designed using ten neural network models, which are trained via the Levenberg-Marquardt Learning Algorithm ([32], [33]). Each neural network is a recurrent multilayer perceptron. The networks have one hidden layer with hyperbolic tangent activation functions and a single neuron with a linear activation function as the output layer. The neural network models are obtained employing the toolbox NNSYSID [34], which runs in MATLAB[1].

[1] MATLAB is a registered trademark of The Math Works, Inc.

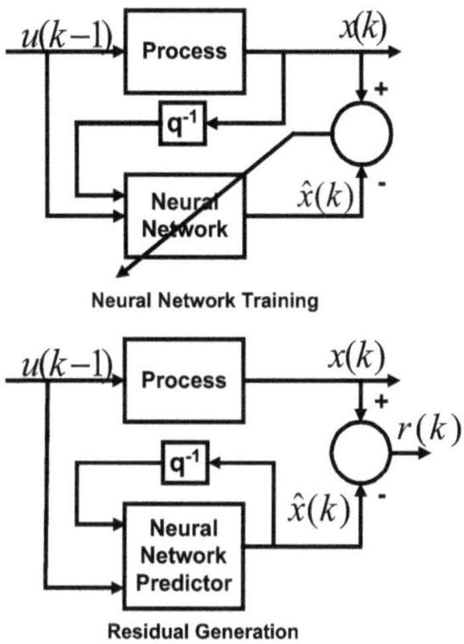

Fig.4. Scheme for training and application of neural networks for residual generation.

All the models have eight input variables and a one output variable with a NNARX[2] structure as:

$$\hat{x}_1(k) = F_1[W_1, x_1(k-1), \cdots, x_1(k-6), u_1(k-1), \cdots, u_1(k-6), \cdots, u_8(k-1), \cdots, u_8(k-6)] \quad (37)$$

$$\hat{x}_2(k) = F_2[W_2, x_2(k-1), \cdots, x_2(k-4), u_1(k-1), \cdots, u_1(k-4), \cdots, u_8(k-1), \cdots, u_8(k-4)] \quad (38)$$

$$\hat{x}_3(k) = F_3[W_3, x_3(k-1), \cdots, x_3(k-6), u_1(k-1), \cdots, u_1(k-6), \cdots, u_8(k-1), \cdots, u_8(k-6)] \quad (39)$$

$$\hat{x}_4(k) = F_4[W_4, x_4(k-1), \cdots, x_4(k-6), u_1(k-1), \cdots, u_1(k-6), \cdots, u_8(k-1), \cdots, u_8(k-6)] \quad (40)$$

[2] Neural Network AutoRegressive, eXternal input.

$$\hat{x}_5(k) = F_5[W_5, x_5(k-1), \cdots, x_5(k-5), u_1(k-1), \cdots, u_1(k-5), \cdots, u_8(k-1), \cdots, u_8(k-5)] \quad (41)$$

$$\hat{x}_6(k) = F_6[W_6, x_6(k-1), \cdots, x_6(k-3), u_1(k-1), \cdots, u_1(k-3), \cdots, u_8(k-1), \cdots, u_8(k-3)] \quad (42)$$

$$\hat{x}_7(k) = F_7[W_7, x_7(k-1), \cdots, x_7(k-6), u_1(k-1), \cdots, u_1(k-6), \cdots, u_8(k-1), \cdots, u_8(k-6)] \quad (43)$$

$$\hat{x}_8(k) = F_8[W_8, x_8(k-1), \cdots, x_8(k-6), u_1(k-1), \cdots, u_1(k-6), \cdots, u_8(k-1), \cdots, u_8(k-6)] \quad (44)$$

$$\hat{x}_9(k) = F_9[W_9, x_9(k-1), \cdots, x_9(k-6), u_1(k-1), \cdots, u_1(k-6), \cdots, u_8(k-1), \cdots, u_8(k-6)] \quad (45)$$

$$\hat{x}_{10}(k) = F_{10}[W_{10}, x_{10}(k-1), \cdots, x_{10}(k-6), u_1(k-1), \cdots, u_1(k-6), \cdots, u_8(k-1), \cdots, u_8(k-6)] \quad (46)$$

where the input variables are

$u_1(.)$ = Fossil oil flow (%).
$u_2(.)$ = Air flow (%).
$u_3(.)$ = Condensed water flow (Litres per minute).
$u_4(.)$ = Water flow for temperation (Kg/s).
$u_5(.)$ = Feedwater flow (T/H).
$u_6(.)$ = Replacement flow to condenser (Litres per second).
$u_7(.)$ = Steam water flow (Litres per minute).
$u_8(.)$ = Burner inclination angle (Degrees)

and the output variables are

$x_1(.)$ = Load power (MW).
$x_2(.)$ = Boiler pressure (Pa).
$x_3(.)$ = Drum level (m).
$x_4(.)$ = Reheated steam temperature (ºK).
$x_5(.)$ = Superheated steam temperature (ºK).

$x_6(.)$ = Reheated steam pressure (Pa).
$x_7(.)$ = Drum pressure (Pa).
$x_8(.)$ = Differential pressure (spray steam – fossil oil flow) (°K).
$x_9(.)$ = Fossil oil temperature to burners (°K).
$x_{10}(.)$ = Feedwater temperature (°K).

W_i represent the weights for each neural network model. The lag structure of each neural network model is determined using the same criterion as in [34]. Once the neural network predictors have been trained, its weights are fixed and used as a parallel scheme for carry out the predictions. Neural networks models are validated with healthy fresh data. Prediction errors close to 1 % are obtained for each model. We display in Fig. 5 a validation test with neural network model given by equation (37) working as a parallel scheme to carry out predictions. This validation test considers load power changes, and it is assumed that initial condition for the neural network model is different to the data acquired $x_1(0)$ from full scale simulator.

The residual generation scheme is implemented for six faults: *waterwall tubes breaking*, *superheater tubes breaking*, *superheated steam temperature control fault*, *dirty regenerative preheater*, *velocity varier of feedwater pumps operating to maximum value* and *blocked fossil oil valve* named as *fault* 1 to *fault* 6, respectively.

For faults 1 to 4, data bases are acquired with a full scale simulator for 75% of initial load power (225 MW), 15 % of severity fault, 2 minutes for inception and 8 minutes of simulation time. Furthermore, for *fault* 5 and *fault* 6 the simulator has only available severity and inception which are chosen as 15 % and 2 minutes, respectively. For these two faults, data bases are acquired for 3 and 4 minutes of simulation time, respectively. The *fault* 5 is very critical because it can activate the drum level alarm and forces the fossil electric power plant out of operation. It is clear that *fault* 6 is detectable when the load power is changed by the operator because the fossil oil valve does not work adequately. For the six cases, residuals are close to zero before fault inception. After this interval, residuals deviate from zero in different ways.

Even if all the described faults are taken into account, in this chapter, due to space limitations, for explaining the process to generate fault patterns, only *fault* 1 and *fault* 5 are considered. The respective residuals are displayed in Fig. 6 and Fig. 7.

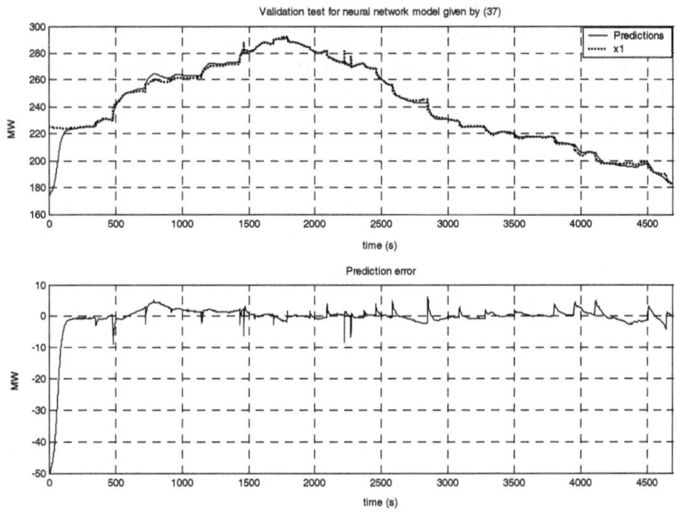

Fig. 5. Validation test for neural network model given by equation (37).

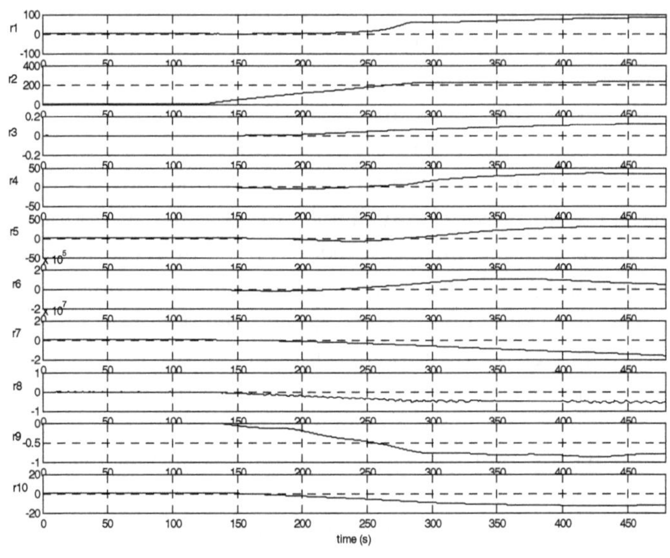

Fig. 6. Residuals for *fault* 1: *waterwall tubes breaking.*

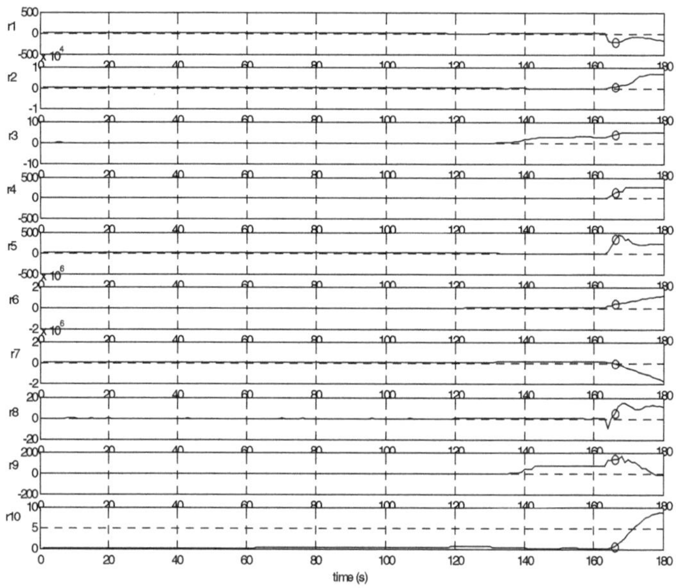

Fig. 7. Residuals for *fault* 5: *velocity varier of feedwater pumps operating to maximum value*. The symbol 'o' denotes that fossil electric power plant is taken out of operation.

5.2.2 Fault Diagnosis

Figure 8 presents a scheme to carry out the fault diagnosis via this associative memory. The previous stage generates a residual vector with ten elements which are evaluated by detection thresholds. Detection thresholds are contained in Table I. This evaluation provides a set of residuals encoded as bipolar vectors $[s_1(k), s_2(k), \ldots, s_{10}(k)]^T$ where

$$s_i(k) = \begin{cases} -1 \text{ if } r_i < \tau_i \\ 1 \text{ if } r_i \geq \tau_i \end{cases}, \quad i = 1, 2, \ldots, 10. \tag{47}$$

Residuals are encoded on-line for every fault. Residuals for *fault* 1 and *fault* 5 are displayed in Fig. 9 and Fig. 10. Once residuals are encoded, it is necessary to analyze them to choose the fault patterns to store in the associative memory. This selection is done in order to discriminate adequately every fault, to reduce false alarms and to isolate fault as soon as possible. The patterns obtained are used, based on the optimal synthesis algorithm proposed in section 4 to train the recurrent neural network and to design the respective associative memory as a way to isolate the faults. Fault patterns, for all the six faults previously mentioned, are contained in Table II, where *fault* 0 pattern is included to denote a normal operation condition.

The optimal training algorithm is programmed in MATLAB. The number of neurons is n=10 (fault pattern length) and the patterns are m=7 (number of fault patterns). The Lagrange multipliers matrix LM= $[\Lambda^1, \Lambda^2, \ldots, \Lambda^n]$, the weight matrix $WM=[W^1, W^2, \ldots, W^{n+1}]$ and the bias vector $BV=[b^1, b^2, \ldots, b^n]$ are obtained as in (48), (49) and (50). The matrices A, T and I are calculated as in (51), (52) and (53). The associative memory is evaluated with these matrices fixed using encoded residuals as input bipolar vectors. Fig. 11 and Fig. 12 illustrate how *fault* 1 and *fault* 5 patterns are retrieved by the associative memory.

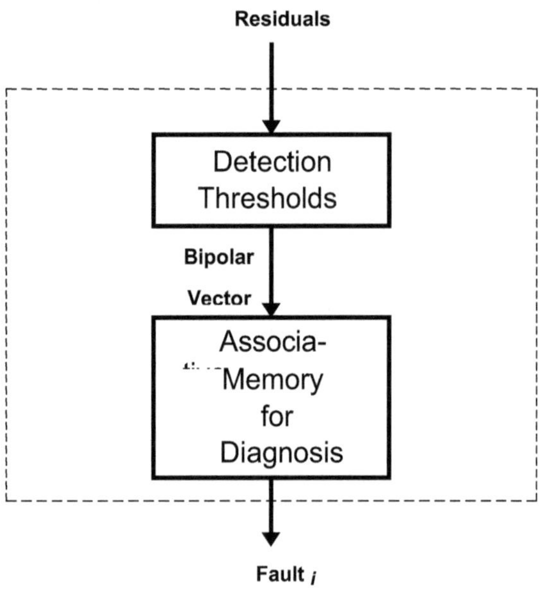

Fig. 8. Scheme for fault diagnosis.

Table 1. Detection thresholds.

i	τ_i
1	$\pm$ 25 MW
2	$\pm$ 30 Pa
3	$\pm$ 0.022 m
4	$\pm$ 4 °K
5	$\pm$ 10 °K
6	$\pm$ 20000 Pa
7	$\pm$ 42000 Pa
8	$\pm$ 0.85 %
9	$\pm$ 0.24 °K
10	$\pm$ 10 °K

$$LM = \begin{bmatrix} 0.00 & 0.00 & 0.30 & 0.00 & 0.00 & 0.05 & 0.17 & 0.00 & 0.09 & 0.00 \\ 0.08 & 0.22 & 0.00 & 0.08 & 0.00 & 0.00 & 0.00 & 0.13 & 0.07 & 0.08 \\ 0.00 & 0.00 & 0.00 & 0.17 & 0.50 & 0.20 & 0.00 & 0.04 & 0.00 & 0.00 \\ 0.09 & 0.09 & 0.00 & 0.24 & 0.00 & 0.18 & 0.01 & 0.00 & 0.14 & 0.09 \\ 0.00 & 0.00 & 0.20 & 0.00 & 0.50 & 0.05 & 0.26 & 0.00 & 0.03 & 0.00 \\ 0.00 & 0.00 & 0.10 & 0.05 & 0.00 & 0.11 & 0.14 & 0.04 & 0.19 & 0.05 \\ 0.07 & 0.13 & 0.00 & 0.09 & 0.00 & 0.01 & 0.03 & 0.22 & 0.00 & 0.07 \end{bmatrix}, \tag{48}$$

$$WM = \begin{bmatrix} 0.51 & 0.37 & -0.07 & 0.22 & 0.22 & 0.22 & 0.07 & 0.37 & 0.37 & 0.51 & -0.22 \\ 0.68 & 0.95 & -0.40 & 0.68 & 0.13 & 0.13 & -0.13 & 0.40 & 0.41 & 0.68 & -0.68 \\ 0.00 & 0.00 & 0.00 & 0.00 & 0.00 & 0.00 & 0.00 & 0.00 & 0.00 & 0.00 & 0.00 \\ 0.27 & 0.46 & -0.09 & 0.64 & 0.27 & 0.27 & 0.09 & 0.09 & 0.09 & 0.27 & -0.27 \\ 0.00 & 0.00 & 0.00 & 0.00 & 0.00 & 0.00 & 0.00 & 0.00 & 0.00 & 0.00 & 0.00 \\ 0.05 & 0.01 & 0.05 & 0.05 & 0.05 & 0.12 & 0.01 & 0.01 & 0.09 & 0.05 & 0.01 \\ 0.03 & -0.03 & 0.16 & 0.03 & 0.16 & 0.03 & 0.23 & -0.03 & -0.03 & 0.03 & 0.10 \\ 1.13 & 0.68 & -0.68 & 0.22 & 0.22 & 0.22 & -0.22 & 1.59 & 0.68 & 1.13 & -1.13 \\ 0.02 & 0.01 & 0.00 & 0.00 & 0.00 & 0.02 & -0.00 & 0.01 & 0.03 & 0.02 & -0.00 \\ 0.05 & 0.37 & -0.07 & 0.22 & 0.22 & 0.22 & 0.07 & 0.37 & 0.37 & 0.51 & -0.22 \end{bmatrix}, \tag{49}$$

$$BV = [-0.07 \quad -0.05 \quad 0.40 \quad -0.40 \quad 0.00 \quad 0.37 \quad 0.04 \quad -0.72 \quad 0.00 \quad -0.07]^T, \tag{50}$$

$$A = \begin{bmatrix} 1 & 0 & 0 & 0 & 0 & 0 & 0 & 0 & 0 & 0 \\ 0 & 1 & 0 & 0 & 0 & 0 & 0 & 0 & 0 & 0 \\ 0 & 0 & 1 & 0 & 0 & 0 & 0 & 0 & 0 & 0 \\ 0 & 0 & 0 & 1 & 0 & 0 & 0 & 0 & 0 & 0 \\ 0 & 0 & 0 & 0 & 1 & 0 & 0 & 0 & 0 & 0 \\ 0 & 0 & 0 & 0 & 0 & 1 & 0 & 0 & 0 & 0 \\ 0 & 0 & 0 & 0 & 0 & 0 & 1 & 0 & 0 & 0 \\ 0 & 0 & 0 & 0 & 0 & 0 & 0 & 1 & 0 & 0 \\ 0 & 0 & 0 & 0 & 0 & 0 & 0 & 0 & 1 & 0 \\ 0 & 0 & 0 & 0 & 0 & 0 & 0 & 0 & 0 & 0 \end{bmatrix}, \tag{51}$$

$$T = \begin{bmatrix} 2.51 & 0.37 & -0.07 & 0.22 & 0.22 & 0.22 & 0.07 & 0.37 & 0.37 & 0.51 \\ 0.68 & 2.95 & -0.40 & 0.68 & 0.13 & 0.13 & -0.13 & 0.40 & 0.40 & 0.68 \\ 0.00 & 0.00 & 2.00 & 0.00 & 0.00 & 0.00 & 0.00 & 0.00 & 0.00 & 0.00 \\ 0.27 & 0.46 & -0.09 & 2.64 & 0.27 & 0.27 & 0.09 & 0.09 & 0.09 & 2.77 \\ 0.00 & 0.00 & 0.00 & 0.00 & 2.00 & 0.00 & 0.00 & 0.00 & 0.00 & 0.00 \\ 0.05 & 0.01 & 0.05 & 0.05 & 0.05 & 2.12 & 0.01 & 0.01 & 0.09 & 0.05 \\ 0.03 & -0.03 & 0.16 & 0.03 & 0.16 & 0.03 & 2.23 & -0.03 & -0.03 & 0.03 \\ 1.13 & 0.68 & -0.68 & 0.22 & 0.22 & 0.22 & -0.22 & 3.59 & 0.68 & 1.13 \\ 0.02 & 0.01 & 0.00 & 0.00 & 0.00 & 0.02 & -0.00 & 0.01 & 2.03 & 0.02 \\ 0.51 & 0.37 & -0.07 & 0.22 & 0.22 & 0.22 & 0.07 & 0.37 & 0.37 & 2.51 \end{bmatrix}, \tag{52}$$

$$I = \begin{bmatrix} -0.22 & -0.68 & 0.00 & -0.27 & 0.00 & 0.01 & 0.10 & -1.13 & 0.00 & -0.22 \end{bmatrix}^T. \tag{53}$$

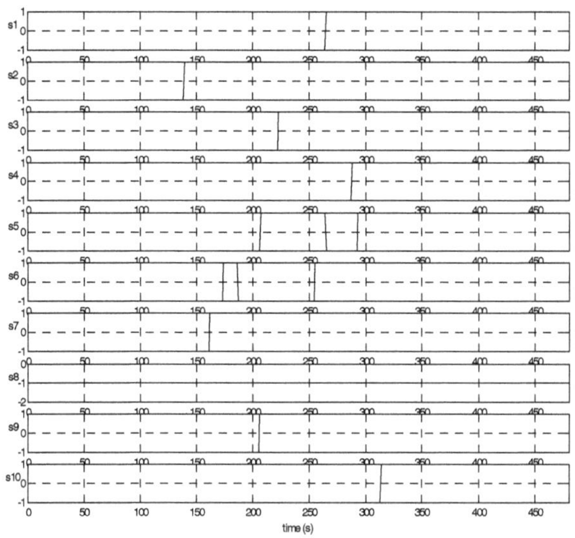

Fig. 9. Encoded residuals for *fault* 1.

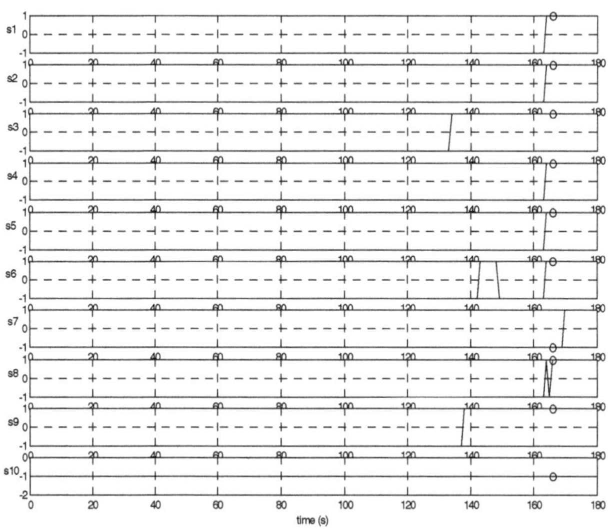

Fig. 10. Encoded residuals for *fault* 5.

Table 2. Fault patterns to store in associative memory.

α^0	α^1	α^2	α^3	α^4	α^5	α^6
-1	1	-1	-1	-1	-1	1
-1	1	-1	-1	-1	-1	-1
-1	1	1	1	1	1	1
-1	1	-1	1	-1	-1	-1
-1	1	1	1	-1	-1	1
-1	1	-1	1	-1	1	1
-1	1	1	1	1	-1	1
-1	-1	-1	-1	-1	-1	1
-1	1	-1	-1	-1	1	1
-1	1	-1	-1	-1	-1	1

When any fault (1 to 6) evolutes, fault patterns which are retrieved by the associative memory can correspond to a wrong pattern. This fact is mainly due to the input transient bipolar vectors which force the associative memory to converge to wrong fault patterns. However, during this interval the encoded residuals do not correspond to true fault pattern and this fact is taken into account to carry out an efficient diagnosis. Taking into account this fact, information to operator is presented as in Fig.13 and Fig. 14, for *fault* 1 and *fault* 5, respectively.

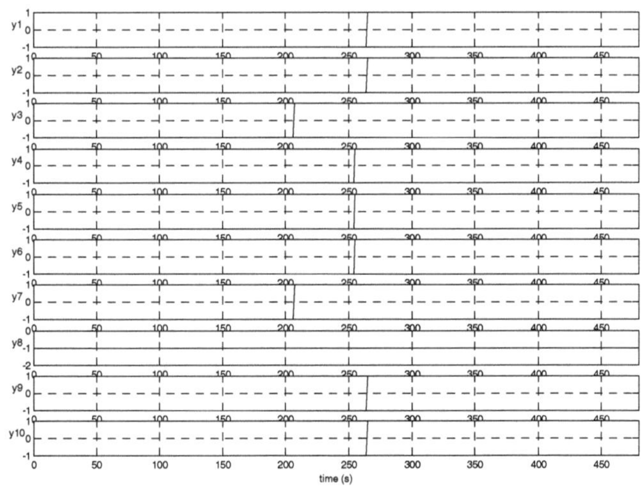

Fig. 11. Fault pattern retrieved by associative memory, *fault* 1.

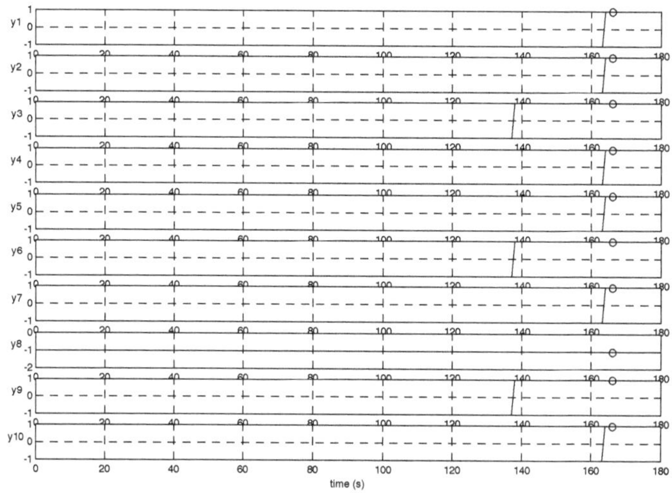

Fig. 12. Fault pattern retrieved by associative memory, *fault* 5.

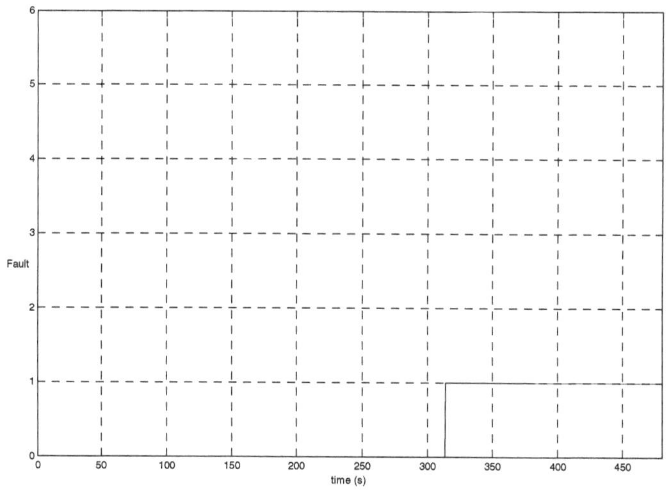

Fig. 13. Information to operator, *fault* 1.

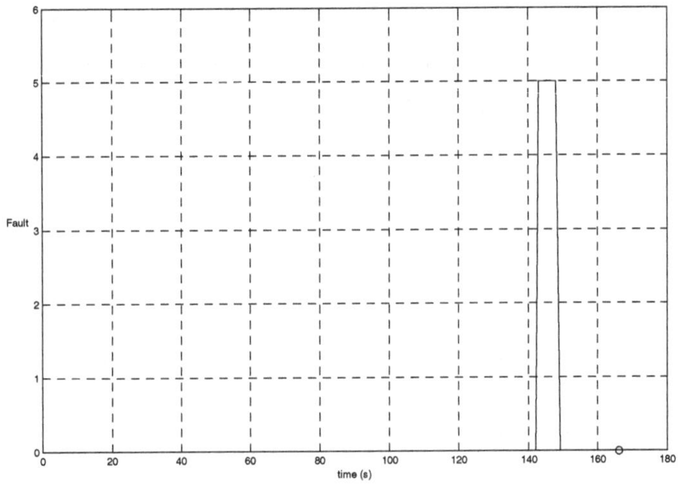

Fig. 14. Information to operator, *fault* 5.

6 Conclusions

Results demonstrate that the optimal training algorithm proposed in this work is adequate to train associative memories based on recurrent neural networks. Based on this approach, an associative memory is designed and applied to fault diagnosis in fossil electric power plants. As future work, it is necessary to analyze convergence for this new algorithm.

Acknowledgment

The authors thank support of the CONACYT, Mexico on project 39866Y. Besides, they also thank the Process Supervision Department of IIE (Instituto de Investigaciones Electricas), Cuernavaca, Mexico for its support. The first author thanks Universidad Autonoma del Carmen, Mexico, for allow him to pursue Ph. D. Studies.

References

1. Liu, D., Lu, Z.: A new synthesis approach for feedback neural networks based on the perceptron training algorithm. IEEE Trans. Neural Networks, Vol. 8 (1997) 1468-1482.
2. Hopfield, J. J.: Neural networks and physical systems with emergent collective computational abilities. Proc. Nat. Academy Sci. USA, Vol. 79 (1982) 2554-2558.
3. Personnaz, L., Guyon, I., Dreyfus, G.: Information storage and retrieval in spin-glass like neural networks, J. Phys. Lett. , Vol.46 (1985), L359-365.
4. Personnaz, L., Guyon, I., Dreyfus, G., Collective computational properties of neural networks: New learning mechanisms. Phys. Rev. A. , Vol. 34 (1986) 4217-4228.
5. Li, J. -H., Michel, A. N., Porod W.: Analysis and synthesis of a class of neural networks: Linear systems operating on a closed hypercube. IEEE Trans. Circuits Syst., Vol. 36 (1989) 1405-1422.
6. Michel, A. N., Si, J., Yen, G.: Analysis and synthesis of a class of discrete-time neural networks described on hypercubes. IEEE Trans. Neural Networks, Vol. 2 (1991) 32-46.
7. Michel, A. N., Farrel, J. A.: Associative memories via artificial neural networks, IEEE Contr. Syst. Mag., Vol. 10 (1990) 6-17.
8. Liu, D., Michel A. N.: Dynamical Systems with Saturations Nonlinearities: Analysis and Design. Springer Verlag, New York, USA, (1994).
9. Liu, D., Michel, A. N.: Cellular neural networks for associative memories. IEEE Trans. Circ. And Syst. II, Vol. 40 (1993) 119-121.
10. Liu, D., Michel, A. N.: Sparselly interconnected neural networks for associative memories. IEEE Trans. Circ. and Syst. II, Vol. 41 (1994) 295-307.
11. Liu, D., Michel A. N.: Robustness analysis and design of a class of neural networks with sparse interconnecting structure, Neurocomputing, Vol. 12 (1996) 59-76.
12. Hassoun, M. H.: Associative Neural Memories: Theory and Implementation, Oxford Univ. Press, Oxford, U. K., (1993).
13. Hassoun, M. H., Youssef, A. M.: Associative Neural Memory capacity and Dynamics. Proc. Int. J. Conf. Neural Networks, San Diego, C A, Vol. 1 (1990) 763-769.
14. Li, J. -H., Michel, A. N., Porod, W.: Qualitative analysis and synthesis of a class of neural networks. IEEE Trans. Circuits Syst., Vol. CAS-35 (1988) 976-986.
15. Seiler, G., Schuler, A. J., Nossek, J. A.: Design of robust cellular neural networks. IEEE Trans. Circuits Syst. I, Vol. 40 (1993) 358-364.
16. Boser, B. E., Gullon, E. M, Vapnik, V. N.: A training algorithm for optimal margin classifiers. Proceedings of the Fifth Annual Workshop of Computational Learning Theory, Pittsburg, Vol.5 (1992) 144-152.

17. Cortes, C., Vapnik, V. N.: Support Vector Networks. Machine Learning, Vol. 20 (1995) 273-297.
18. Luemberger D.: Linear and Non Linear Programming. Addison Wesley Publishing Company, USA, 1984.
19. Hopfield, J. J.: Neurons with graded response have collective computational abilities. Proc. Nat. Academy Sci. USA, Vol. 81 (1984) 3088-3092.
20. Mynski, M. L., Papert S. A. , Perceptrons: An introduction to Computational Geometry. MA: MIT Press, Cambridge (1969) and (1988 expanded version).
21. Rosemblatt, F., Principles of Neurodynamics. NY: Spartan, New York, 1962.
22. Ruz-Hernandez, J. A., Sánchez, E. N., Suarez D. A.: Neural Networks-based Scheme for Fault Diagnosis in Fossil Electric Power Plants. Proc. of International Joint Conference on Neural Networks, Montreal, Canada, (2005) 1740-1745.
23. Vapnik V. N.: Estimation of Dependences Based in Empirical Data, Addendum I. Springer Verlag, New York, USA (1982).
24. Kuhn, H. W., Tucker, A. W., : Nonlinear Programming. Proceedings of the Second Berkeley Symposium on Mathematical, Statistics and Probability, University of California Press, Berkeley and Los Angeles, Calif., USA, (1961) 481-492.
25. Frank, P. M.: Diagnostic Procedure in the Automatic Control Engineering, Survey paper at- Automatic Control Engineering, Vol. 2 (1994) 47-63.
26. Patton, R. J., Frank, P. M., Clark, R. N.: Fault Diagnosis in Dynamic Systems: Theory and Application, Prentice Hall, New York, USA (1989).
27. Chen, J., Patton, R. J.: Robust Model Based Fault Diagnosis for Dynamic Systems, Kluwer Academic Publishers, Norwell, Massachusetts, USA (1999).
28. Köppen-Seliger, B., Frank, P. M.: Fault detection and isolation in technical processes with neural networks. Proceedings of the 34th Conference on Decision & Control, New Orleans, USA, (1995) 2414-2419.
29. Comisión Federal de Electricidad. Manual del Centro de Adiestramiento de Operadores Ixtapantongo, Módulo III, Unidad 1, México (1997).
30. Ruz-Hernandez, J. A., Suarez, D. A., Shelomov, E., Villavicencio, A.: Predictive Control Based on an Auto-Regressive Neuro-Fuzzy Model Applied to the Steam Generator Startup Process at a Fossil Power Plant, Revista de Computacion y Sistemas, Vol. 6 (2003) 204-212.
31. Noorgard, M., Ravn, O., Poulsen, N. K., Hansen, L. K.: Neural Networks for Modelling and Control of Dynamic Systems, Springer Verlag, London, Great Britain (2000).
32. Marquardt, D.: An algorithm for least-squares estimation of nonlinear parameters, SIAM Journal Appl. Mathematics, Vol. 11 (1963).
33. Levenberg, K.: A method for solution of certain nonlinear problems in least squares. Quart. Appl. Mathematics, Vol. 2 (1944) 164-168.
34. Nøorgard M., Neural Network based system identification toolbox (NNSYSID TOOLBOX). Tech. Report 97 E-851, Department of Automation, DTU Press, Lyngby, Denmark (1997).

Acceleration Output Prediction of Buildings Using a Polynomial Artificial Neural Network

Francisco J. Rivero-Angeles, and Eduardo Gomez-Ramirez

[1] Gerencia de Instrumentación Sísmica. CANDE Ingenieros, S.A. de C.V., Clemente Orozco 18, Col. Ciudad de los Deportes, México, D. F., 03710, frivero@candeingenieros.com, [2] Laboratorio de Investigación y Desarrollo de Tecnología Avanzada (LIDETEA). Universidad La Salle. Benjamín Franklin 47, Col. Condesa, México, D. F., 06140, egr@ci.ulsa.mx

Abstract. Severe earthquake motions could make civil structures to undergo hysteretic cycles and crack or yield their resistant elements. The present research proposes the use of a polynomial artificial neural network to identify and predict, on-line, the behavior of such nonlinear systems. Predictions are carried out first on theoretical hysteretic models and later using two real seismic records acquired on a 24-story concrete building in Mexico City. Only two cycles of movement are needed for the identification process and the results show fair prediction of the acceleration output.

1 Introduction

Civil structures, such as buildings or bridges, are instrumented to acquire acceleration, velocity and displacement output data due to lateral loads, which could arise from severe wind or strong earthquake motions. The data is later analyzed to assess the lateral resistant capacity of the structure and to check output maximums against those allowed by construction codes. In some instances, wind or earthquakes induce lateral loads such that energy may dissipate through hysteretic phenomena, thus reducing their resistant capacity [10]. Many buildings have been instrumented around the world in order to monitor their structural health. The identification of such nonlinear systems is therefore an important task for engineers who work in those areas affected by these natural hazards, and thus, the subject of the present paper.

To cope with this problem, a polynomial artificial neural network (PANN) [8] is proposed, so a model could be identified to represent such nonlinear systems. The proposed algorithm is able to estimate with good

F.J. Rivero-Angeles and E. Gomez-Ramirez : *Acceleration Output Prediction of Buildings Using a Polynomial Artificial Neural Network*, StudFuzz **208**, 357–383 (2007)
www.springerlink.com

accuracy the acceleration output. Training is done on-line with only a limited amount of data. It is worth noting that on-line algorithms are favored in closed loop control applications, decision making strategies and fault detection, evidently in real-time. Thus, the present research implements such algorithms.

Also, in structural dynamics, many parametric models are proposed to simulate the behavior of the structure, nonetheless, the PANN model seems easier to implement with real data and with the following advantages: (a) the external driving force is assumed unknown and not needed for the identification scheme, which makes it interesting because it does not limit its application to ground acceleration records, but opens its use in other accidental loading such as high wind or explosions, in which the actual loads are unknown; and (b) the physical properties of the structure, such as mass, stiffness or damping, are not needed, which in turn makes the proposed algorithm advantageous when modeling actual structures only by using acceleration records.

Additionally, from the training view point, usual neural networks are trained following two criteria: (a) by minimizing the error assuming an established threshold, or (b) until a given number of epochs is reached. In this case, the proposed PANN is trained with a weight variance criterion, that is, during on-line training, the weights reach a "constant" value when the variance is close to zero. Besides the variance analysis, a moving average was used, as a tendency indicator, to complete the revision.

To sum up, the present research proposes the use of a PANN with a superior training speed, with a reduced number of samples for the on-line identification process, with no *a priori* knowledge of the structural parameters, the type of nonlinear system, or the driving force.

2 Contents

The present research shows the use of a polynomial artificial neural network (PANN) [10] to identify and predict the acceleration output of four theoretical single-degree-of-freedom systems (SDOF). Two of them were modeled with a linear-elastic behavior. The other two were modeled as nonlinear, hysteretic systems with the Bouc-Wen model [24]. These systems were subject to the Loma Prieta earthquake, recorded at the Treasure Island station in Santa Cruz, California, USA, 1989, and to the Mexico City earthquake, recorded at the Communications Secretary (SCT), in 1985. The results of the simulations show that the PANN is capable to identify both type of systems (linear and nonlinear) and predict with good accuracy the acceleration output after training.

Later, one actual seismic record, attained at the roof of a real 24-story concrete structure in Mexico City, acquired in the year 2002, is used to identify the behavior of the building. The identified model is then used to predict the acceleration motion of the same real building, subjected to another actual record acquired ten months later, in the year 2003, and the results show that this simple model predicts with very good accuracy the behavior of the system.

A long term aim of the present research is to develop a technique that could be used in conjunction with fault detection analysis, structural health monitoring, and structural control.

3 State of the Art

In this section, the state of the art in forecasting time series is shown, and of course, their use in the identification of civil structures. Forecasting time series has been solved with a broad range of algorithms such as ARMAX [3], NARMAX [6], Fuzzy Logic [23], Neural Networks [8], etc.

In civil structures, some researchers have successfully identified nonlinear systems with a wide variety of proposed algorithms. The techniques could be divided into on-line and off-line algorithms. In [22] a statistical technique is proposed, which considers the typical vibration signature of nonlinear mechanical systems, using linear techniques. One of the conclusions drawn is that linear models are not able to fully predict the structural behavior, even nonlinear characteristics might modify the structural damping of the first mode. Thus, nonlinear techniques have been developed to cope with this problem. In [9] a thorough literature review is given. In [11] a NARMAX orthogonal model is used for parametric identification. In [21] a spectral method is shown, and in [13] a sequential regression analysis, Gauss-Newton optimization, least squares and extended Kalman filter is used. Also, least squares methods have been used, with some modifications, by [16], [19], and [26]. Another example is the use of ERA-OKID, Subspace and Least Squares algorithms to estimate linear parameters of structures [15]. All papers so far have in common the use of off-line algorithms, where the complete acceleration, velocity and displacement records, or a combination of them, are used.

Nonetheless, as it was mentioned earlier, for structural control or decision making strategies, on-line algorithms are suitable. Evidently, some researchers have worked with such type of algorithms. That is the case of [5], in which an adaptable least squares is used to see the evolution of stiffness changes in time of linear systems. In [25], besides parameter estimation, the partly unknown excitation is estimated, also with a least squares algorithm.

Although artificial neural networks have not been widely used in civil and structural engineering, some researchers have successfully applied them; that is the case of [12], [17], and [18]. Nonetheless, the models and architectures of those networks seem complex and are computer time consuming. Also, a partial knowledge of the structure and the nonlinear dynamics of the restoring force are needed.

4 Basic Structural Dynamics

This section contains some of the concepts shown in any structural dynamics text book for civil engineers. A good starting point is [2] and the reader is referred to it for a more extensive revision. For the case of the proposed PANN, the following structural parameters are not needed, yet they are used to simulate the theoretical models shown herein, and their acceleration output is used for comparison purposes with the PANN.

For civil structures, the aim is to build useful infrastructure for our societies. Building codes give some rules such that a structure satisfies the following: (a) it will not suffer any damage under low intensity earthquake motions, (b) damage in nonstructural elements, such as architectural walls, plaster, windows and partitions, is limited and reparable under medium intensity earthquakes, and (c) the structure will not collapse under high intensity earthquakes. Also, interstory drifts are limited to ensure comfort and safety of the inhabitants. One type of structural analysis and design of buildings under lateral loads is by using response spectra, thus, it is necessary to instrument the ground and the structure to acquire acceleration, velocity and displacement records. Those records allow to define instrumental measurements of the earthquake intensity.

On the other hand, from the structure's point of view, its response will be a function of the excitation and the building characteristics. The differential equation shown in (1) describes the dynamics of a SDOF system in continuous time and subject to external driving force:

$$\ddot{x} + \frac{c}{m}\dot{x} + \frac{k}{m}x = \frac{f(t)}{m}. \tag{1}$$

In this case, the implicit relation which maps the external driving force $f(t)$ with the displacement $x(t)$ is known. The structural parameters are mass m, damping c and stiffness k. In a typical seismic instrumentation, accelerometric sensors are adopted, thus velocities and displacements are obtained by numeric integration. The quotient k/m is, by definition, the circular frequency ϖ squared, and period T could be computed with $2\pi/\varpi$. If the response of the system depends on its dynamic characteristics, it is no surprise that building codes give design spectra which depend on the period of the structure and consider reduction factors to take into account the energy dissipated through hysteretic phenomena (nonlinear). The vibration periods of a structure are function of the mass and the stiffness, yet, if these parameters are unknown, (2) or (3) could be used.

$$T = 0.1N; \tag{2}$$

$$T = 0.075H^{\frac{3}{4}}; \tag{3}$$

where$N$ is the number of stories in a building, and H in meters represents the total height of the building. The vibration period physically represents the number of seconds needed for a building to oscillate a complete cycle.

Another influential characteristic in the response of a building is viscous damping c, which is function of a fraction ς of the critical damping. It could be noted that the magnitude of the spectral ordinates diminishes when ς gets larger (the larger the damping, the lesser the spectral acceleration). This kind of damping considers dissipating energy arising from internal friction, friction at the bearings or with nonstructural elements, etc., thus, the magnitude of the effects depends on the level of damage and is difficult to quantify with certainty. To deal with this problem, viscous damping is introduced as a numerical aid due to the fact that 5% of critical damping (usual in civil concrete structures) leads to good results. If a different formulation is required, the reader is referred to the work in [14].

Earthquakes produce alternating loading conditions (cyclic), and construction codes admit the occurrence of structural damage, thus, load-deformation characteristics are of interest. The load-deformation curve of a structure depends, among other factors, on the curves of the elements, the materials and the area properties of the sections. If the stiffness of the structure changes with time on cyclic loading, it leads to hysteretic cycles. In some instances, the stiffness reduces considerably from one cycle to the other, such phenomenon is known as deterioration.

A structural system is ductile if is able to hold important deformations under almost constant loads, without reaching excessive damage levels, which is opposed to a fragile behavior. A common measure is the ductility factor μ, defined for elastoplastic structures as the deformation required for collapse, divided by the corresponding elastic deformation. The energy dissipating capacity in a loading cycle is computed by the area enclosed in a hysteresis cycle. In deteriorating systems, the area enclosed in every cycle is reduced, and consequently, the dissipating capacity is diminished. In this sense, this kind of systems are less effective to resist cyclic loads.

A formal way to define hysteresis is to consider it as a special constitutive relation with memory. In other words, hysteresis arises when the displacement output could not be determined at any given instant t, but it depends on the evolution of the displacements in the interval $[0,t]$ and possibly on the initial conditions [20].

Many references to differential hysteretic models could be found in the literature. Nevertheless, the Bouc-Wen differential hysteretic model [24] is widely used in structural engineering. There are two interesting features of this model: (a) it is capable to represent different shapes, and (b) correlates well the numerical results to experimental data derived from tests.

Several hysteretic friction models exist, such as the Dahl [7], LuGre [4] or Duhem Operator [14]. Nonetheless, the most popular remains the Bouc-Wen model, modified to consider deterioration [1]. According to Wen, the restoring force in a nonlinear hysteretic system could be decomposed into the two parts shown in (4):

$$Q(x,\dot{x}) = g(x,\dot{x}) + z(x); \tag{4}$$

where $g(x,\dot{x})$ is a non-hysteretic component, generally nonlinear, which depends on the instantaneous displacement and velocity, and $z(x)$ is the hysteretic component, which depends on the displacement output. The behavior of $z(x)$ depends on the material characteristics, amplitude of the response and the structural characteristics. For civil SDOF structures, subject to ground motion, the following equation applies:

$$m\ddot{x} + c\dot{x} + kx = f(t) = -m\ddot{x}_g; \tag{5}$$

where the SDOF has mass m, damping c, stiffness k, ground acceleration $\ddot{x}_g$. Variables x, $\dot{x}$ and $\ddot{x}$ represent the displacement, velocity and acceleration, respectively. Baber and Wen [1] proposed the following set of nonlinear differential simultaneous equations to describe the motion of a hysteretic SDOF.

$$m\ddot{x} + c\dot{x} + \alpha kx + (1-\alpha)kz = f(t); \tag{6}$$

$$\dot{z} = \frac{A\dot{x} - \nu\left(\beta|\dot{x}||z|^{n-1}z + \gamma\dot{x}|z|^n\right)}{\eta} \tag{7}$$

Equation (6) represents the motion of the system and (7) models the time change of the hysteretic component (see [1]). This set of equations allow to model hysteretic degrading and nondegrading systems. From this revision, it is evident the necessity to select materials, designs and structural details to avoid fragile failure and deterioration.

5 Polynomial Artificial Neural Network

An interesting feature of the neural networks is the ability to identify nonlinear systems, thus, for the case of nonlinear hysteretic structures, neural networks may identify more accurately the dynamic behavior of a non-parametric model. The model of a polynomial artificial neural network (PANN) is described in (8).

$$\hat{y}_k = [\phi(x_{1,k}, x_{2,k}, \quad \ldots x_{n_i,k}, \quad x_{1,k-1}, \quad x_{2,k-1}, \quad \ldots, \quad x_{n_i,k-n_1} \quad \ldots\ldots \quad y_{k-1}, \quad y_{k-2}, \quad \ldots y_{k-n_2})]_{\phi_{\min}}^{\phi_{\max}}; \tag{8}$$

where $\hat{y}_k \in \Re$ is the estimated function, $\phi(x,y) \in \Re$ is a nonlinear function, $x_i \in X$ are the inputs with $i = 1,\ldots,n_i$; and n_i is the number of inputs. $y_{k-j} \in Y$ are the previous output values, for $j = 1,\ldots,n_2$; n_1 is the number of delays or previous values in the input, n_2 is the number of delays in the output, X and Y are compact subsets of $\Re$. To simplify the notation, (9) is given.

$$z = (x_{1,k}, x_{2,k}, \ldots, x_{n_1,k}, \ldots, y_{k-1}, y_{k-2}, \ldots, y_{k-n_2}); \tag{9}$$
$$z = \left(z_1, z_2, z_3, \ldots, z_{n_v}\right);$$

where n_v is the total number of elements in description z, and $n_v = n_i + n_1 n_i + n_2$. The nonlinear function $\phi(z) \in \Phi_p$ belongs to a family Φ_p of polynomials which could be represented by (10).

$$\Phi_p\left(z_1, z_2, \ldots, z_{n_v}\right) = \phi(z); \tag{10}$$
$$\phi(z) = a_0\left(z_1, z_2, \ldots, z_{n_v}\right) + a_1\left(z_1, z_2, \ldots, z_{n_v}\right) + \ldots + a_p\left(z_1, z_2, \ldots, z_{n_v}\right).$$

Sub index p is the maximum power of the polynomial expression. Polynomials $a_i\left(z_1, z_2, \ldots, z_{n_v}\right)$ are homogeneous polynomials of total degree i, for $i = 0,\ldots,p$. Each homogeneous polynomial could be written as shown in (11).

$$\begin{aligned}
&a_0\left(z_1,z_2,\ldots,z_{n_v}\right)=w_0;\\
&a_1\left(z_1,z_2,\ldots,z_{n_v}\right)=w_{1,1}z_1+w_{1,2}z_2+\ldots+w_{1,n_v}z_{n_v};\\
&a_2\left(z_1,z_2,\ldots,z_{n_v}\right)=w_{2,1}z_1^2+w_{2,2}z_1z_2+\ldots+w_{2,N_2}z_{n_v}^2;\\
&\vdots\\
&a_p\left(z_1,z_2,\ldots,z_{n_v}\right)=w_{p,1}z_1^p+w_{p,2}z_1^{p-1}z_2+\ldots+w_{p,N_p}z_{n_v}^p;
\end{aligned} \tag{11}$$

where $w_{i,j}$ is the associated weight of the PANN. Term w_0 corresponds to the input bias of the network, homogeneous polynomial $a_1(z)$ is equivalent to weight the inputs, polynomials $a_2(z)$ to $a_p(z)$ represent the modulation between the inputs and the power of each polynomial. N_i is the number of terms of each polynomial with

$$N_0=1;N_1=n_v;N_2=\sum_{i=1}^{n_v}i;N_3=\sum_{s_1=0}^{n_v-1}\sum_{i=1}^{n_v-s_1}i;\ldots;N_p=\sum_{s_p-2=0}^{n_v-1}\cdots\sum_{s_2=0}^{n_v-s_3}\sum_{s_1=0}^{n_v-s_2}\sum_{i=1}^{n_v-s_1}i. \tag{12}$$

Dimension N_Φ of each family Φ_p could be computed with $N_\Phi=\sum_{i=0}^{p}N_i$. Activation function of the network is given by (13).

$$\left[\phi(z)\right]_{\phi_{\min}}^{\phi_{\max}}=\begin{cases}\phi_{\max} & \phi(z)\geq\phi_{\max}\\ \phi(z) & \phi_{\min}<\phi(z)<\phi_{\max}\\ \phi_{\min} & \phi(z)\leq\phi_{\min}\end{cases}. \tag{13}$$

The architecture of the PANN is shown in fig. 1. Learning of the PANN is done by minimizing an error function, thus, the approximation error in the PANN is defined as:

$$\begin{aligned}
err_n\left(Y_n,\phi(z)\right)&=\frac{1}{n}\sum_{k=1}^{n}\left(y_k-\phi(z)_k\right)^2;\\
&=\frac{1}{n}\sum_{k=1}^{n}\left(y_k-\hat{y}_k\right)^2;\\
Y_n&=\left(y_1,y_2,\ldots,y_n\right);
\end{aligned} \tag{14}$$

where Y_n is the desired output, $\phi(z)_k\in\Phi_p(z)$, and n is the number of previous values of the output. The optimal error is defined as:

$$opterr_n\left(Y_n,\phi(z)\right)=err_n\left(Y_n,\phi_n^*(z)\right); \tag{15}$$

where $\phi_n^*(z)\in\Phi_p$ is the optimal estimation of Y_n. Thus, the PANN $\phi(z)\in\Phi_p$ learns the desired output uniformly with precision ε if

$$err_n\left(Y_n,\phi(z)\right)-err\left(Y_n,\phi_n^*(z)\right)\le\varepsilon;\varepsilon>0 \tag{16}$$

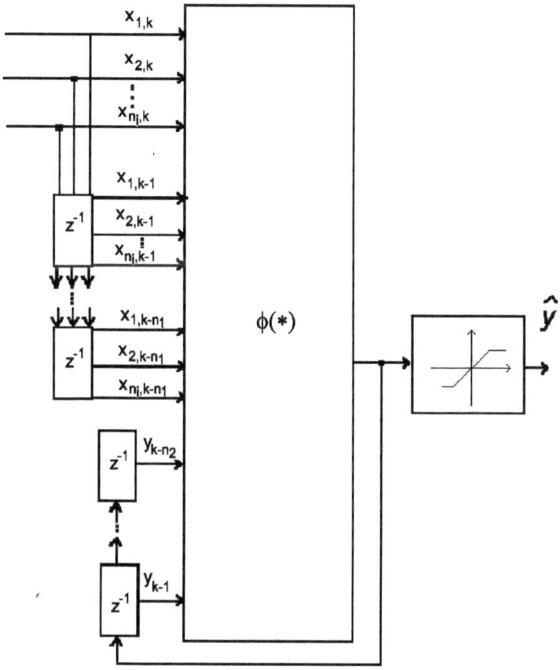

Fig. 1. Architecture of the PANN

From the latter, the learning problem of the PANN consists on finding the structure of $\phi\in\Phi_p(z)$ which verifies (16). With this architecture, the weights of the PANN could be found with a recursive least squares algorithm during training, for example (17).

$$P_{k+1}=P_k-\frac{P_k\phi(z)_{k+1}^T\phi(z)_{k+1}P_k}{1+\phi(z)_{k+1}P_k\phi(z)_{k+1}^T}; \tag{17}$$

$$\hat{w}_{k+1}=\hat{w}_k+P_{k+1}\phi(z)_{k+1}^T\left(y_{k+1}-\phi(z)_{k+1}\hat{w}_k\right);$$

where $\phi(z)_{k+1}$ is a row vector, $k+1$ is the identification instant, $\hat{w}$ are the weights to be identified and P is a matrix, approximately to the covariance of the parameters with $P(0) > 0$. Also, a forgetting factor could be used, nonetheless, this simple algorithm lead to good results, thus it was kept as a simpler approximation. If a more complex least squares algorithm is desired, it could be changed easily since the PANN does not require this specific identification scheme.

6 Simulations

Four SDOF theoretical systems were modeled. SDOF systems act like shear buildings with only lateral movement of the mass. No mass rotations are considered. In this case, two SDOFs were simulated with the Loma Prieta base acceleration record and the last two with the Mexico City earthquake record. In both cases, simulations were run with linear elastic and nonlinear hysteretic behavior. The simulations with the SDOF subject to the excitation of Loma Prieta earthquake are presented next.

6.1. SDOF Subject to Loma Prieta Record

The first SDOF has the following structural properties: mass $m = 1kg$, damping $c = 1.2566 kgf \cdot s / cm$, and stiffness $k = 157.9137 kgf / cm$, subject to the base ground acceleration record acquired during the Loma Prieta earthquake. Simulations consisted on the following: (a) fully linear elastic structure, and (b) hysteretic nonlinear structure with Bouc-Wen parameters $\alpha = 1/21$, $A = \eta = \nu = n = 1.0$, $\beta = \gamma = 0.5$. In every single case the PANN was trained with the acceleration output corrupted with 2% random noise and parameters $p = 2$, $n_i = n_1 = 0$, and $n_2 = 4$, which leads to a second order polynomial, no input values considered and 4 previous values of the output.

These values were obtained by trial and error, such that the results were acceptable and still obtain a simple model. Additionally, the PANN was trained with the first 100 samples, two cycles of movement, from the record at 50 samples per second. Note that two cycles of movement could be related to $2T$, with T the vibration period of the structure, in this case, 1 second. Next the simulations with SDOF (a), fully linear elastic response, are shown.

Figure 2 contains the samples used during the training process of the proposed PANN. Note that the values have been normalized, such that the values are bounded between zero and one.

Figure 3 shows the weight variation. It could be noted that the recursive least squares algorithm quickly identifies the weights of the PANN model. As it was stated previously, training termination was done with a variance analysis. In this case, the variance of the weights could be considered as a time series, in which variance at instant k results from a set of three previous variance values. It is apparent that the variance reduces until it reaches a value of 0.0008303, small enough to assume they have reached an almost constant value of the weights.

Another way to observe the tendency of the variance of the weights is through a moving average of sets of five previous values. The moving average "softens" the variance time series, and gives a clearer indication of the tendency. As an analogy, the slope of the weights should tend to zero to assume constant values. In that same figure, the slope of the first weight (as an example) is shown with a value of –0.0021, assumed very close to zero, and thus, the absolute variation clearly indicates a tendency to zero.

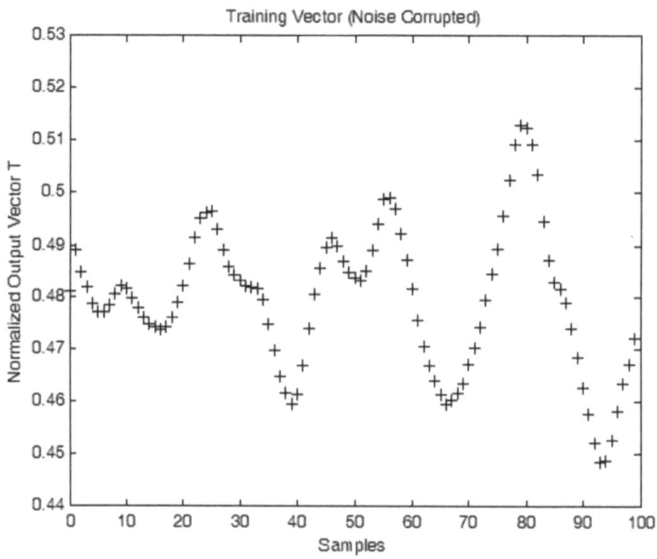

Fig. 2. Training vector, SDOF (a), Loma Prieta earthquake

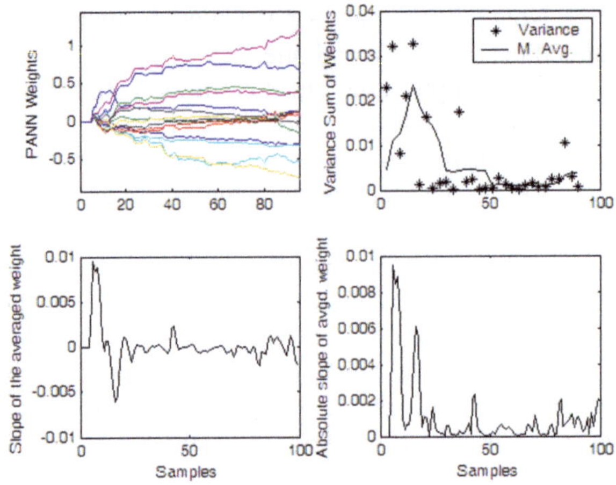

Fig. 3. Weight variation, SDOF (a), Loma Prieta earthquake

Figure 4 shows the training interval with 100 samples and the acceleration output prediction with 1,990 samples, divided by a vertical line in the figure. It is worth noting that during training, the PANN identifies the weights with only 2 seconds of motion and the prediction closely follows the theoretical output. It could also be noted that during the intense phase of the motion, the estimation error increases a bit. That figure also contains the prediction errors. Figure 5 shows a window with the acceleration output prediction of the intense part (from 12 to 15 seconds of motion). Note that the model is able to adapt to unknown conditions with no *a priori* knowledge of the type of structure or excitation.

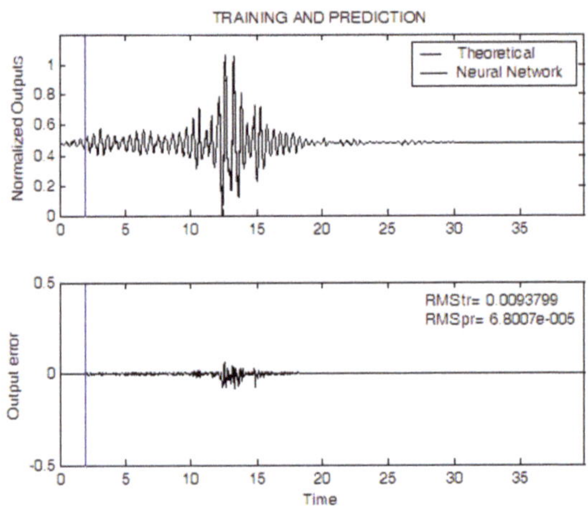

Fig. 4. Training and prediction, SDOF (a), Loma Prieta earthquake

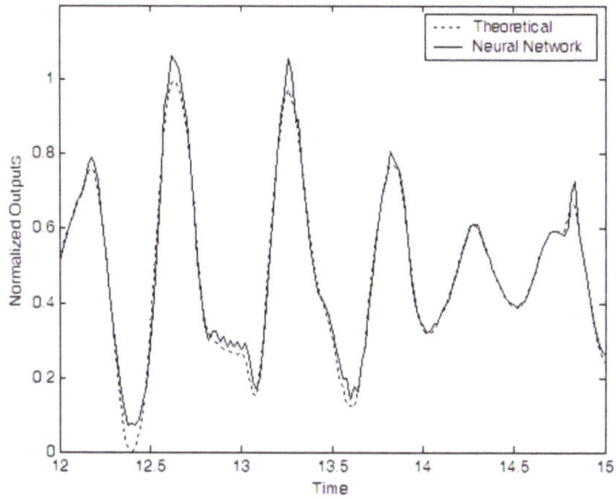

Fig. 5. Acceleration output prediction, SDOF (a), Loma Prieta earthquake

Next, the simulation of SDOF (b) hysteretic structure is shown. Figure 6 contains the associated hysteresis of the SDOF. Figure 7 contains the samples used during the training process. Fig. 8 shows the weight variation, and it could be noted a quick identification. Also, the weight variance reduces to a value of 0.00065185. The moving average also tends to zero in this interval. Again, the slope of the first weight reaches a value of 0.0032, assumed close to zero, such that the absolute variation is assumed to tend to zero. Figure 9 contains the training interval with 100 samples and prediction with 1,990 samples divided by a vertical line. Remember that two cycles of motion are related to $2T$.

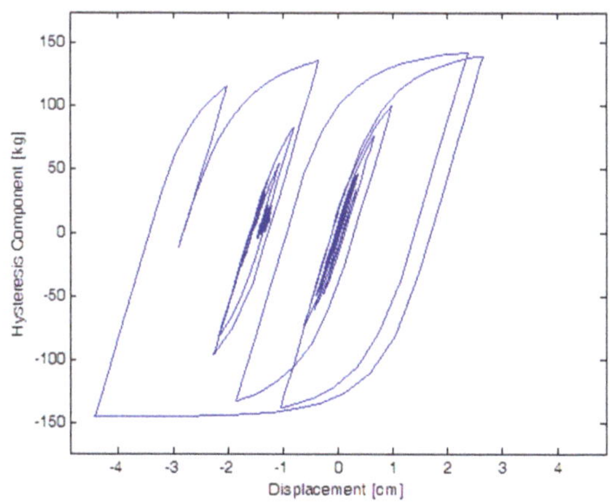

Fig. 6. Associated hysteresis, SDOF (b), Loma Prieta earthquake

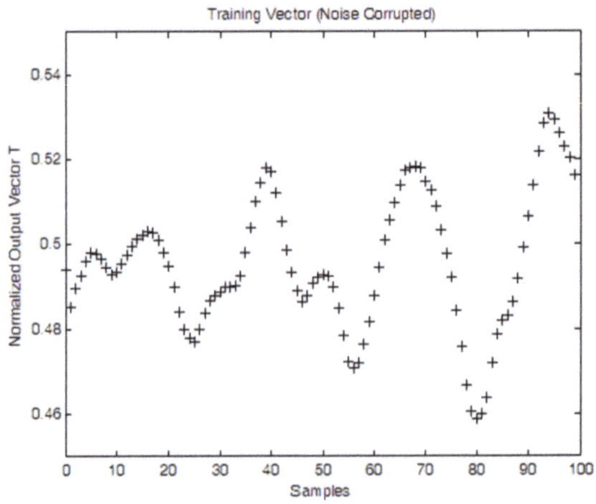

Fig. 7. Training vector, SDOF (b), Loma Prieta earthquake

During the acceleration output prediction, the weights of the model were kept constant, and it could be noted that the prediction is still quite acceptable, nonetheless, when energy dissipation through hysteretic phenomena takes place, the estimation error increases. Even though, the PANN is able to adapt to this nonlinear system. In that figure the estimation error is also shown. Figure 10 shows a window from 12 to 15 seconds of prediction. Note that during the first two seconds of motion, the system is basically linear and elastic, therefore, if the system would have behaved nonlinearly, the PANN would have identified it better. This statement could be verified with the results of the simulations run with the SDOF subject to Mexico City earthquake.

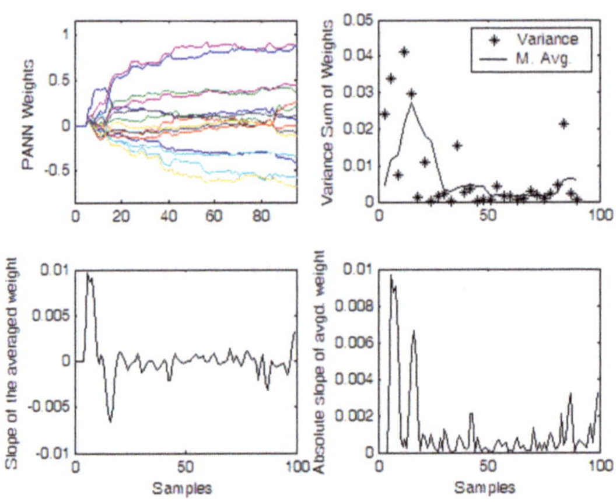

Fig. 8. Weight variation, SDOF (b), Loma Prieta earthquake

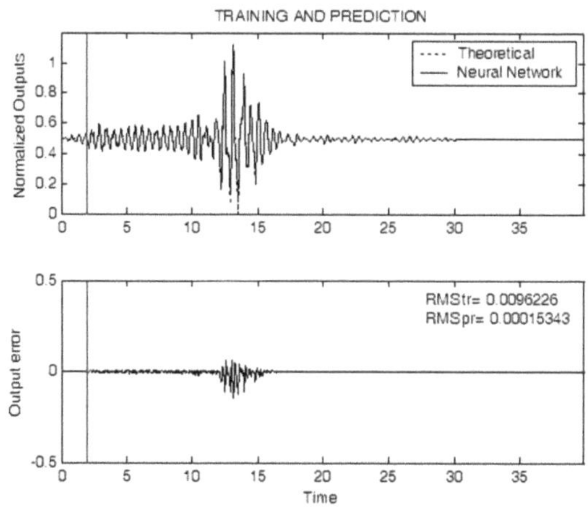

Fig. 9. Training and prediction, SDOF (b), Loma Prieta earthquake

6.2. SDOF Subject to Mexico City Record

The second SDOF has the following structural properties: mass $m = 1kg$, damping $c = 0.3142 kgf \cdot s / cm$, and stiffness $k = 9.8696 kgf / cm$, subject to the base ground acceleration record acquired during the Mexico City earthquake, reduced to 30% amplitude to develop compact and stable hysteresis cycles. Again, simulations consisted on the following: (a) fully linear elastic structure, and (b) hysteretic nonlinear structure with Bouc-Wen parameters $\alpha = 1/21$, $A = \eta = \nu = n = 1.0$, $\beta = \gamma = 0.5$. In every single case the PANN was trained with the acceleration output corrupted with 2% random noise and parameters $p = 2$, $n_i = n_1 = 0$, and $n_2 = 4$, which leads to a second order polynomial, no input values considered and 4 previous values of the output.

Additionally, the PANN was trained with the first 200 samples, two cycles of movement, from the record at 50 samples per second. In this case, the vibration period of the structure is of 2 seconds, in resonance with the driving force. Next the simulations with SDOF (a), fully linear elastic response, are shown.

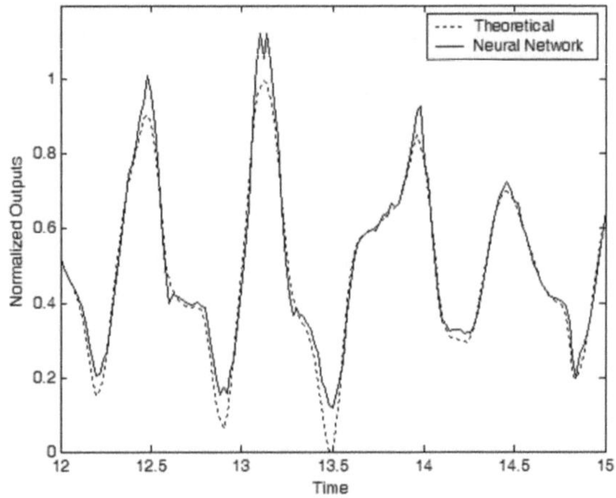

Fig. 10. Acceleration output prediction, SDOF (b), Loma Prieta earthquake

Figure 11 contains the samples used during training. Fig. 12 shows the weight variation. The variance of the weights reduces to a value of 4.2677×10^{-6}, small enough to assume constant values of the weights. The moving average clearly shows a tendency to zero, and the slope of the first weight reaches a value of 2.6754×10^{-5}, therefore, the absolute slope tends to zero. Note that in this case, the identification results are better due to the fact that response of the SDOF is larger. This could be explained because the vibration period of the structure is in resonance to the vibration period of the driving base acceleration earthquake record.

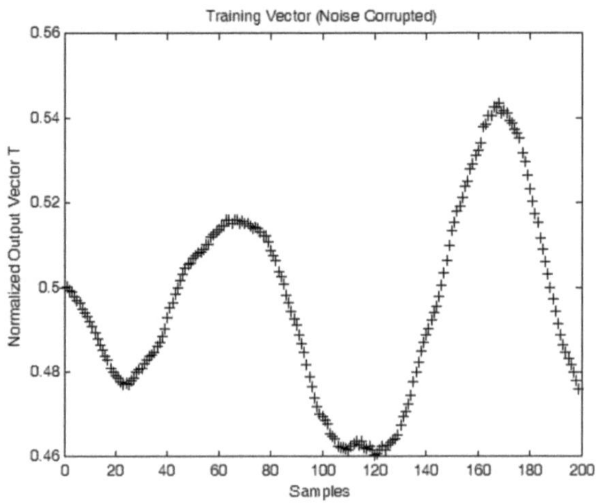

Fig. 11. Training vector, SDOF (a), Mexico City earthquake

Figure 13 shows the training interval with 200 samples and the prediction with 2,300 samples divided by a vertical line. It is worth noting that training is done with only 4 seconds of motion. During the intense phase, the estimation error is kept small. The latter could be noted in the figure.

Figure 14 contains a window of the intense part for added resolution, between 25 and 35 seconds of motion. Note that the model is able to adapt to the unknown external conditions.

Next, the simulations with the hysteretic SDOF (b) are shown.

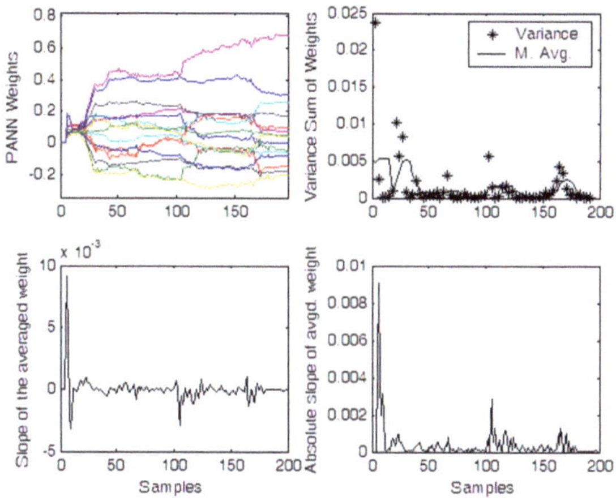

Fig. 12. Weight variation, SDOF (a), Mexico City earthquake

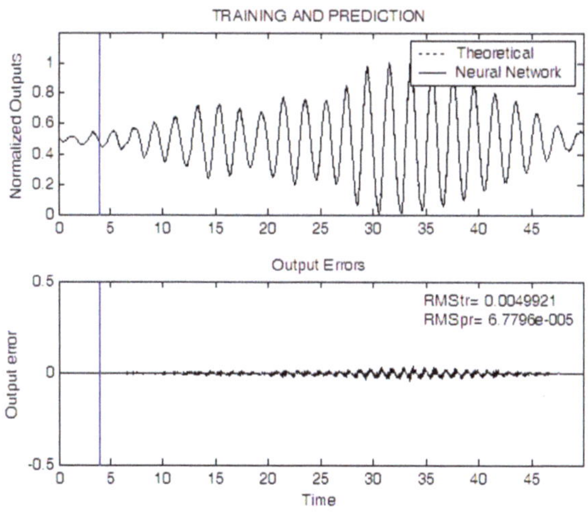

Fig. 13. Training and prediction, SDOF (a), Mexico City earthquake

Figure 15 shows the hysteretic behavior associated to the SDOF. Figure 16 contains the samples used during training of the PANN. Figure 17 shows the weight variation reducing to a value of 0.0002141, small enough to assume that the weights reached an almost constant value. The latter is easily noted with the moving average as an indicator of the tendency of the variance time series. Clearly the moving average shows a tendency to zero. Also, the slope of the first weight reaches a value of 1.3556×10^{-5}, and therefore, the absolute slope variation tends to zero.

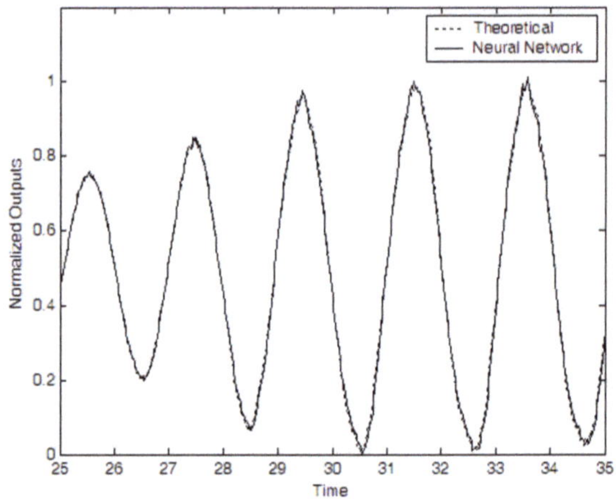

Fig. 14. Acceleration output prediction, SDOF (a), Mexico City earthquake

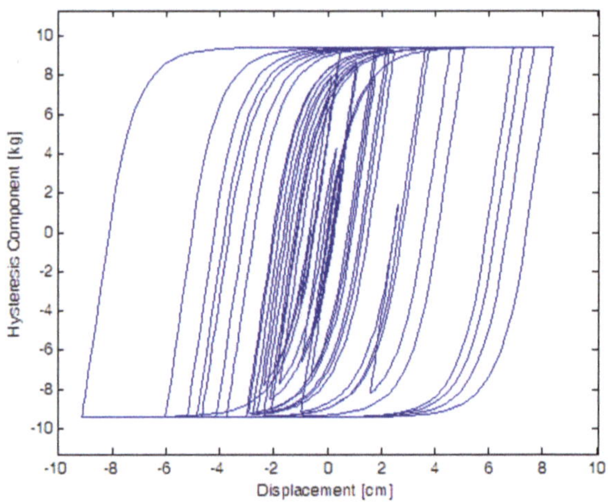

Fig. 15. Associated hysteresis, SDOF (b), Mexico City earthquake

Figure 18 shows the training interval. In this case, training was performed with 200 samples and prediction with the rest of the samples in the record (2,300 samples), divided by a vertical line in the figure. Again, training was done with the first 4 seconds of motion. During the acceleration output prediction phase, the weights of the network were kept constant, and it shows that the prediction closely follows the theoretical output. It is apparent that even when the energy dissipation is maximum, the estimation error is still close to zero. This fact could be explained because the nonlinear behavior takes place practically from the beginning of the motion, and the PANN is able to identify it.

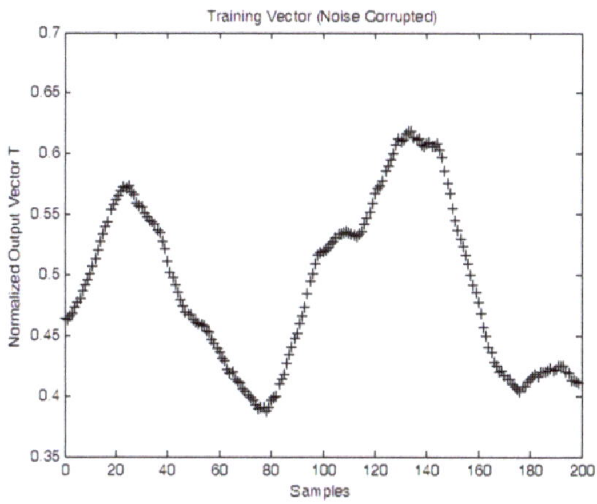

Fig. 16. Training vector, SDOF (b), Mexico City earthquake

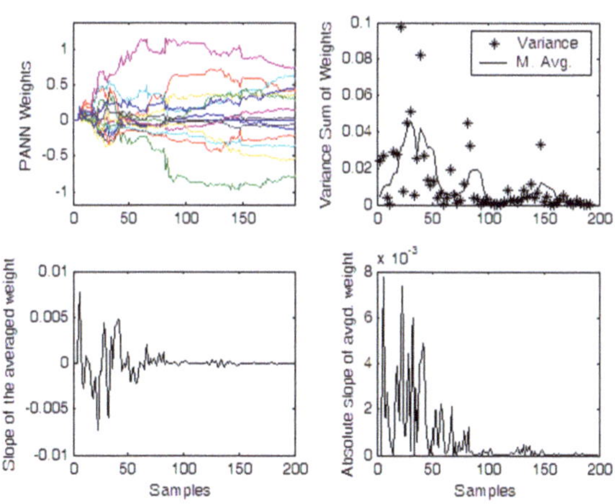

Fig. 17. Weight variation, SDOF (b), Mexico City earthquake

Figure 19 shows a window of the intense part of the motion for added resolution. The window is between 25 and 35 seconds of the acceleration output of the mass. Note that the model accurately reconstructs the output acceleration. It is worth noting that the system behaves nonlinearly and the PANN is able to identify it with only two cycles of motion.

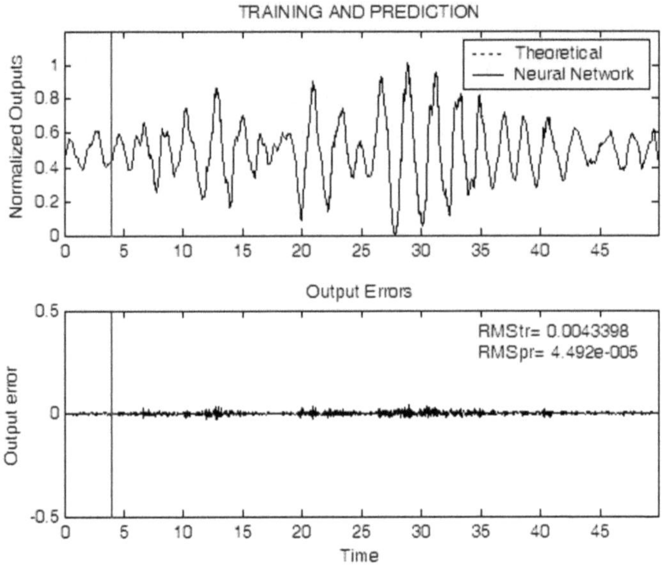

Fig. 18. Training and prediction, SDOF (b), Mexico City earthquake

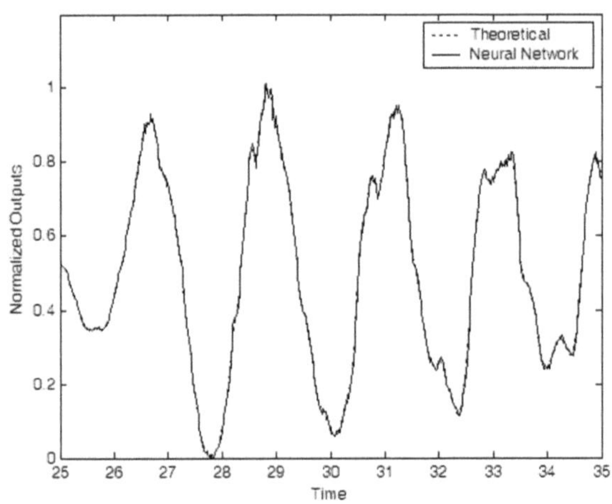

Fig. 19. Acceleration output prediction, SDOF (b), Mexico City earthquake

7 Identification Using Real Data

This section uses the actual seismic acceleration record acquired at the roof of a real structure to prove the effectiveness of the proposed algorithm. This instrumented construction is located in Mexico City. It is a reinforced concrete, 24-story, office building; 8 stories are used for parking, and the rest are administrative offices. The instrumentation has been active since the year 2000, and to date, several earthquakes of low intensity have been acquired.

The earthquake record acquired in April 18th 2002 is used for training, and only the acceleration output of the roof, parallel to the street, is used. The building has a period of 2.95 seconds, and almost 6 seconds of motions are used for training. The PANN was trained again with a second order polynomial model, no input values and 4 previous output values.

Additionally, the PANN was trained with the first 290 samples, also obtained from a record at 50 samples per second. Figure 20 contains the samples used during training of the PANN.

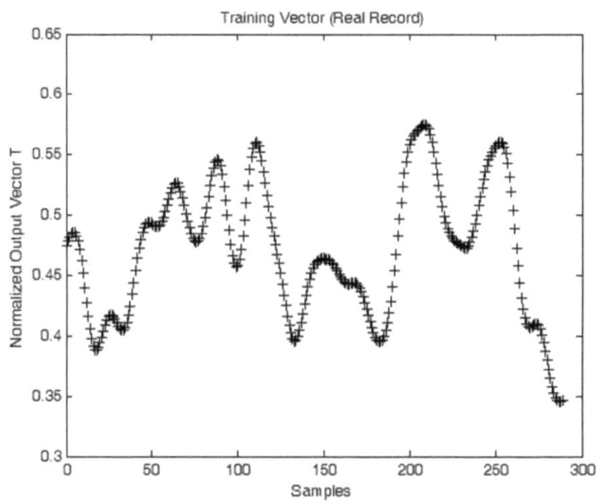

Fig. 20. Training vector, actual data, April 22, 2002 earthquake

Figure 21 shows the weight variation. The weight variance, in sets of three previous values, quickly reduces to a constant value of 0.00010439, thus training is attained. Also in that figure, the moving average of five previous values indicates a tendency to zero of the weight variance.

It could also be noted that the slope reaches a value of -1.5994×10^{-5} and therefore, the absolute variation is assumed to tend to zero. This way, a training interval is ensured with a fairly small amount of samples.

Figure 22 shows the training interval with 290 samples, along with the prediction interval with 9,956 samples, separated by a vertical line. As it would be expected, during training, the PANN is quick enough to identify the weights with only 6 seconds of motion, of a total of 200 seconds.

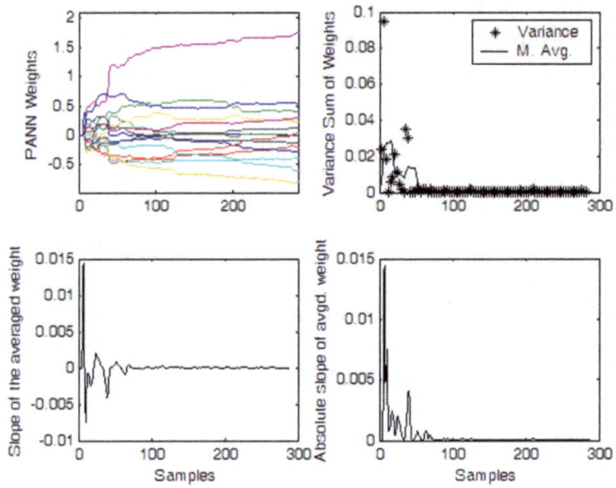

Fig. 21. Weight variation, actual data, April 22, 2002 earthquake

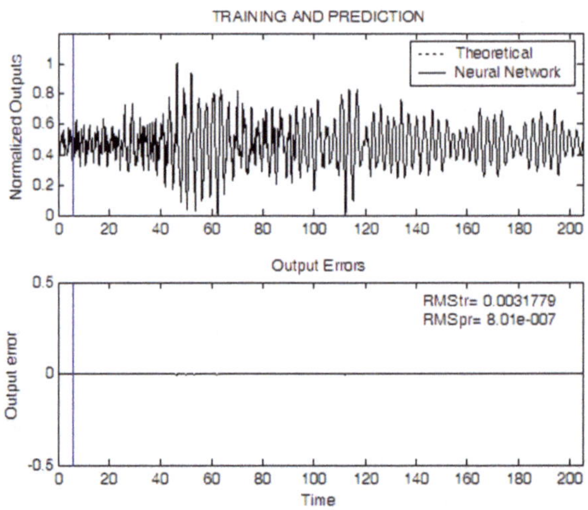

Fig. 22. Training and prediction, actual data, April 22, 2002 earthquake

In order to predict the acceleration output, the weights of the model have been kept constant. It could be noted that the prediction output is quite good. Even at the intense part of the record, the estimation error practically does not change. In that figure, the error has also been plotted along with the training and prediction. The RMS errors are also shown for training and prediction.

Figure 23 shows a window for added resolution. This window contains the intense part from 40 to 60 seconds of motion. It is worth mentioning that the acceleration output prediction at the roof of the real building is quite accurate for a model with no *a priori* knowledge of the structure, the type of nonlinear behavior it might show, or the intensity of the earthquake that hit it.

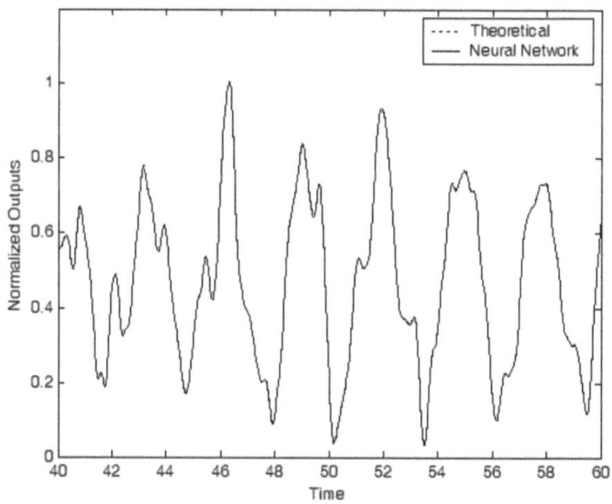

Fig. 23. Acceleration output prediction, actual data, April 22, 2002 earthquake

Additionally, and as a final test, once the model was identified, another seismic record acquired 10 months later by the instrumentation, January 21st, 2003, was used to predict the acceleration output of the roof.

Using the weights computed previously, the acceleration output prediction is performed with this new and totally different excitation. Fig. 24 shows the prediction with 19,360 samples. During the intense part of the motion, the estimation error practically does not change. The error has been plotted along with the identification.

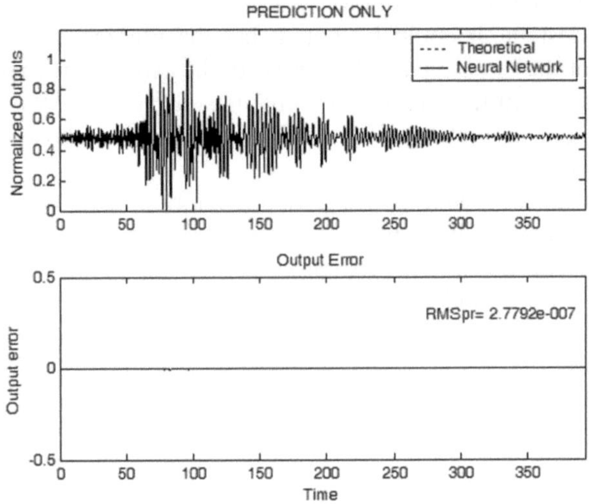

Fig. 24. Acceleration output prediction, actual data, January 21, 2003 earthquake

Figure 25 shows a window for added resolution, from 94 to 100 seconds of the total motion. Note that in this case, the model was identified 10 months earlier with another earthquake record, with totally different characteristics, and the model still predicts quite accurately the acceleration output.

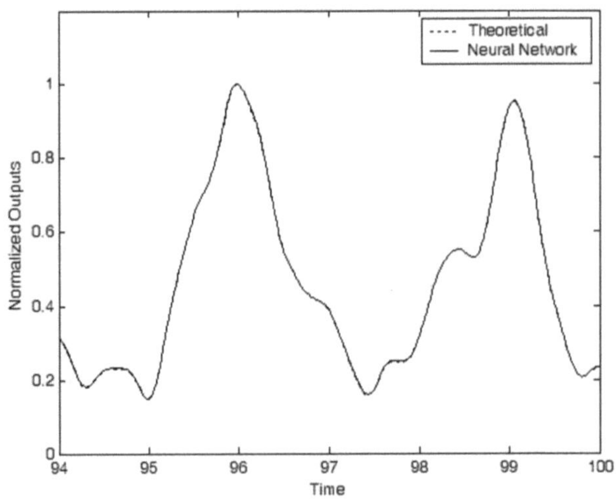

Fig. 25. Acceleration output prediction, actual data, January 21, 2003 earthquake

8 Conclusions

In the last two decades, several buildings have been instrumented to monitor the resistant lateral capacity of the structure through the analysis of the acceleration, velocity and displacement records, acquired by that instrumentation when subject to earthquake or high wind.

The present research proposes the use of a polynomial artificial neural network (PANN) to identify the behavior of such nonlinear systems and predict the acceleration output. The PANN is trained on-line, only with the first two cycles of motion. Usually, seismic records are acquired at 50 samples per second, which means that only a small amount of samples is needed for training. The latter was revised with a variance analysis of the weights during training. The variance, in conjunction with a moving average, could suggest a tendency to zero, and thus could be interpreted as the weights reaching a constant value.

In order to verify the effectiveness of the proposed algorithm, simulations of four theoretical structures are shown. Two of them are simulated with linear elastic behavior, and the last two are simulated with nonlinear hysteretic behavior modeled with Bouc-Wen. Two theoretical models were excited with the Loma Prieta seismic record, and the other two with the Mexico City. The results of the simulations show fair convergence speed of the weights and good accuracy in the acceleration output prediction. Finally, the PANN identified a model of a real 24-story concrete building, located in Mexico City. The structure was identified with a record acquired in April, 2002, and the prediction results are adequate. Additionally, as a last test, once the model was identified, a seismic record acquired in January, 2003, was used to predict the acceleration output of the roof. Once again, the results show that the identified model predicts particularly well the output of the system.

The prediction is achieved with a model with no *a priori* knowledge of the structural physical parameters, the type of nonlinear behavior, or the type of excitation. A long term aim of the present research is the development of algorithms which allow its use, in conjunction with fault detection techniques, structural health monitoring, decision making strategies, and structural control, obviously in real-time.

References

1. Baber, T., and Wen, Y. K., "Random Vibration of Hysteretic, Degrading Systems", *J. of Engineering Mechanics Division*, ASCE, 107(6), (1981)
2. Bazán Zurita, E., and Meli Piralla, R., *Manual de Diseño Sísmico de Edificios*, Editorial Limusa, Grupo Noriega Editores, México, D.F., (1995)

3. Box, G. E. P., and Jenkin, G. M., *Time Series Analysis: Forecasting and Control*, San Francisco, CA, Holden-Day, (1970)
4. Canudas de Wit, C., Olsson, H., Aström, K. J., and Lischinsky, P., "A New Model for Control of Systems with Friction", *IEEE Transactions on Automatic Control*, Vol. 40, No. 3, (1995)
5. Chase, J. G., Barroso, L. R., and Hwang, K. S., "LMS-based Structural Health Monitoring Methods for the ASCE Benchmark Problem", *Proceedings of the American Control Conference ACC*, Boston, MA, June 30 - July 2, (2004)
6. Chen, S., and Billings, A., "Representations of Nonlinear Systems: the NARMAX model", *Int. J. of Control*, Vol. 49, No. 3, (1989)
7. Dahl, P. R., "Solid Friction Damping of Mechanical Vibrations", *AIAA Journal*, Vol. 14, (1976) pp. 1675-1682
8. Gomez-Ramirez, E., Pozniak, A., Gonzalez-Yunes, A., and Avila-Alvarez, M., "Adaptive Architecture of Polynomial Artificial Neural Network to Forecast Nonlinear Time Series", *Congress on Evolutionary Computation, CEC '99*, Mayflower, Washington, D.C., USA, July 6 - 9, (1999)
9. Housner, G. W., Bergman, L. A., Caughey, T. K., Chassiakos, A. G., Claus, R. O., Masri, S. F., Skelton, R. E., Soong, T. T., Spencer, B. F., and Yao, J. T. P., "Structural Control: Past, Present and Future", *Journal of Engineering Mechanics*, Vol. 123, No. 9, Sep., (1997)
10. Humar, J. L., *Dynamics of Structures*, A. A. Balkema Publishers, 2nd Edition, (2001)
11. Korenberg, M., Billings, S. A., Liu, Y. P., and McIlroy, P. J., "Orthogonal Parameter Estimation Algorithm for Non-Linear Stochastic Systems", *International Journal of Control*, Vol. 48, No. 1, (1988)
12. Kosmatopoulos, E. B., Smyth, A. W., Masri, S. F., and Chassiakos, A. G., "Robust Adaptive Neural Estimation of Restoring Forces in Nonlinear Structures", *Transactions of the ASME, Journal of Applied Mechanics*, Vol. 68, November, (2001)
13. Loh, C. H., and Chung, S. T., "A Three-Stage Identification Approach for Hysteretic Systems", *Earthquake Engineering and Structural Dynamics*, Vol. 22, (1993) 129-150
14. Macki, J. W., Nistri, P., and Zecca, P., "Mathematical Models for Hysteresis", *SIAM Rev.*, Vol. 35, (1993) 94-123
15. Martinez-Garcia, J. C., Gomez-Gonzalez, B., Martinez-Guerra, R., and Rivero-Angeles, F. J., "Parameter Identification of Civil Structures Using Partial Seismic Instrumentation", *in 5th Asian Control Conference, ASCC*, Melbourne, Australia, July 20-23, (2004)
16. Masri, S. F., Miller, R. K., Saud, A. F., and Caughey, T. K., "Identification of Nonlinear Vibrating Structures: Part I – Formulation", *Transactions of the ASME, J. of Applied Mechanics*, Vol. 57, Dec., (1987)
17. Masri, S. F., Chassiakos, A. G., and Caughey, T. K., "Structure-unknown nonlinear dynamic systems: identification through neural networks", *Smart Mater. Struct.*, 1, (1992) 45-56
18. Masri, S. F., Chassiakos, A. G., and Caughey, T. K., "Identification of nonlinear dynamic systems using neural networks", *J. of Applied Mechanics*, 60, (1993) 123-133

19. Mohammad, K. S., Worden, K., and Tomlinson, G. R., "Direct Parameter Estimation for Linear and Non-linear Structures", *Journal of Sound and Vibration*, 152 (3), (1992)
20. Ni, Y. Q., Ko, J. M., and Wong, C. W., "Nonparametric Identification of Nonlinear Hysteretic Systems", *Journal of Engineering Mechanics*, Vol. 125, No. 2, February, (1999)
21. Rice, H. J., and Fitzpatrick, J. A., "A Procedure for the Identification of Linear and Non-linear Multi-Degree-of-Freedom Systems", *J. of Sound and Vibration*, 149 (3), (1991)
22. Smyth, A. W., Masri, S. F., Caughey, T. K., and Hunter, N. F., "Surveillance of Mechanical Systems on the Basis of Vibration Signature Analysis", *Transactions of the ASME, J. of Applied Mechanics*, Vol. 67, Sept., (2000)
23. Sugeno, M., *Industrial Applications of Fuzzy Control*, Elsevier Science Pub. Co., (1985)
24. Wen, Y. K., "Method for Random Vibration of Hysteretic Systems", *Journal of Engineering Mechanics, ASCE*, 102(2), (1976) 249-263
25. Yang, J. N., Pan, S., and Lin, S., "Identification and Tracking of Structural Parameters with Unknown Excitations", *Proceedings of the American Control Conference ACC*, Boston, MA, June 30 - July 2, (2004)
26. Yar, M., and Hammond, J. K., "Parameter Estimation for Hysteretic Systems", *J. of Sound and Vibration*, 117 (1), (1987)

Time Series Forecasting of Tomato Prices in Mexico Using Modular Neural Networks and Processing in Parallel

Ileana Leal, Patricia Melin

Division of Graduate Studies, Tijuana Institute of Technology, Mexico, pmelin@tectijuana.mx

Abstract. In this paper we describe the concepts of Time Series, Neural Networks, Modular Neural Networks, and Parallelism. Modular Neural Networks and Parallel Processing for Time Series Forecasting of Tomato Prices in Mexico are described in this paper. A particular modular neural network architecture implemented in parallel was used. Simulation results with the modular neural network approach for this application are very good.

1 Introduction

At this time all companies make use of forecasts of product prices or sales, which are used for planning of their production. If in the process of forecasting a significant error is obtained, this could leave a company without the raw material or supplies necessary for the production or in the case of agricultural products, the loss of a great amount of land without harvesting because of low prices of the product.

In these cases, the erroneous forecasting directly affects the profits of the company. Diverse factors exist so that the forecasting is erroneous, for example, the validity and availability of historical data, the desired precision of the prognosis, the benefits of the result, the future periods that are desired to foretell, among others [18].

Neural Networks are not more than an artificial model, which is a simplification of the human brain. The artificial model has the ability to learn from data in a similar way as humans do. The human brain is the most perfect example that we have of a system that is able to acquire knowledge through experience [25]. For this reason an artificial model of this system can be used for solving many problems.

I. Leal and P. Melin : *Time Series Forecasting of Tomato Prices in Mexico Using Modular Neural Networks and Processing in Parallel*, StudFuzz **208**, 385–402 (2007)
www.springerlink.com

The use of Modular Neural Networks is considered when the applications are very complex. However, if a higher level of efficiency is required and a smaller processing time is required then no longer a Modular Neural Network in a single machine is sufficient, it is necessary to distribute that Network in several processors [18].

In order to carry out this distributed processing, a method of parallelism will be implemented, which defines the division of processing work between multiple processors that operate simultaneously. The achieved results of parallel processing are more efficient when compared to sequential processing.

2 Time Series

When we speak of a sequence of values observed throughout time, and therefore ordered chronological, we called that, in a general sense, a time series. It is difficult to imagine a branch of science in which they do not appear. The data of many problems can be considered as time series [17].

If we know the past values of the series, and it is not possible to predict with total certainty the next value of the variable, we say that the series is nondeterministic or random, and logically this forms the body of knowledge called "time series analysis".

2.1. Prediction

Prediction is the estimation of future values of the variable based on the past behavior of the time series [17].

Evidently although the future value of a time series is not predictable with complete accuracy, this area of study has great interest, the result can either be completely random, or with some existing regularity as far as their behavior in the time [17].

2.2. Price

The prices are the key of the income as well as the profits of a company. It is considered that the determination of prices is the key activity within the capitalist system of the free company [32].

3 Neural Networks

One of the multiple branches by which the investigation of the Artificial intelligence has been developed, are the development of the so called "Neural Networks". We can define a Neural Network as a mathematical model inspired by biological systems, which are simulated in conventional computers [14].

3.1. Operation

Neural networks try to give a different approach from the traditional way to solve the problems. The training consists of presenting/displaying to the network a set to him of input-output patterns that of some form are known a priori. This set of input-output patterns is called training set. It is desired that the neural network infers the rule that governs these patterns and is able to generalize. That is, the goal is to obtain the correct output for input patterns that do not belong to the training set.

3.2. Methods of Learning

The learning methods are classified as follows:

3.2.1. Supervised Learning

This learning method consists of presenting the input patterns next to the desired output patterns for each input pattern, and for this reason it is called supervised learning.

3.2.2. Learning Supervised

In this type of learning the desired output patterns are not presented to the network, since it is not indicated to the network the results that it must give, but are left to the network to organize the data.

3.2.3. Reinforced Learning

In this case the supervisor is limited to indicate if the output offered by the network is correct or incorrect, but does not indicate the answer that must be given [14].

3.3. Elements

The elements of a neural network are the following ones:

3.3.1. Aggregation Function

This function calculates the base value or total input to the unit, generally using a simple weighed sum of all the received inputs, that is to say, of the inputs multiplied by the weight or value of the connections. It is equivalent to the combination of the excitatory and inhibitory signals of the biological neurons.

3.3.2. Function of Activation

It is perhaps the basic characteristic of the neurons, the one that better defines their behavior. This function is in charge of calculating the level or state of activation of the neuron based on the total input.

3.3.3. Weighted Connections

Play the role of the synaptic connections, the weight of the connection is equivalent to the force or effectiveness of the synapse. The existence of connections determines if it is possible that a unit influences another one, they define the type (excitatory/inhibitory) and the intensity of the influence.

3.3.4. Output

Calculates the output of the neuron based on the activation of the same one, although normally it is not applied more than the function identity, and the value of activation is taken as the output. The value of the output would act the as of the rate of firing in the biological neurons.

4 Modular Neural Networks

A modular neural network is based on the idea of "divides and conquer": a complex problem can be divided in a series of simpler sub-problems that can efficiently be solved by smaller networks.

Modular networks solve a problem by fragmenting it into simpler sub-problems, each one is solved entirely by a module [24].

4.1 Architecture of a Modular Neural Network

A modular architecture is a network using several modules in which each module is a monolithic neural network. The task of each module is simpler that the one of the complete network [18]. The characteristics of the modular neural networks are:

- The modules are connected like neurons in a network
- Independent Modules = Work in parallel
- Scheme of control = together works of useful way

In figure 1, we show the Architecture of a Modular Neural Net-work.

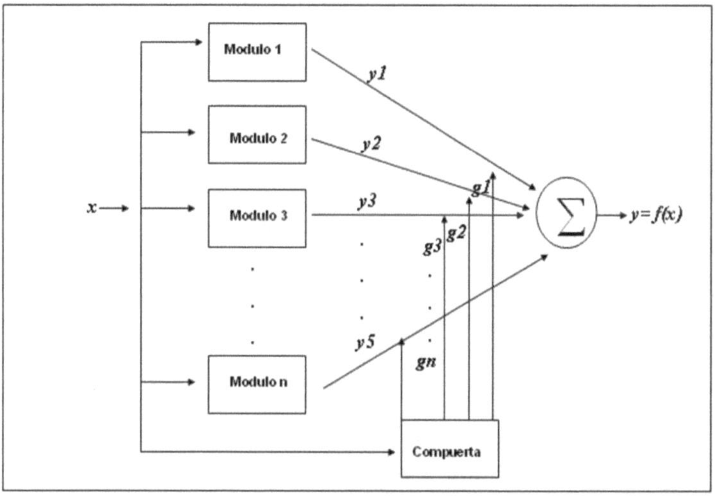

Fig. 1. Architecture of Modular Neural Network

5 Parallelism

Parallel programming defines the division of processing work between multiple processors that operate simultaneously. The goal is to make such processing in a more efficient form, compared with its execution in a sequential system [25].

5.1. Use of Clusters

A cluster parallel system combines a group of calculation systems, connected by means of a network of high speed and a specific scheme of programming to obtain a power of calculation similar to a supercomputer [25].

The calculation systems work together to complete a common processing task, by means of the division of such processing among them, and generating a result as if it was a unique system of calculation. Normally, such processing is an intensive operation on data or numerical values. The success of clusters of computers is due to its ability to combine a group of calculation systems, in such a way that they provide with processing capacities that a single computer could not solve.

The use of clusters to develop, and to execute parallel applications is gaining popularity, becoming a great alternative to the use of specialized architectures, more due to the great value of these systems. In figure 2, the structure of cluster is shown.

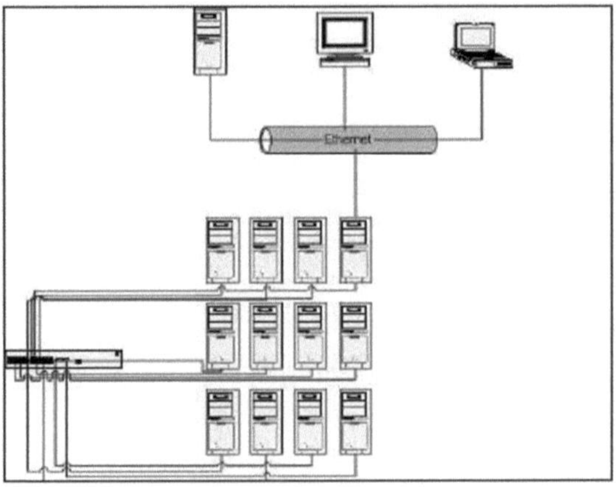

Fig. 2 Structure of a Cluster

5.2. Master-Slave Method

The Master-Slave method works in such a way that it has a single administrator of the tasks of all the processors. The processors are the slaves and perform their processing and return their results to the administrator also called Butler or Teacher [25]. In this method it is possible to have from one to n enslaved processors as it shown in figure 3.

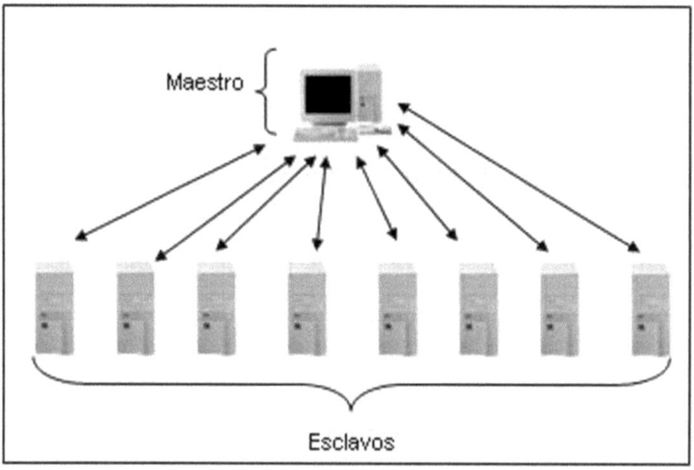

Fig. 3 Cluster with many processors

6 Proposed Architectures

For achieving the goals of this research we compiled tables of real data of the price (in pesos per kg.) of the Saladette Tomato, this price is the one that is offered to the final consumer, the tables contain data from 5 cities of Mexico (Distrito Federal, Guadalajara, Mérida, Monterrey and Torreón), which include the years from 1998 to 2004. The data are given per week, which were obtained from the web page of the National System of Information and Integration of Markets [27]. The design of the modular neural network and the one with parallelism were based on experience and on the references.

6.1 Architecture of the Modular Neural Network

The modular network has as input the weekly period and as the output the predicted price, and consists of 5 modules, which represent each one of the cities mentioned before. Each module counts on 5 sub-modules to do the work.

In table 1, we show the parameters of the modular neural network.

Table 1. Architecture of the Modular Neural Network for all the modules and their respective submodules.

MODULE				
SUBMODULE	LAYERS	NEURONS	EPOCHS	ERROR GOAL
1	2	14,14	2000	0.0000002
2	2	14,12	2000	0.0000002
3	2	15,14	2000	0.0000002
4	2	15,12	2000	0.0000002
5	2	12,12	2000	0.0000002

6.2. Architecture in Parallel: Master-Slave

For this research six processors were used, which constitute the Master-Slave architecture, and it was used as follows: one processor as the master and five processors as the slaves.

Each of the five slaves corresponds to each one of the above mentioned cities, which represent each of the 5 modules of the modular architecture before mentioned. The Master is the one in charge of distributing the tasks for each one of the slaves. In figure 5, the graphical representation used for the Master-Slave architecture is shown.

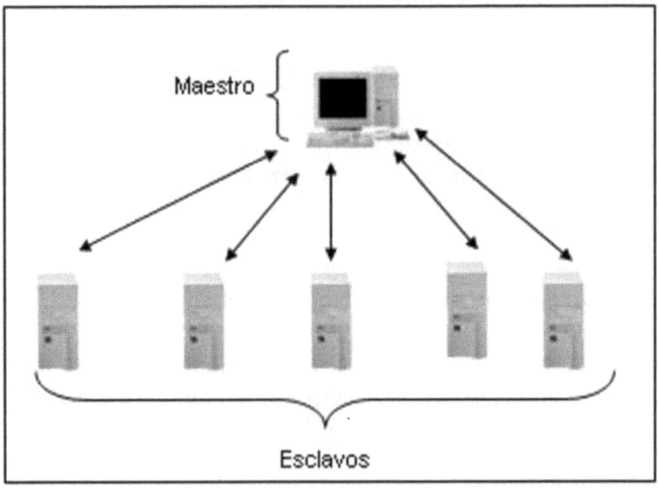

Fig. 5 Master-Slave architecture.

6.3. Integration

For the integration of the five modules that represent each of the cities used, a decision module was used. For the integration of the results coming from the sub-modules an average operator was applied, which consisted of adding each of the sub-modules and dividing the result between the number of sub-modules.

7 Results

In this section, the results obtained from each of the cities are shown. The simulation results are for time series forecasting of the tomato price.

7.1. Module 1: Distrito Federal

In figure 6, we show the graphical representation of the training for Module 1; in which the real data and the trained data is represented.

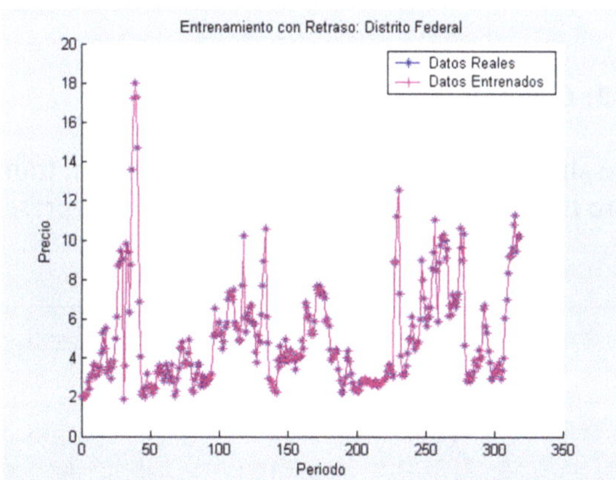

Fig. 6 Training of Module 1: Distrito Federal.

In figure 7, we have the graphical representation of the forecasting for Module 1.

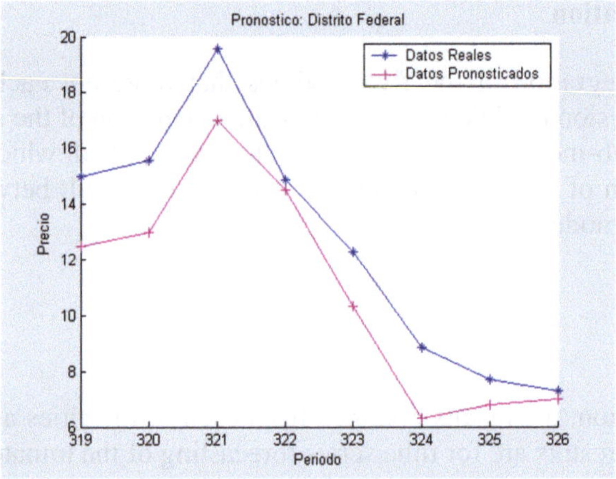

Fig. 7 Forecasting of Module 1: Distrito Federal.

The training for this module was obtained with the modular neural network in parallel and the forecasting is acceptable, since an average error of 1.59 pesos was obtained, which indicates us that the foretold prices are very near the real prices.

7.2. Module 2: Guadalajara

In figure 8 we show the graphical representation of the training for Module 2; in which are the real data and the trained data are represented.

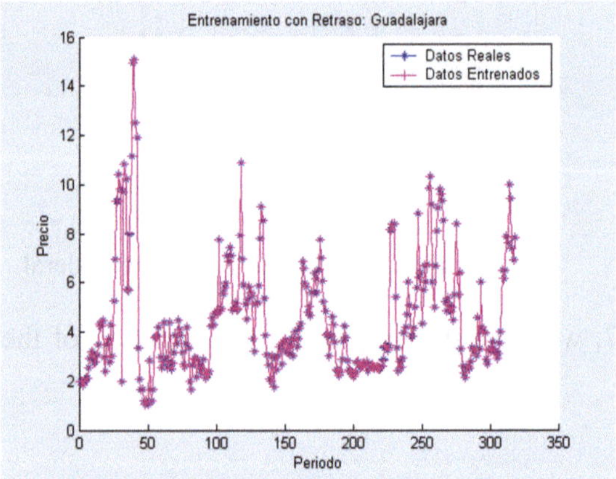

Fig. 8 Training of Module 2: Guadalajara.

In figure 9, we have the graphical representation of the forecasting for Module 2.

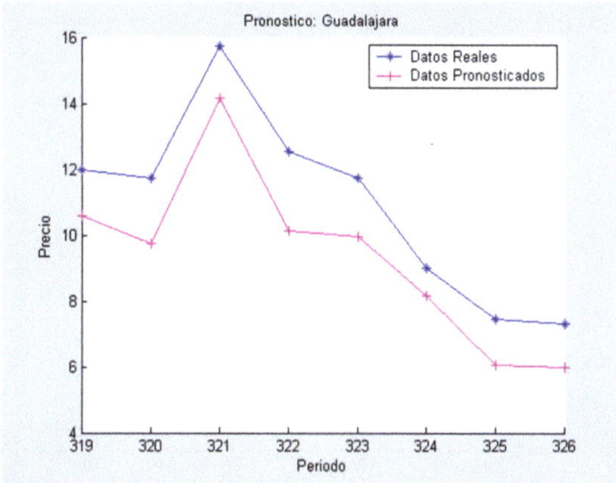

Fig. 9 Forecasting of Module 2: Guadalajara.

The training for this module was obtained with the modular Neural network in parallel and the forecasting is acceptable, since an average error of 1.80 pesos was achieved, which indicates us that the foretold prices are very near the real prices.

7.3. Module 3: Mérida

In figure 10, we have the graphical representation of the training for Module 3; in which the real data and the trained data are shown.

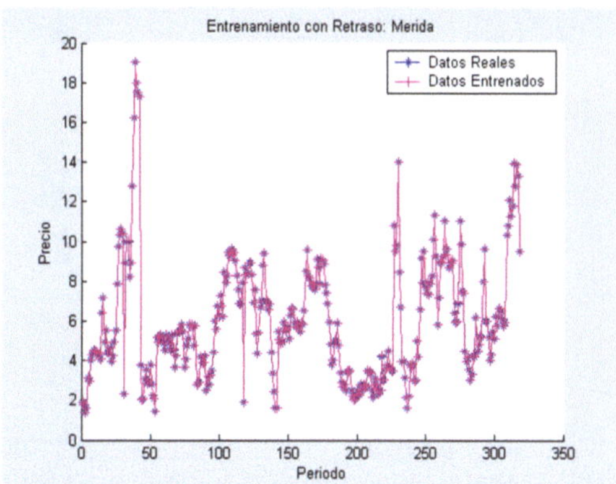

Fig. 10 Training of Module 3 for Mérida.

In figure 11, we have the graphical representation of the forecasting for Module 3.

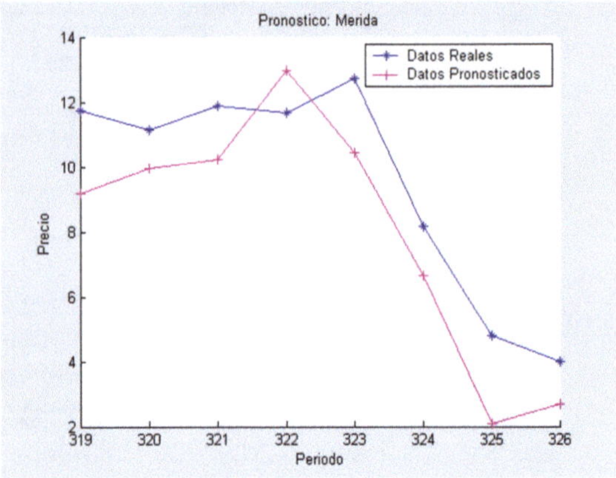

Fig. 11 Forecasting of Module 3 for Mérida.

The forecasting for module 3 of Mérida is acceptable, since an average error of 1.69 pesos was obtained, which indicates us that the foretold prices are very near the real prices.

7.4. Module 4: Monterrey

In figure 12, we show the graphical representation of the training for Module 4; in which are to the real data and the trained data.

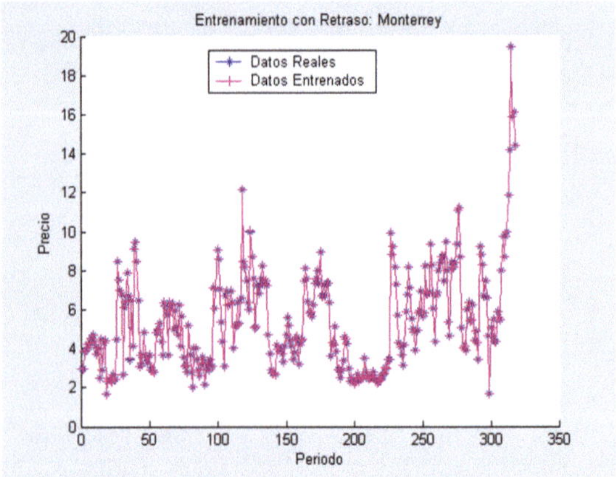

Fig. 12 Training of Module 4: Monterrey.

In figure 13, we have the graphical representation of the forecasting for Module 4.

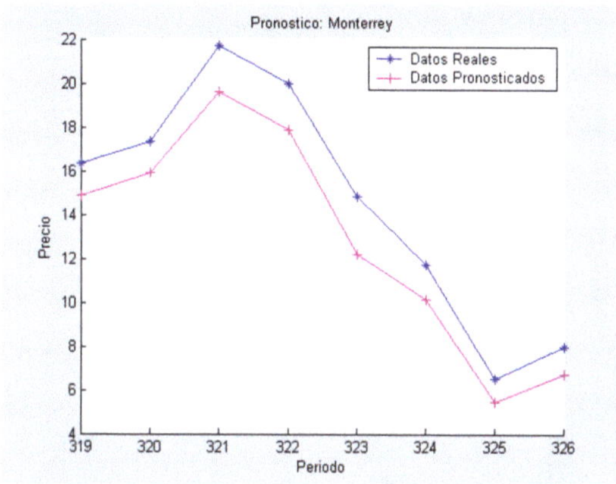

Fig. 13 Forecasting of Module 4: Monterrey.

The training for this module was obtained with the modular neural network in parallel and the forecasting is acceptable, since an average error of 1.71 pesos was obtained, which indicates us that the foretold prices are very near the real prices.

7.5 Module 5: Torreón

In figure 14, is the graphical representation of the training for Module 5; in which are to the real data and the trained data.

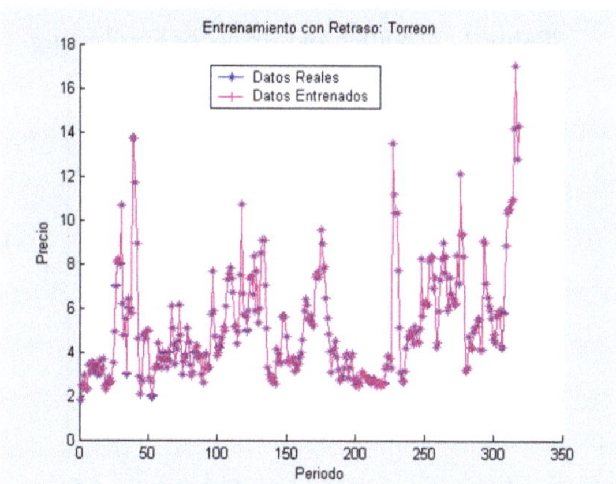

Fig. 14 Training of Module 5: Torreón.

In figure 15, we have the graphical representation of the forecasting for Module 5.

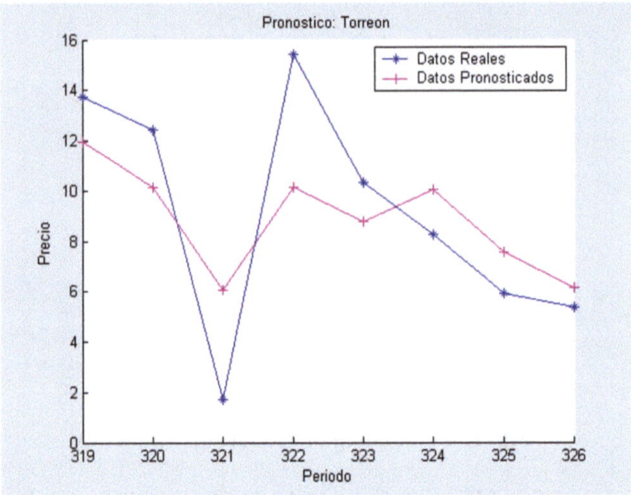

Fig. 15 Forecasting of the Module 5: Torreón.

The training for this module was obtained with the modular neural network in parallel and the forecasting is acceptable, since an average error of 2.42 pesos was obtained, which indicates us that the foretold prices are very near the real prices.

In table 2, they are each one of the modules with its respective forecasting errors.

Table 2. Modules and Errors of Forecasting.

MODULES AND ERRORS OF FORECASTING	
MODULE	ERROR (Pesos)
1	1.69
2	1.59
3	1.80
4	1.71
5	2.42

In the previous table we can observe that the best forecasting is the one of Module 2: Guadalajara, since an average error of 1.59 pesos was obtained and the forecasting with a greater error is the one of Module 5: Torreón, of 2.42 pesos, both errors are acceptable because the real data are very near the forecasted data. The difference between the errors of the modules is because the real data from the cities are very different.

In table 3, we show the training time for each of the Modules

Table 3 Training time of the Modules.

TRAINING TIME OF THE MODULES	
MODULE	**TIME**
1	2 hr and 41min approx
2	2 hr and 53 min approx
3	1 hr and 25 min approx
4	1 hr and 20 min approx
5	1 hr and 16 min approx

The Processing in Parallel (Cluster) approximately had a total training time of 2 hours and 53 minutes.

8 Conclusions

Time Series Forecasting for the price of the tomato in Mexico is very complicated, since a great variation between the prices exists. The use of modular neural networks allowed an acceptable prediction and the use of parallelism contributed to the results in that a smaller time of processing was obtained. It is possible to mention that the information that exists about the price of the Tomato in Mexico and the one of parallelism is very limited. The Master-Slave method allowed a good distribution of the tasks and contributed to obtaining good results for this type of application. As future work, a genetic algorithm can be used to optimize the architecture of the modular neural network.

References

1. **Álvarez D.**, "Tutorial de Redes Neuronales", Departamento de Señales y Radiocomunicaciones (SSR), México. 1997. Available on the following web page: http://www.gc.ssr.upm.es/inves/neural/ann2/anntutorial.html
2. **Andrade L.** "Simulación distribuida de arquitecturas multi-DSP". Reporte Técnico. Laboratorio en investigación y desarrollo en informática, Universidad de la Plata. 1997.
3. **Andrade L.** "Parmatlab (Toolbox para procesamiento en paralelo en Matlab)" 2004. Available from: http://www.cdsp.neu.edu/info/students/landrade/
4. **Arellano M.**, "Introducción al Análisis Clásico de Series de Tiempo", Tecnológico de Monterrey en línea, 2004. Available from the web page: http://ciberconta.unizar.es/LECCION/seriest/100.HTM#_Toc523661809
5. **Arroyo L.** "RedesNeuronales". 2004. Available from the web page: http://www.fortunecity.com/skyscraper/chaos/279/articulos/redesneuronales.htm
6. **Ayala M, Castillo R**, "Un modelo de predicción para el valor TRM: Un acercamiento desde las redes neuronales artificiales", Escuela de Matemáticas, Universidad Sergio Arboleda, Cali, Colombia. 2002. Available from: http://www.usergioarboleda.edu.co/civilizar/matematicas/Pdfs/mayala.pdf
7. **Bolella A.**, "Redes Neuronales", Ilustrados.com. 2003. Available from: http://www.ilustrados.com/publicaciones/EpyVyyEAFyqhTOudht.php
8. **Coello C.**, "Uso de Técnicas de Inteligencia Artificial para Aplicaciones Financieras", *LANIA*, Año 7, Vol. 26 y 27, Otoño-Invierno.
9. **Consejo Superior de Investigaciones**, "The Parallel Virtual File System Project", Peru. Available from: http://csi.unmsm.edu.pe/paralelo/pvfs/
10. **Cortés G.** "Redes Neuronales y Algoritmos de Primer Orden", Universidad Católica de Valparaíso, Chile. 1997. Available from the web page: http://html.wanadoo.com/redes-neuronales-y-algoritmos-de-primer-orden.html
11. **Duarte R.**, "Series Temporales". 2002. Available from http://www.bccr.fi.cr/ndie/Documentos/DIE-02-2002-NT-ASPECTOS%20 CONCEPTUALES %20SOBRE% 20SEATS.pdf
12. **Fernández L., Vasconcelos P.**, "El Comportamiento de las series de tiempo de los precios de los activos", México, Instituto Tecnológico Autónomo de México. Available from the following web page: http://cursos.itam.mx/mhess/download/studentprojects/fe2003/Martin.doc
13. **García I.**, "Análisis Y Predicción De La Serie De Tiempo Del Precio Externo Del Café Colombiano Utilizando Redes Neuronales Artificiales", Universitas Scientiarum, *Revista De La Facultad De Ciencias, Pontificia Universidad Javeriana* Vol. 8, 45-50. http://www.javeriana.edu.co/universitas_scientiarum/MATEMATICAS/7-analisis.pdf
14. **Gómez M.**, "Redes Neuronales Artificiales (Tutorial)", IIIA-Instituto de Investigación en Inteligencia Artificial, CSIC-Spanish Scientific Research

Council, Barcelona, Spain, 2004. Available from the web page: http://www.iiia.csic.es/%7Emario/Tutorial/RNA_marcos.htm
15. **González M.**, "Series de Tiempo". 2000. Available from the web page: http://g.unsa.edu.ar/asades/actas2000/11-31.html
16. **Gonzáles S.**, "Predicciones De Variables Energéticas Mediante Análisis De Series Temporales", Facultad de Ingeniería Mecánica, Universidad Nacional de Ingeniería Lima –Perú, *TECNIA*, Vol. 8 No.2, Págs.81-87, 1998.
17. **Jiménez J.**, "Series Temporales". 1997. Available from the web page: http://ciberconta.unizar.es/LECCION/seriest/100.HTM#_Toc523661810
18. **Jiménez M.**, " Redes Neuronales Modulares ". 2001. Available from http://strix.ciens.ucv.ve/~iartific/Material/PP_RNAs_Modulares.ppt.
19. **Lizarraga C.**, "Clusters de Linux", Departamento de Física, Universidad de Sonora, Hermosillo, Sonora, 2002. Available from the web page: http://www.fisica.uson.mx/carlos/LinuxClusters/
20. **López I.**, "Redes Modulares", Grupo de Tratamiento Avanzado de Señal (GTAS), Universidad de Cantabria. 2001. Available from the web page: http://www.gtas.dicom.unican.es/miembros/Nacho/Doctorado/NNs/Modular.ppt
21. **Mtz. de Lejarza I**. "Redes Neuronales". Universidad de Valencia, 1998. Available from: http://www.uv.es/~mlejarza/redes.htm
22. **Mallo C.** "Predicción de la demanda eléctrica horaria mediante redes neuronales artificiales", Departamento de Economía Cuantitativa. Universidad de Oviedo. 2003. Available from the web page: http://www.uv.es/asepuma/recta/ordinarios/5/5-1.pdf
23. **Medina D.**, "Series de Tiempo". Available from the web page: http://es.wikipedia.org/wiki/Series_de_tiempo
24. **Melín P., Bravo D.** "Estudio Comparativo de Métodos de Procesamiento en Paralelo para Redes Neuronales Modulares". International Seminar on Computational Intelligence 2004, December 1-3 2004, pags 71-79.
25. Web page of the Supercomputing Department of UNAM University. 2004. http://clusters.unam.mx/index.php
26. Web page of the Institution in charge of design and coordinate the operation of the "Sistema Nacional de Información del Sector Agroalimentario y Pesquero", 2005. http://www.siap.sagarpa.gob.mx
27. Web of the "Sistema Nacional de Información e Integración de Mercados", 2005. http://www.economia-sniim.gob.mx/e_default.asp?
28. **Pérez M., Hidalgo H., Ocampo F.**, "Time series prediction using artificial neural networks", Instituto de Investigaciones Oceanológicas, Universidad Autónoma de Baja California. 2002. Available from the web page: http://rcmarinas.ens.uabc.mx/~pagina/resumenes/2002/28-1/f0999resfinal.pdf
29. **Pérez R F.** "Modelos TAR para detectar cambios de régimen en series financieras y económicas", Universidad de Medellín, julio 2004. Available from: http://sigma.poligran.edu.co/politecnico/apoyo/Decisiones/simposio/documentos/Modelos-TAR-fp.pdf
30. **Plummer E. A.**, "Time Series Forecasting With Feed-Forward Neural Networks: Guidelines And Limitations", Laramie, Wyoming, July, 2000.

http://www.karlbranting.net/papers/plummer/Paper_7_12_00.htm#_Toc48807 6541

31. **Rodríguez M.**, "Predicción en los Mercados de Valores con Redes Neuronales", 5 campus.com. 2002. Available from the web page: http://ciberconta.unizar.es/LECCION/REDES/180.HTM
32. **Salas J.**, "Mercadotecnia del Precio". Available from the web page: http://www.uady.mx/sitios/contadur/material/MK10_Materialdeclase2.ppt.
33. **Sterling T.** "Una introducción a los Clusters de PC's para cómputo de alto rendimiento", California Institute of Technology, 2001.
34. **Storti M.** "Cálculo Paralelo". 2003. Available from the web page: http://venus.arcride.edu.ar/calculoparalelo/slides_MPI
35. **Wilkinson B.** "Parallel Programming: Techniques and applications using networked workstations and parallel computers", Prentice-Hall 1998.

Modular Neural Networks with Fuzzy Sugeno Integration Applied to Time Series Prediction

Patricia Melin, Valente Ochoa, Luis Valenzuela, Gabriela Torres, Daniel Clemente

Dept. of Computer Science, Tijuana Institute of Technology, Mexico

Abstract. We describe in this paper the application of several neural network architectures to the problem of simulating and predicting the dynamic behavior of complex economic time series. We use several neural network models and training algorithms to compare the results and decide at the end, which one is best for this application. We also compare the simulation results with the traditional approach of using a statistical model. In this case, we use real time series of prices of consumer goods to test our models. Real prices of tomato and green onion in the U.S. show complex fluctuations in time and are very complicated to predict with traditional statistical approaches.

1 Introduction

Forecasting refers to a process by which the future behavior of a dynamical system is estimated based on our understanding and characterization of the system. If the dynamical system is not stable, the initial conditions become one of the most important parameters of the time series response, i.e. small differences in the start position can lead to a completely different time evolution. This is what is called sensitive dependence on initial conditions, and is associated with chaotic behavior [2, 16] for the dynamical system.

The financial markets are well known for wide variations in prices over short and long terms. These fluctuations are due to a large number of deals produced by agents that act independently from each other. However, even in the middle of the apparently chaotic world, there are opportunities for making good predictions [4,5]. Traditionally, brokers have relied on technical analysis, based mainly on looking at trends, moving averages, and certain graphical patterns, for performing predictions and subsequently making deals. Most of these linear approaches, such as the well-known Box-Jenkins method, have disadvantages [9].

P. Melin et al.: *Modular Neural Networks with Fuzzy Sugeno Integration Applied to Time Series Prediction*, StudFuzz **208**, 403–413 (2007)
www.springerlink.com

More recently, soft computing [10] methodologies, such as neural networks, fuzzy logic, and genetic algorithms, have been applied to the problem of forecasting complex time series. These methods have shown clear advantages over the traditional statistical ones [12]. The main advantage of soft computing methodologies is that, we do not need to specify the structure of a model a-priori, which is clearly needed in the classical regression analysis [3]. Also, soft computing models are non-linear in nature and they can approximate more easily complex dynamical systems, than simple linear statistical models. Of course, there are also disadvantages in using soft computing models instead of statistical ones. In classical regression models, we can use the information given by the parameters to understand the process, i.e. the coefficients of the model can represent the elasticity of price for a certain good in the market. However, if the main objective if to forecast as closely as possible the time series, then the use of soft computing methodologies for prediction is clearly justified.

2 Monolithic Neural Network Models

A neural network model takes an input vector X and produces and output vector Y. The relationship between X and Y is determined by the network architecture. There are many forms of network architecture (inspired by the neural architecture of the brain). The neural network generally consists of at least three layers: one input layer, one output layer, and one or more hidden layers. Figure 1 illustrates a neural network with p neurons in the input layer, one hidden layer with q neurons, and one output layer with one neuron.

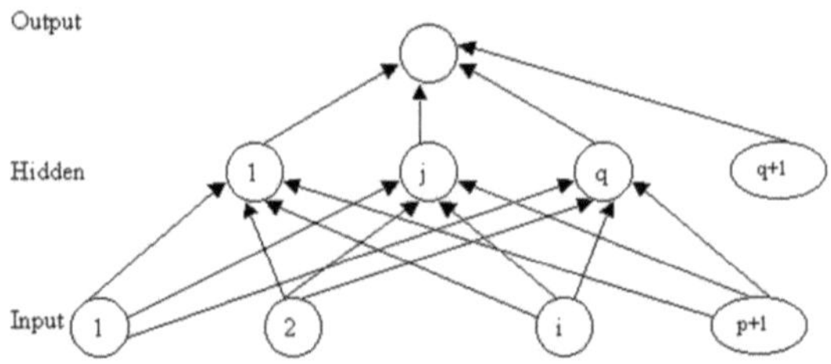

Fig. 1. Single hidden layer feedforward network.

In the neural network we will be using, the input layer with p+1 processing elements, i.e., one for each predictor variable plus a processing element for the bias. The bias element always has an input of one, $X_{p+1}=1$. Each processing element in the input layer sends signals Xi (i=1,…,p+1) to each of the q processing elements in the hidden layer. The q processing elements in the hidden layer (indexed by j=1,…,q) produce an "activation" $a_j=F(\Sigma w_{ij}X_i)$ where w_{ij} are the weights associated with the connections between the p+1 processing elements of the input layer and the jth processing element of the hidden layer. Once again, processing element q+1 of the hidden layer is a bias element and always has an activation of one, i.e. $a_{q+1}=1$. Assuming that the processing element in the output layer is linear, the network model will be

$$Y_t = \sum_{j=1}^{p+1} \pi_\iota \, x_{it} + \sum_{j=1}^{p+1} \theta_j F\left(\sum_{i=1}^{p+1} w_{ij} \, x_{it} \right) \tag{1}$$

Here π_ι are the weights for the connections between the input layer and the output layer, and θ_j are the weights for the connections between the hidden layer and the output layer. The main requirement to be satisfied by the activation function F(·) is that it be nonlinear and differentiable. Typical functions used are the sigmoid, hyperbolic tangent, and the sine functions.

The weights in the neural network can be adjusted to minimize some criterion such as the sum of squared error (SSE) function:

$$E_1 = \tfrac{1}{2} \sum_{1=1}^{n} (d_1 - y_1)^2 \tag{2}$$

Thus, the weights in the neural network are similar to the regression coefficients in a linear regression model. In fact, if the hidden layer is eliminated, (1) reduces to the well-known linear regression function. It has been shown [22] that, given sufficiently many hidden units, (1) is capable of approximating any measurable function to any accuracy. In fact F(.) can be an arbitrary sigmoid function without any loss of flexibility.

The most popular algorithm for training feedforward neural networks is the backpropagation algorithm [14,18]. As the name suggests, the error computed from the output layer is backpropagated through the network, and the weights are modified according to their contribution to the error function. Essentially, backpropagation performs a local gradient search, and hence its implementation does not guarantee reaching a global minimum. A number of heuristics are available to partly address this problem, some of which are presented below. Instead of distinguishing between the weights of the different layers as in Equation (1), we refer to them generically as w_{ij} in the following.

After some mathematical simplification the weight change equation suggested by backpropagation can be expressed as follows:

$$w_{ij} = -\eta \frac{\partial E_1}{\partial w_{ij}} + \theta \Delta wij \tag{3}$$

Here, η is the learning coefficient and θ is the momentum term. One heuristic that is used to prevent the neural network from getting stuck at a local minimum is the random presentation of the training data.

3 Modular Neural Networks

There exists a lot of neural network architectures in the literature that work well when the number of inputs is relatively small, but when the complexity of the problem grows or the number of inputs increases, their performance decreases very quickly. For this reason, there has also been research work in compensating in some way the problems in learning of a single neural network over high dimensional spaces.

In the work of Sharkey [20], the use of multiple neural systems (Multi-Nets) is described. It is claimed that multi-nets have better performance or even solve problems that monolithic neural networks are not able to solve. It is also claimed that multi-nets or modular systems have also the advantage of being easier to understand or modify, if necessary.

In the literature there is also mention of the terms "ensemble" and "modular" for this type of neural network. The term "ensemble" is used when a redundant set of neural networks is utilized, as described in Hansen and Salamon [8]. In this case, each of the neural networks is redundant because it is providing a solution for the same task, as it is shown in Figure 2.

On the other hand, in the modular approach, one task or problem is decompose in subtasks, and the complete solution requires the contribution of all the modules, as it is shown in Figure 3.

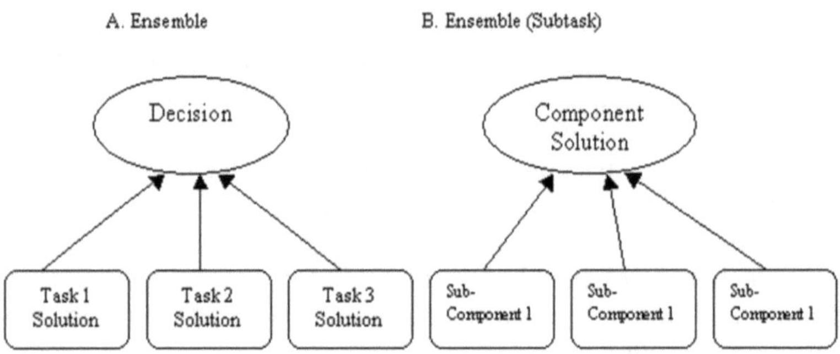

Fig. 2. Ensembles for one task and subtask.

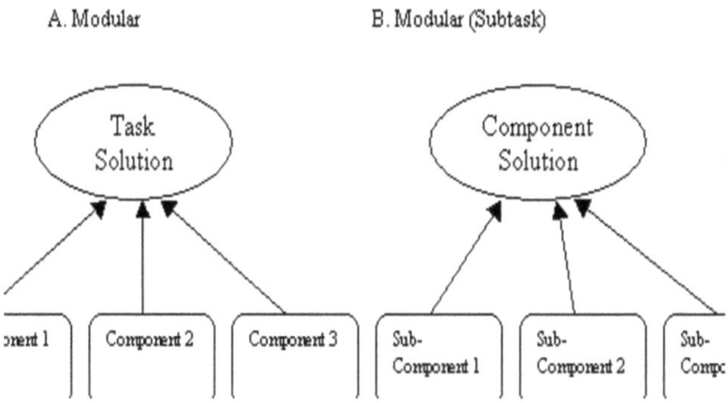

Fig. 3. Modular approach for task and subtask.

4 Methods for Response Integration

In the literature we can find several methods for response integration, that have been researched extensively, which in many cases are based on statistical decision methods. We will mention briefly some of these methods of response integration, in particular the ones based on fuzzy logic. The idea of using these types of methods, is that the final decision takes into account all of the different kinds of information available about the time series. In particular, we consider aggregation operators, and the fuzzy Sugeno integral [21].

Yager [23] mentions in his work, that fuzzy measures for the aggregation criteria of two important classes of problems. In the first type of problems, we have a set $Z=\{z_1,z_2,\ldots,z_n\}$ of objects, and it is desired to select one or more of these objects based on the satisfaction of certain criteria. In this case, for each $z_i \in Z$, it is evaluated $D(z_i)=G(A_i(z_i),\ldots,A_j(z_i))$, and then an object or objects are selected based on the value of G. The problems that fall within this structure are the multi-criteria decision problems, search in databases and retrieving of documents.

In the second type of problems, we have a set $G=\{G_1,G_2,\ldots,G_q\}$ of aggregation functions and object z. Here, each G_k corresponds to different possible identifications of object z, and our goal is to find out the correct identification of z. For achieving this, for each aggregation function G, we obtain a result for each z, $D_k(z)=G_k(A_1(z), A_2(z), \ldots ,A_n(z))$. Then we associate to z the identification corresponding to the larger value of the aggregation function.

A typical example of this type of problems is pattern recognition. Where A_j corresponds to the attributes and $A_j(z)$ measures the compatibility of z with the attribute. Medical applications and fault diagnosis fall into this type of problems. In diagnostic problems, the A_j corresponds to symptoms associated with a particular fault, and G_k captures the relations between these faults.

Fuzzy integrals can be viewed as non-linear functions defined with respect to fuzzy measures. In particular, the "gλ-fuzzy measure" introduced by Sugeno [21] can be used to define fuzzy integrals. The ability of fuzzy integrals to combine the results of multiple information sources has been mentioned in previous works.

Definition 1. A function of sets $g:2^x$-(0.1) is called a fuzzy measure if:

1. g(0)=0 g(x)=1
2. $g(A) \leq g(B)$ if $A \subset B$
3. if $\{A_i\}i^\alpha = 1$ is a sequence of increments of the measurable set then

$$\lim_{i \to \infty} g(A_i) = g(\lim_{i \to \infty} A_i) \tag{4}$$

From the above it can be deduced that g is not necessarily additive, this property is replaced by the additive property of the conventional measure.

From the general definition of the fuzzy measure, Sugeno introduced what is called "gλ-fuzzy measure", which satisfies the following additive property: For every $A, B \subset X$ and $A \cap B = \theta$,

$$g(A \cup B) = g(A) + g(B) + \lambda\, g(A)g(B), \tag{5}$$

for some value of $\lambda > -1$.

This property says that the measure of the union of two disjunct sets can be obtained directly from the individual measures. Using the concept of fuzzy measures, Sugeno [21] developed the concept of fuzzy integrals, which are non-linear functions defined with respect to fuzzy measures like the gλ-fuzzy measure.

Definition 2 let X be a finite set and $h:X \to [0,1]$ be a fuzzy subset of X, the fuzzy integral over X of function h with respect to the fuzzy measure g is defined in the following way,

$$\begin{aligned} h(x) \circ g(x) &= \max_{E \subseteq X} [\min(\min_{x \in E} h(x), g(E))] \\ &= \sup_{\alpha \in [0,1]} [\min(\alpha, g(h_\alpha))] \end{aligned} \tag{6}$$

where h_α is the level set α of h,

$$h_\alpha = \{ x \mid h(x) \geq \alpha \}. \tag{7}$$

We will explain in more detail the above definition: h(x) measures the degree to which concept h is satisfied by x. The term $\min(h_x)$ measures the degree to which concept h is satisfied by all the elements in E. The value g(E) is the degree to which the subset of objects E satifies the concept measure by g. As a consequence, the obtained value of comparing these two quantities in terms of operator min indicates the degree to which E satifies both criteria g and min(h_x). Finally, operator max takes the greatest of these terms.

5 Simulation and Forecasting Prices in the U.S. Market

We will consider the problem forecasting the prices of tomato in the U.S. market. The time series for the prices of this consumer good show very complicated dynamic behavior, and for this reason it is interesting to analyze and predict the future prices for this good. We show in Figure 4 the time series of monthly tomato prices in the period of 1960 to 1999, to give an idea of the complex dynamic behavior of this time series.

We will apply both the modular and monolithic neural network approach and also the linear regression method to the problem of forecasting the time series of tomato prices. Then, we will compare the results of these approaches to select the best one for forecasting.

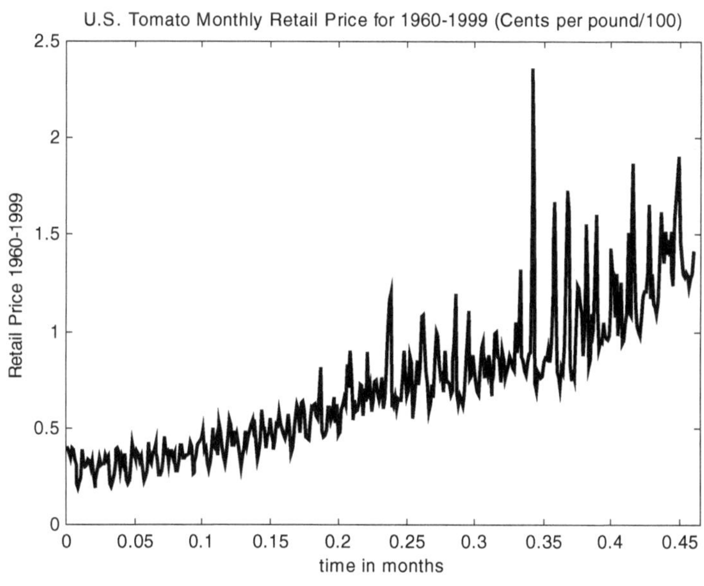

Fig. 4. Prices in US Dollars of tomato from January 1960 to December 1999.

6 Experimental Results

We describe, in this section, the experimental results obtained by using neural networks to the problem of forecasting tomato prices in the U.S. Market. We show results of the application of several architectures and different learning algorithms to decide on the best one for this problem. We also compare at the end the results of the neural network approach with the results of linear regression models, to measure the difference in forecasting power of both methodologies.

First, we will describe the results of applying modular neural networks to the time series of tomato prices. We used the monthly data from 1960 to 1999 for training a Modular Neural Network with four Modules, each of the modules with 80 neurons and one hidden layer. We show in Figure 5 the result of training the modular neural network with this data. In Figure 5, we can appreciate how the modular neural network approximates very well the real time series of tomato prices over the relevant period of time.

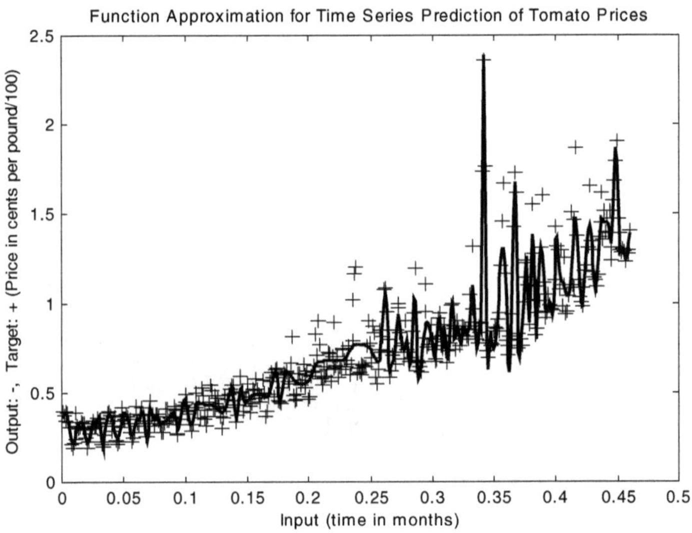

Fig. 5. Modular network for tomato prices with Levenberg-Marquardt algorithm.

We have to mention that the results shown in Figure 5 are for the best modular neural network that we were able to find for this problem. We show in Figure 6 the comparison between several of the modular neural networks that we tried in our experiments. From Figure 6 we can appreciate that the modular neural network with one time delay and Leverberg-Marquardt (LM) training algorithm is the one that fits best the data and for this reason is the one selected.

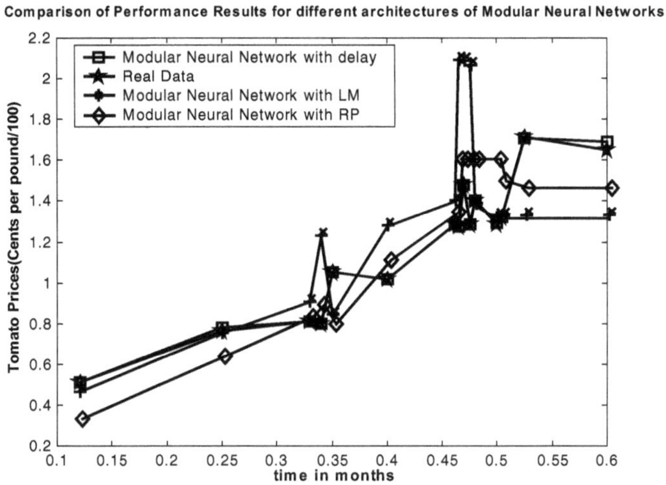

Fig. 6. Comparison of performance results for several modular neural networks.

We show in Figure 7 the comparison of the best monolithic network against the best modular neural network. The modular network clearly fits better the real data of the problem.

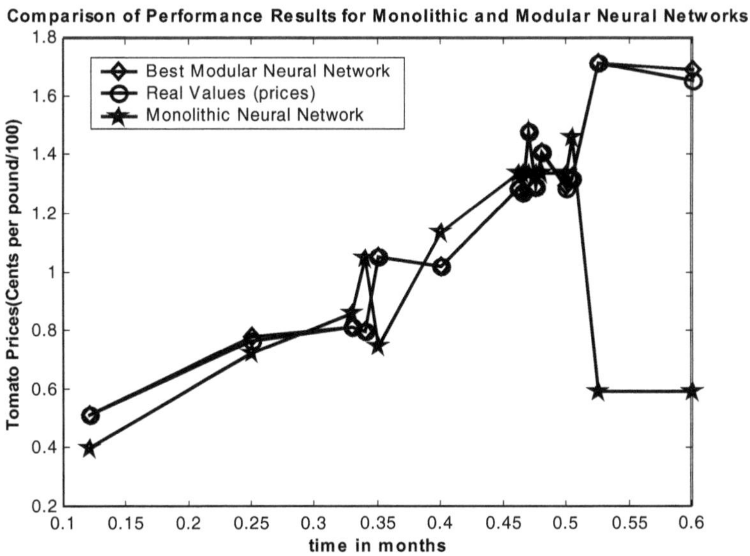

Fig. 7. Comparison of monolithic and modular neural networks.

7 Conclusions

We described in this paper the use of modular neural networks for simulation and forecasting time series of consumer goods in the U.S. Market. We have considered a real case to test our approach, which is the problem of time series prediction of tomato prices in the U.S. market. We have applied monolithic and modular neural networks with different training algorithms to compare the results and decide which is the best option. The Levenberg-Marquardt learning algorithm gave the best results. The performance of the modular neural networks was also compared with monolithic neural networks. The forecasting ability of modular networks was clearly superior.

References

1. Boers, E. and Kuiper, H. (1992) Biological Metaphors and the Design of Modular Artificial Neural Networks. Departments of Computer Science and Experimental and Theoretical Psychology at Leiden University, the Netherlands.
2. Brock, W.A., Hsieh, D.A., and LeBaron, B. (1991). "Nonlinear Dynamics, Chaos and Instability", MIT Press, Cambridge, MA, USA.
3. Castillo, O. and Melin, P. (1996). "Automated Mathematical Modelling for Financial Time Series Prediction using Fuzzy Logic, Dynamical System Theory and Fractal Theory", Proceedings of CIFEr'96, IEEE Press, New York, NY, USA, pp. 120-126.
4. Castillo, O. and Melin P. (1998). "A New Fuzzy-Genetic Approach for the Simulation and Forecasting of International Trade Non-Linear Dynamics", Proceedings of CIFEr'98, IEEE Press, New York, USA, pp. 189-196.
5. Castillo, O. and Melin, P. (1999). "Automated Mathematical Modelling for Financial Time Series Prediction Combining Fuzzy Logic and Fractal Theory", Edited Book "Soft Computing for Financial Engineering", Springer-Verlag, Germany, pp. 93-106.
6. O. Castillo and P. Melin, Soft Computing and Fractal Theory for Intelligent Manufacturing. Springer-Verlag, Heidelberg, Germany, 2003.
7. Fu, H.-C., Lee, Y.-P., Chiang, C.-C., and Pao, H.-T. (2001). Divide-and-Conquer Learning and Modular Perceptron Networks in IEEE Transaction on Neural Networks, vol. 12, No. 2, pp. 250-263.
8. Hansen, L. K. and Salamon P. (1990). Neural Network Ensembles, IEEE Transactions on Pattern Analysis and Machine Intelligence, Vol. 12, No. 10, pp. 993-1001.
9. Haykin, S. (1996). "Adaptive Filter Theory", Third Edition, Prentice Hall.
10. Jang, J.-S. R., Sun, C.-T., and Mizutani, E. (1997). "Neuro-fuzzy and Soft Computing: A Computational Approach to Learning and Machine Intelligence", Prentice Hall.

11. Lu, B. and Ito, M. (1998). Task Decomposition and module combination based on class relations: modular neural network for pattern classification. Technical Report, Nagoya Japan, 1998.
12. Maddala, G.S. (1996). "Introduction to Econometrics", Prentice Hall.
13. Murray-Smith, R. and Johansen, T. A. (1997). Multiple Model Approaches to Modeling and Control. Taylor and Francis, UK.
14. Parker, D.B. (1982). " Learning Logic", Invention Report 581-64, Stanford University.
15. Quezada, A. (2004). Reconocimiento de Huellas Digitales Utilizando Redes Neuronales Modulares y Algoritmos Geneticos. Thesis of Computer Science, Tijuana Institute of Technology, Mexico.
16. Rasband, S.N. (1990). "Chaotic Dynamics of Non-Linear Systems", Wiley.
17. Ronco, E. and Gawthrop, P. J. (1995). Modular neural networks: A State of the Art. Technical Report, Center for System and Control. University of Glasgow, Glasgow, UK, 1995.
18. Rumelhart, D.E., Hinton, G.E., and Williams, R.J. (1986). "Learning Internal Representations by Error Propagation", in "Parallel Distributed Processing: Explorations in the Microstructures of Cognition", MIT Press, Cambridge, MA, USA, Vol. 1, pp. 318-362.
19. Schdmit, A. and Bandar, Z. (1997). A Modular Neural Network Architecture with Additional Generalization Abilities for High Dimensional Input Vectors, Proceedings of ICANNGA'97, Norwich, England.
20. Sharkey, A. (1999). Combining Artificial Neural Nets: Ensemble and Modular Multi-Nets Systems, Ed. Springer-Verlag, London, England.
21. Sugeno, M. (1974). Theory of fuzzy integrals and its application, Doctoral Thesis, Tokyo Institute of Technology, Japan.
22. White, H. (1989). "An Additional Hidden Unit Test for Neglected Nonlinearity in Multilayer Feedforward Networks", Proceedings of IJCNN'89, Washington, D.C., IEEE Press, pp. 451-455.
23. Yager, R. R. (1999). Criteria Aggregations Functions Using Fuzzy Measures and the Choquet Integral, International Journal of Fuzzy Systems, Vol. 1, No. 2.

On Linguistic Summaries of Time Series Using a Fuzzy Quantifier Based Aggregation via the Sugeno Integral

Janusz Kacprzyk, Sławomir Zadrożny and Anna Wilbik

Systems Research Institute, Polish Academy of Sciences
Warsaw, Poland
[kacprzyk,zadrozny,wilbik@ibspan.waw.pl]

Abstract. We propose and advocate the use of linguistic summaries as descriptions of trends in time series data. We consider two general types of such summaries: summaries based on frequence and summaries based on duration. We employ the concept of a linguistic database summary due to Yager. To account for a specificity of time series data summarization we employ the Sugeno integrals for linguistic quantifier based aggregation.

1. Introduction

A linguistic summary is meant as a concise, human-consistent description of a data set that captures the very essence of data in the sense of their values, variability, etc. The concept of a linguistic data(base) summary has been introduced by Yager [13] and further developed by Kacprzyk and Yager [8], and Kacprzyk, Yager and Zadrożny [9]. In this approach the content of a database is summarized via a natural language like expression semantics of which is provided via Zadeh's calculus of linguistically quantified propositions [14].

In this paper we consider a specific type of data, namely time series that can be viewed as a certain real valued function of time. For a manager, stock exchange players, etc. it might be very useful to obtain a brief, natural language like description of trends present in the data on, e.g., a company performance, stock exchange quotations, etc. over a certain period of time. This is not meant as a replacement for a classical statistical analysis but rather as an additional form of data description that exhibits a remarkably by its high human consistency because for a human being the only fully natural means for the articulation, communication, etc. is natural language.

Technically, the summaries we propose refer to trends identified here with straight line segments of the piece-wise linear approximation of time

J. Kacprzyk et al.: *On Linguistic Summaries of Time Series Using a Fuzzy Quantifier Based Aggregation via the Sugeno Integral*, StudFuzz **208**, 415–433 (2007)
www.springerlink.com

series. Thus, the first step is the construction of such an approximation. For this purpose we use a modified version of the simple, easy to use Sklansky and Gonzalez's algorithm [12]. Then we employ a set of features (attributes) to characterize the trends such as the slope of the line, a goodness of approximation of the original data points by the line segment and the length of time period over which the trend may occur.

Basically, the summaries proposed by Yager are interpreted in terms of a number or proportion of elements possessing a certain property. In the framework considered here a summary might look like: "Most of the trends are short" or in a more sophisticated form: "Most long trends are increasing". Such expressions are easily interpreted using Zadeh's [14] calculus of linguistically quantified propositions. The most important element of this interpretation is a linguistic quantifier exemplified by "most". In Zadeh's approach it is interpreted in terms of a proportion of elements possessing a certain property (e.g., a length of a trend) among all the elements considered (e.g., all trends). In Kacprzyk, Wilbik and Zadrożny [7] we have proposed to use Yager's linguistic summaries, interpreted through Zadeh's calculus, for the summarization of time series.

Another type of summaries we propose here do not use the linguistic quantifier based aggregation over the number of trends but over the time instants they take altogether. For example, an interesting summary may take the following form: "Trends taking most of time are increasing" or "Increasing trends taking most of the time are of a low variability". Such summaries do not directly fit the framework of the original Yager's [13] approach. In order to overcome this difficulty we generalize our previous approach (cf. Kacprzyk, Wilbik and Zadrożny [7]), modelling the linguistic quantifier based aggregation both over the number of trends as well over the time they take with the use of the Sugeno integral.

In this paper, first, we describe the way the trends are extracted from time series and characterized using a set of attributes. Then we briefly remind the basics of the original Yager's approach to linguistic summarization and discuss how it may be used to describe a set of trends. In the next section we show how these summaries might be interpreted using the concept of fuzzy measure and the Sugeno integral. Finally we present some simple examples of linguistic summaries of an artificial data set.

2. Characterization of Time Series

In our approach time series data $\{(x_i, y_i)\}$ are approximated by a piece-wise linear function f such that for a given $\varepsilon > 0$, there holds

$$\forall i: \; |f(x_i) - y_i| \leq \varepsilon \tag{1}$$

There exist many algorithms that find such approximations (cf. [5, 6]). Our starting point is the Sklansky and Gonzalez algorithm [12] that seems to be a good choice due to its simplicity and efficiency. We modified it in the following way. The algorithm constructs the intersection of cones starting from a point p_i of the time series and including the circle of radius ε around the subsequent data points p_{i+j}, $j = 1, \dots$ until this intersection becomes empty. If for p_{i+k} the intersection is empty, then the points $p_i, p_{i+1}, \dots, p_{i+k-1}$ are approximated by a straight line segment and to approximate the remaining points we construct a new cone starting at p_{i+k-1}. Figure 1 presents the idea of the algorithm. The family of possible solutions, i.e., straight line segments to approximate points p_1 and p_2, is indicated with a dark gray area.

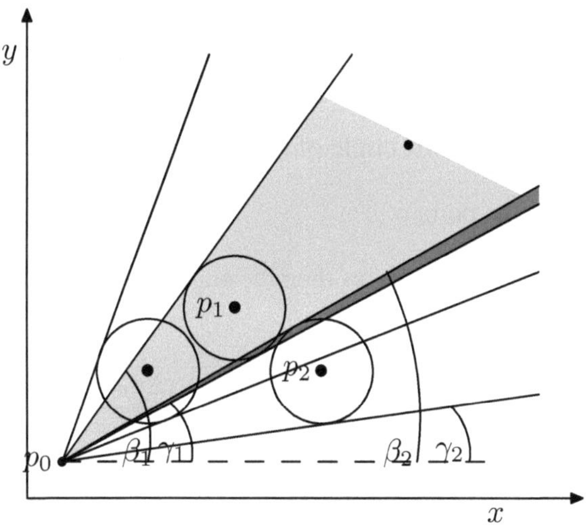

Fig. 1. An illustration of the algorithm [12] for an uniform ε-approximation

To make it more intuitively appealing we will now present the algorithm in the form of a pseudocode. First, denote by:

- `p_0` – the current starting point,
- `p_1` – the last point successfully verified, i.e. for it the intersection of cones starting at `p_0` is non-empty,
- `p_2` – the next point to be checked
- `Alpha_01` – a pair of angles (γ_1, β_1), meant as an interval that defines the current cone, as shown in Figure 1 (indicated by light gray and dark gray area)
- `Alpha_02` – a pair of angles defining the cone constructed to check `p_2` (i.e., the cone starting at point `p_0` and inscribing the circle of radius ε around the point `p_2` (cf. (γ_2, β_2) in Figure 1))
- function `read_point()` fetches the next data point,
- function `find()` finds a pair of angles defining the cone starting at point `p_0` and inscribing the circle of radius ε around of the point `p_2`

The pseudocode of the procedure that extracts the trends is depicted in Figure 2.

The bounding values of `Alpha_02` (γ_2, β_2), computed by function `find()`, are the slopes of two lines such that:

- they are tangent to the circle of radius ε around point `p_2`
- they start at the point `p_0`

Let $\Delta x = x_0 - x_2$ and $\Delta y = y_0 - y_2$ then the angles γ_2, β_2 can be expressed by the formulas:

$$\gamma_2 = arctg\left(\frac{(\Delta x)(\Delta y) - \varepsilon\sqrt{(\Delta x)^2 + (\Delta y)^2 - \varepsilon^2}}{(\Delta x)^2 - \varepsilon^2}\right)$$

$$\beta_2 = arctg\left(\frac{(\Delta x)(\Delta y) + \varepsilon\sqrt{(\Delta x)^2 + (\Delta y)^2 - \varepsilon^2}}{(\Delta x)^2 - \varepsilon^2}\right)$$

Then, as an approximation of points $p_0, \ldots, p_1$ we assume either a single straight line segment, chosen as, e.g. a bisector, or one that minimizes the distance (e.g. assumed as sum of squared errors, SSE) from the approximated points, or the whole family of possible solutions, i.e. the segments of the rays of the cone.

```
read_point(p_0);
read_point(p_1);
do
{
  p_2 = p_1;
  Alpha_02 = find();
  Alpha_01 = Alpha_02;
  do
  {

    Alpha_01 = Alpha_01 ∩ Alpha_02;

    p_1=p_2;
    read_point(p_2);
    Alpha_02 = find();

  } while(Alpha_01 ∩ Alpha_02 ≠ ∅);

  save_found_trend();
  p_0 = p_1;
  p_1 = p_2;
}
```

Fig. 2. Pseudocode of the procedure for extracting trends. For technical reasons we do not use Greek letters to denote variables.

This method is fast as it requires only a single pass through the data.

We characterize the trends, meant as the straight line segments of the above described uniform ε-approximation, using the following three features:

- dynamics of change,
- duration, and
- variability.

In what follows we will briefly discuss these factors.

2.1. Dynamics of Change

Under the term *dynamics of change* we understand the speed of changes. It can be described by the slope of a line representing the trend, (cf. any angle η from the interval $\langle\gamma,\beta\rangle$ in Figure 1). Thus, to quantify dynamics of change we may use the interval of possible angles $\eta \in \langle -90;90\rangle$ or their trigonometrical transformation.

However it might be impractical to use such a scale directly while describing trends. Therefore we may use a fuzzy granulation in order to meet the users' needs and the task specificity. The user may construct a scale of linguistic terms corresponding to various directions of a trend line as, e.g.:

- quickly decreasing,
- decreasing,
- slowly decreasing,
- constant,
- slowly increasing,
- increasing, and
- quickly increasing.

Notice that the number of different linguistic values, i.e. 7, does follow a very well-known psychological rule, the so-called Miller's magic number 7 ± 2, that basically states how many distinct (linguistic) values a human being can distinguish.

Figure 3 illustrates the lines corresponding to the particular linguistic terms.

In fact, each term represents a fuzzy granule of directions. In batyrshin et al. [1, 2] there are presented many methods of constructing such a fuzzy granulation. The user may define a membership functions of particular linguistic terms depending on his or her needs.

We map a single value η (or the whole interval of the angles corresponding to the gray area in Figure 1) characterizing the dynamics of change of a trend identified using the algorithm shown in Figure 2, into a fuzzy set best matching a given angle. Then we will say that a given trend is, e.g., "decreasing to a degree 0.8", if $\mu_{decreasing}(\eta) = 0.8$, where $\mu_{decreasing}$ is the membership function of a fuzzy set representing the linguistic term "decreasing" that is a best match for the angle η characterizing the trend under consideration.

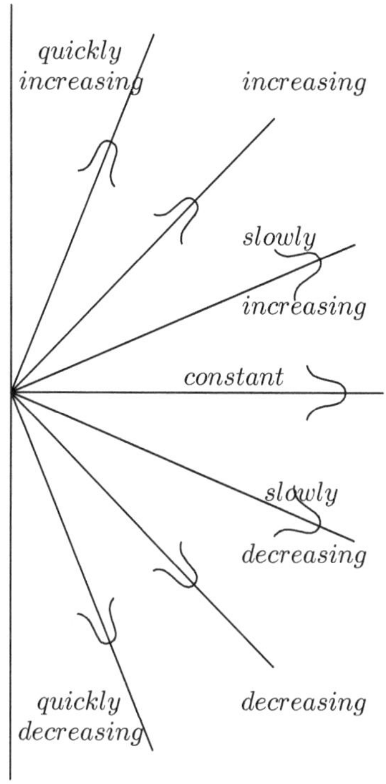

Fig. 3. A visual representation of angle granules defining dynamics of change

2.2. Duration

Duration describes the length of a single trend. Again we will treat it as a linguistic variable. An example of its linguistic labels is "long" defined as a fuzzy set whose membership function might be assumed as in Figure 4, where OX is the axis of time measured with units that are used in the time series data under consideration.

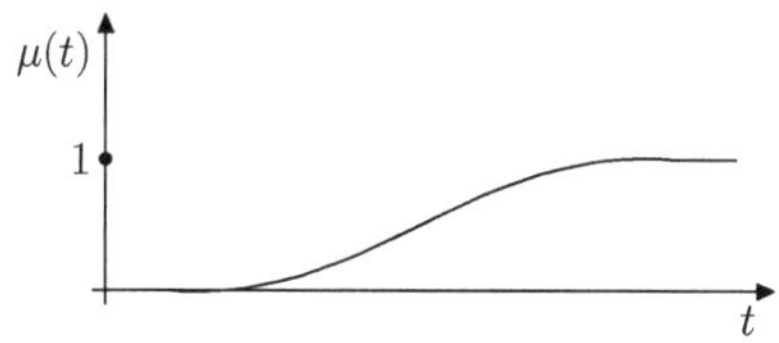

Fig. 4. Example of member ship function describing the term "long" concerning the trend duration

The actual definitions of linguistic terms describing the duration depends on the perspective assumed by the user. He or she, analyzing the data, may adopt this or another time horizon implied by his or her needs. The analysis may be a part of a policy, strategic or tactical planning, and thus, may require a global or local look, respectively.

2.3. Variability

Variability refers to how "spread out" (in the sense of values taken on) a group of data is. There are five frequently used statistical measures of variability:

- the range (maximum - minimum). Although this range is computationally the easiest measure of variability, it is not widely used as it is only based on two extreme data points. This make it very vulnerable to outliers and therefore may not adequately describe real variability.
- the interquartile range (IQR) calculated as the third quartile[1] minus the first quartile[2] that may be interpreted as representing the middle 50% of the data. It is resistant to outliers and is computationally as easy as the range.
- the variance is calculated as $1/n\sum_i(x_i - \bar{x})^2$, where $\bar{x}$ is the mean value.
- the standard deviation – a square root of the variance. Both the variance and the standard deviation are affected by extreme values.
- the mean absolute deviation (MAD), calculated as $1/n\sum_i|x_i - \bar{x}|$. While it has a natural intuitive definition as the "mean deviation from the mean", the introduction of the absolute value makes analytical calculations using this statistics much more complicated.

We propose to measure the variability of a trend as the distance of data points covered by this trend from a linear uniform ε-approximation that represents a given trend. For this purpose we propose to employ a distance between a point and a family of possible solutions, indicated as a dark gray cone in Figure 1. Equation (1) assures that the distance is definitely smaller than ε. We may use this information for the normalization. The normalized distance equals 0 if the point lays in the dark gray area. In the opposite case

[1] third quartile is the 75th percentile

[2] first quartile is the 25th percentile

it is equal to the distance to the nearest point belonging to the cone, divided by ε.

Alternatively, we may bisect the cone and then compute the distance between the point and this ray.

Again the measure of variability is treated as a linguistic variable whose values are linguistic terms (labels) modeled by fuzzy sets defined by the user.

3. Linguistic Summaries and Their Application to Trend Summarization

A linguistic summary, as presented in Kacprzyk and Zadrożny [10], [11] is meant as a natural language-like sentence that subsumes the very essence of a set of data. This set is assumed to be numeric and is usually large, not comprehensible in its original form by the human being. In Yager's approach (cf. Yager [13], Kacprzyk and Yager [8], and Kacprzyk, Yager and Zadrożny [9]) the following context for linguistic summaries mining is assumed:

- $Y = \{y_1, \ldots, y_n\}$ is a set of objects (records) in a database, e.g., the set of workers;
- $A = \{A_1, \ldots, A_m\}$ is a set of attributes characterizing objects from Y, e.g., salary, age, etc. in a database of workers, and $A_j(y_i)$ denotes a value of attribute A_j for object y_i.

A linguistic summary of a data set D consists of:

- a summarizer P, i.e. an attribute together with a linguistic value (fuzzy predicate) defined on the domain of attribute A_j (e.g. "low salary" for attribute "salary");
- a quantity in agreement Q, i.e. a linguistic quantifier (e.g. most);
- truth (validity) T of the summary, meant as a proposition of Zadeh's calculus of linguisitically quantified propositions (e.g. 0.7), i.e. a number from the interval $[0, 1]$ assessing the truth (validity) of the summary (e.g. 0.7); usually, only summaries with a high value of T are interesting;
- optionally, a qualifier R, i.e. i.e. another attribute A_k together with a linguistic value (fuzzy predicate) defined on the domain of attribute A_k determining a (fuzzy subset) of Y (e.g. "young" for attribute "age").

In what follows we will often for brevity identify summarizers and qualifiers with the linguistic terms they contain. In particular we will refer to the membership function μ_P or μ_R of the summarizer or qualifier to be meant as the membership functions of respective linguistic terms.

Thus, a linguistic summary may be exemplified by

$$T(\text{most of employees earn low salary}) = 0.7 \tag{2}$$

A richer form of a linguistic summary may include a qualifier (e.g. young) as in, e.g.,

$$T(\text{most of young employees earn low salary}) = 0.9 \tag{3}$$

Thus, basically, the core of a linguistic summary is a *linguistically quantified proposition* in the sense of Zadeh [14]. A linguistically quantified proposition corresponding to (2) may be written as

$$Qy\text{'s are } P \tag{4}$$

and the one corresponding to (3) may be written as

$$QRy\text{'s are } P \tag{5}$$

Then, the component of a linguistic summary, T, i.e., its truth (validity), directly corresponds to the truth value of (4) or (5). This may be calculated by using either the original Zadeh's calculus of linguistically quantified propositions (cf. [14]), or via other interpretations of linguistic quantifiers. The truth values (from $[0,1]$) of (4) and (5) are calculated, respectively, as

$$T(Qy\text{'s are } P) = \mu_Q\left(\frac{1}{n}\sum_{i=1}^{n}\mu_P(y_i)\right) \tag{6}$$

$$T(QRy\text{'s are } P) = \mu_Q\left(\frac{\sum_{i=1}^{n}(\mu_R(y_i)\wedge\mu_P(y_i))}{\sum_{i=1}^{n}\mu_R(y_i)}\right) \tag{7}$$

where Q is a fuzzy set representing the linguistic quantifier in the sense of Zadeh [14].

In order to characterize the summaries of trends we will refer to Zadeh's concept of a protoform (cf., Zadeh [15]). Basically, a protoform is defined as a more or less abstract prototype (template) of a linguistically quantified proposition. Then, summaries mentioned above might be represented by two types of the protoforms of the following forms. We may consider *frequency based summaries* and we obtain:

- a simple form:

$$Q \text{ trends are } P \tag{8}$$

exemplified by:

Most of trends have *a large variability*

- an extended form:

$$QR \text{ trends are } P \tag{9}$$

exemplified by:

Most of *slowly decreasing trends* have *a large variability*

However it should be noticed that in some cases the summaries of the above types might not properly grasp the character of time series. For example, assuming there are many very short trends of high variability and a few long terms of very low variability we may obtain a summary stating that "Most of trends have a large variability". This might be perceived as somehow incomplete or inaccurate summarization on its own as in fact the trends taking *most of time* have very low variability. Thus we propose to complement the above types of summaries with two more types of *duration based summaries*. These may be represented by the following schemes:

- a simple form:

$$\text{The trends that took } Q \text{ time are } P \tag{10}$$

exemplified by:

The trends that took *most* time have *a large variability*

- an extended form:

$$R \text{ trends that took } Q \text{ time are } P \tag{11}$$

exemplified by:

Slowly decreasing trends that took *most* time have *a large variability*

The truth degrees T of the frequency based summaries (8)-(9) can be directly computed using Zadeh's calculus of linguistically quantified propositions, in particular the formulae (6) and (7) are of use. However this is not the case when we consider duration based summaries. The reason is that in case of (8)-(9) a linguistic quantifier aggregates over the number of trends possessing a certain property while in case of (10)-(11) this aggregation goes over time taken by the trends. Thus, in the former case the *count* (number) of the trends matters that is properly accounted for with the use of the Σ-Count cardinality in formulae (6) and (7). In the latter case however another mode of aggregation is required. In order to to secure a unified solution in both cases we propose to employ the Sugeno integral as explained in the next section.

4. Linguistic Summary Interpretation via the Sugeno Integral

As we explained in the previous section, duration based linguistic summaries do not fit well to the interpretation of a linguistically quantified proposition employed in Zadeh's calculus. Thus we propose here to use the Sugeno integral for that purpose.

Let us start with a brief recall of the basics of the Sugeno integral. Let $X = \{x_1, \ldots, x_n\}$ be a finite set. Then, (cf., e.g., [4]) a *fuzzy measure* on X is a set function $\mu : \mathcal{P}(X) \longrightarrow [0,1]$ such that:

$$\begin{array}{l} \mu(\emptyset) = 0, \mu(X) = 1 \\ \text{if } A \subseteq B \text{ then } \mu(A) \leq \mu(B), \forall A, B \in \mathcal{P}(X) \end{array} \tag{12}$$

where $\mathcal{P}(X)$ denotes a set of all subsets of X.

Let μ is a fuzzy measure on X. The *discrete Sugeno integral* of function $f : X \longrightarrow [0,1]$, $f(x_i) = a_i$, with respect to μ is a function $S_\mu : [0,1]^n \longrightarrow [0,1]$ such that

$$S_\mu(a_1, \ldots, a_n) = \max_{i=1,\ldots,n} (a_{\sigma(i)} \wedge \mu(B_i)) \tag{13}$$

where $\wedge$ stands for the minimum, σ is such a permutation of $\{1, \ldots, n\}$ that $a_{\sigma(i)}$ is the i-th smallest element from among the a_i's and $B_i = \{x_{\sigma(i)}, \ldots, x_{\sigma(n)}\}$.

We can treat function f as a membership function of a fuzzy set $F \in \mathcal{F}(X)$, where $\mathcal{F}(X)$ denotes a family of fuzzy sets defined in X. Then the

Sugeno integral can be equivalently defined as a function $S_\mu : \mathcal{F}(X) \longrightarrow [0,1]$ such that

$$S_\mu(F) = \max_{\alpha_i \in \{a_1,\ldots,a_n\}} (\alpha_i \wedge \mu(F_{\alpha_i})) \tag{14}$$

where F_{α_i} is the α-cut of F and the meaning of other symbols is as in (13).

The fuzzy measure and the Sugeno integral may be intuitively interpreted in the context of multicriteria decision making (MCDM) where we have a set of criteria and some options (decisions) characterized by the degree of satisfaction of particular criteria. In such a setting X is a set of criteria and μ expresses the importance of each subset of criteria, i.e., how the satisfaction of a given subset of criteria contributes to the overall evaluation of the option. Then the properties of the fuzzy measure (12) properly require that the satisfaction of all criteria makes an option fully satisfactory and that the more criteria are satisfied by an option the better its overall evaluation. Finally the set F represents an option and $\mu_F(x)$ defines the degree to which it satisfies the criterion x. Then the Sugeno integral may be interpreted as an aggregation operator yielding an overall evaluation of option F in terms of its satisfaction of the set of criteria X. In such a context the formula (14) may interpreted as follows:

> find a subset of criteria of the highest possible importance (expressed by μ) such that at the same time minimal satisfaction degree of all these criteria by the option F is as high as possible (expressed by α)
> and take the minimum of these two degrees as the overall evaluation of the option F. (15)

Now we will explain how various linguistic summaries discussed in the previous section may be interpreted using the Sugeno integral. The linguistic quantifier Q is still defined as in Zadeh's calculus as a fuzzy set in [0,1], exemplified by (18). We will assume that Q is a regular monotone and non-decreasing quantifier:

$$\mu(0) = 0, \quad \mu(1) = 1 \tag{16}$$

$$x_1 \leq x_2 \Rightarrow \mu_Q(x_1) \leq \mu_Q(x_2) \tag{17}$$

exemplified by

$$\mu_Q(x) = \begin{cases} 1 & \text{for } x \geq 0.8 \\ 2x - 0.6 & \text{for } 0.3 < x < 0.8 \\ 0 & \text{for } x \leq 0.3 \end{cases} \tag{18}$$

The truth value of particular summaries is computed using the Sugeno integral (14). For simple types of summaries we are in a position to provide an interpretation similar to the one given above for the MCDM. For this purpose we will identify the set of criteria X with a set of trends while an option F will be the whole time series under consideration characterized in terms of how well its trends satisfy P.

In what follows $|A|$ denotes the cardinality of set A, summarizers P and qualifiers R are identified with fuzzy sets modelling the linguistic terms they contain, X is the set of all trends extracted from time series and $\text{time}(x_i)$ denotes duration of the trend x_i.

Simple frequency based summaries defined by (8) The truth value of this type of summary may be expressed as $S_\mu(P)$ where

$$\mu(P_\alpha) = \mu_Q\left(\frac{|P_\alpha|}{|X|}\right) \tag{19}$$

Thus, referring to (15), the truth value is determined by looking for a subset of trends of the cardinality high enough as required by the semantics of the quantifier Q and such that all these trends "are P" to the highest possible degree.

Extended frequency based summaries defined by (9) The truth value of this type of summary may be expressed as $S_\mu(P)$ where

$$\mu(P_\alpha) = \mu_Q\left(\frac{|(P \cap R)_\alpha|}{|R_\alpha|}\right) \tag{20}$$

Simple duration based summaries defined with (10) The truth value of this type of summary may be expressed as $S_\mu(P)$ where

$$\mu(P_\alpha) = \mu_Q\left(\frac{\sum_{i:x_i \in P_\alpha} \text{time}(x_i)}{\sum_{i:x_i \in X} \text{time}(x_i)}\right) \tag{21}$$

Thus, referring to (15) the truth value is determined by looking for a subset of trends such that their total duration with respect to the duration of the whole time series is long enough as required by the semantics of the quantifier Q and such that all these trends "are P" to the highest possible degree.

Extended duration based summaries defined with (11) The truth of this type of summary may be expressed as $S_\mu(P)$ where

$$\mu(P_\alpha) = \mu_Q\left(\frac{\sum_{i:x_i \in (P\cap R)_\alpha} \text{time}(x_i)}{\sum_{i:x_i \in R_\alpha} \text{time}(x_i)}\right) \tag{22}$$

Due to the properties (16)-(17) of the quantifiers employed it is obvious that all μ's defined above for particular types of summaries satisfy the axioms (12) of the fuzzy measure.

5. Example

Let us assume that from some given data we have extracted trends listed in Table 1, e.g. using the algorithm shown in Figure 2. We assume the granulation of dynamics of change presented in Section 2.1..

Table 1. Trends extracted

id	dynamics of change (α in degrees)	duration (time units)	variability ([0,1])
1	25	15	0.2
2	-45	1	0.3
3	75	2	0.8
4	-40	1	0.1
5	-55	1	0.7
6	50	2	0.3
7	-52	1	0.5
8	-37	2	0.9
9	15	5	0.0

We can consider the following simple frequency based trend summary:

$$\textit{Most} \text{ of trends are } \textit{decreasing} \tag{23}$$

In this summary *most* is the linguistic quantifier Q. The membership function is as in (18).

"Trends are decreasing" is a summarizer P with the membership function of the "decreasing" term given as in (24). Let us recall, that for brevity we identify summarizers and qualifiers with the linguistic terms they contain.

$$\mu_P(\alpha) = \begin{cases} 0 & \text{for } \alpha \leq -65 \\ 0,066\alpha + 4.333 & \text{for } -65 < \alpha < -50 \\ 1 & \text{for } -50 \leq \alpha \leq -40 \\ -0.01\alpha - 1 & \text{for } -40 < \alpha < -20 \\ 0 & \text{for } \alpha \geq -20 \end{cases} \tag{24}$$

n is the number of all trends, i.e., in this example $n = |X|$=9.
The truth value of (23) is computed according to (14) and (19) that yields:

$$T(\mathit{Most} \text{ of the trends are } \mathit{decreasing}) = \\ = \max_{\alpha_i \in \{a_1,\dots,a_n\}} \left(\alpha_i \wedge \mu_Q \left(\frac{|P_\alpha|}{|X|} \right) \right) = 0.511$$

If we assume the extended form, we may have the following summary:

$$\mathit{Most} \text{ of } \mathit{short} \text{ trends are } \mathit{decreasing} \tag{25}$$

Again, *most* is the linguistic quantifier Q with its membership function given as (18). *"Trends are decreasing"* is a summarizer P as in the previous example. *"Trend is short"* is the qualifier R. We define the membership function $\mu_R(t)$ as follows:

$$\mu_R(t) = \begin{cases} 1 & \text{for } t \leq 1 \\ -\frac{1}{2}t + \frac{3}{2} & \text{for } 1 < t < 3 \\ 0 & \text{for } t \geq 3 \end{cases} \tag{26}$$

The truth value of (25) is computed using the formula (14) and (20):

$$T(\mathit{Most} \text{ of } \mathit{short} \text{ trends are } \mathit{decreasing}) \\ = \max_{\alpha_i \in \{a_1,\dots,a_n\}} \left(\alpha_i \wedge \mu_Q \left(\frac{|(P \cap R)_\alpha|}{|R_\alpha|} \right) \right) = 0.9$$

On the other hand, we may have the following simple duration based linguistic summary:

$$\text{Trends that took } \mathit{most} \text{ time are } \mathit{slowly\ increasing} \tag{27}$$

"Trends are slowly increasing" is the summarizer P with the membership function $\mu_P(\alpha)$ defined as follows:

$$\mu_P(\alpha) = \begin{cases} 0 & \text{for } \alpha \leq 5 \\ 0.1\alpha - 0.5 & \text{for } 5 < \alpha < 15 \\ 1 & \text{for } 15 \leq \alpha \leq 20 \\ -0.05\alpha + 2 & \text{for } 20 < \alpha < 40 \\ 0 & \text{for } \alpha \geq 40 \end{cases} \tag{28}$$

The linguistic quantifier *most* is defined as previously. The truth value of (27) is computed via the formula (14) and (21) and we obtain:

$$\begin{aligned} &T(\text{Trends that took } \textit{most} \text{ time are } \textit{slowly increasing}) \\ &= \max_{\alpha_i \in \{a_1,\dots,a_n\}} \left(\alpha_i \wedge \mu_Q \left(\frac{\sum_{x_i \in P_\alpha} \text{time}(x_i)}{\sum_{i:x_i \in X} \text{time}(x_i)} \right) \right) \\ &= 0.733 \end{aligned}$$

Finally, we may consider an extended form of duration based summaries, here exemplified by:

$$\text{Trends } \textit{with a low variability} \text{ that took } \textit{most} \text{ of the time are } \textit{slowly increasing} \tag{29}$$

Again, *most* is the linguistic quantifier and *"trends are slowly increasing"* is summarizer P, with a membership function defined as in the previous example. *"Trends have a low variability"* is the qualifier R. The membership function $\mu_R(v)$ is given as follows:

$$\mu_R(v) = \begin{cases} 1 & \text{for } v \leq 0.2 \\ -5v + 2 & \text{for } 0.2 < v < 0.4 \\ 0 & \text{for } v \geq 0.4 \end{cases} \tag{30}$$

The truth value of (29) is computed according to the formula (14) and (22) and we obtain:

$$\begin{aligned} &T(\text{Trends } \textit{with low variability} \text{ that took } \textit{most} \text{ of the time are } \textit{slowly increasing}) \\ &= \max_{\alpha_i \in \{a_1,\dots,a_n\}} \left(\alpha_i \wedge \mu_Q \left(\frac{\sum_{i:x_i \in (P \cap R)_\alpha} \text{time}(x_i)}{\sum_{i:x_i \in R_\alpha} \text{time}(x_i)} \right) \right) \\ &= 0.75 \end{aligned}$$

6. Concluding Remarks

We have proposed a new approach to the linguistic summarization of time series data. The basic idea boils down to the identification of trends in time series that are characterized by a set of attributes. Then such a set of trends is directly amenable to the linguistic summarization. The specificity of time series calls for a new type of summaries that cannot be easily cast in the original framework of linguistic summaries as proposed by Yager. As opposed to Zadeh's calculus of linguistically quantified propositions used for the linguistic quantifier based aggregation, we employ here a new method based on the use of the Sugeno integrals. The basic idea is very similar and inspired by the work of Bosc et al.[3]. It seems that the results obtained are more intuitively appealing.

References

[1] I. Batyrshin (2002). On granular Derivatives and the solution of a Granular Initial Value Problem. In *International Journal Applied Mathematics and Computer Science*, 12(3):403-410.

[2] I. Batyrshin, L. Sheremetov. Perception Based Functions in Qualitative Forecasting. (to appear)

[3] P. Bosc P., L. Lietard, O. Pivert (2003). Sugeno fuzzy integral as a basis for the interpretation of flexible queries involving monotonic aggregates. In *Information Processing and Management*, 39(2):287–306.

[4] M. Grabisch (1998). Fuzzy integral as a flexible and interpretable tool of aggregation. In Bouchon-Meunier B. (ed.) *Aggregation and Fusion of Imperfect Information*, Studies in Fuzziness and Soft Computing, Heidelberg, New York: Physica–Verlag, 51–72.

[5] J. Colomer, J. Melendez, J. L. de la Rosa, J. Augilar (1997). A qualitative/quantitative representation of signals for supervision of continuous systems. In *Proceedings of the European Control Conference -ECC97*, Brussels.

[6] B. Hugeney, B. Bouchon-Meunier (2001). Time-Series Segmentation and Symbolic Representation, from Process-Monitoring to Data-Mining. In *LNCS* 2206:118-123.

[7] J. Kacprzyk, A. Wilbik, S. Zadrożny (2006). Linguistic summarization of trends: an approach. (in press).

[8] J. Kacprzyk and R.R. Yager (2001). Linguistic summaries of data using fuzzy logic. In *International Journal of General Systems*, 30:33-154.

[9] J. Kacprzyk, R.R. Yager and S. Zadrożny (2000). A fuzzy logic based approach to linguistic summaries of databases. In*International Journal of Applied Mathematics and Computer Science*, 10:813-834.

[10] J. Kacprzyk, S. Zadrożny (2005). Linguistic database summaries and their protoforms: toward natural language based knowledge discovery tools. In *Information Sciences* 173:281-304.

[11] J. Kacprzyk, S. Zadrożny (2005). Fuzzy linguistic data summaries as a human consistent, user adaptable solution to data mining. In B. Gabrys, K. Leiviska, J. Strackeljan (Eds.) *Do Smart Adaptive Systems Exist?* Springer, Berlin Heidelberg New York, 321-339.

[12] J. Sklansky and V. Gonzalez (1980) Fast polygonal approximation of digitized curves. In *Pattern Recognition* 12(5):327-331.

[13] R.R. Yager (1982). A new approach to the summarization of data. In *Information Sciences*, 28:69-86.

[14] L.A. Zadeh (1983). A computational approach to fuzzy quantifiers in natural languages. In *Computers and Mathematics with Applications*, 9, 149-184.

[15] L.A. Zadeh (2002). A prototype-centered approach to adding deduction capabilities to search engines – the concept of a protoform. BISC Seminar, University of California, Berkeley.